HYDROGENATED AMORPHOUS SILICON ALLOY DEPOSITION PROCESSES

APPLIED PHYSICS

A Series of Professional Reference Books

Series Editor

ALLEN M. HERMANN

University of Colorado at Boulder
Boulder, Colorado

1. Hydrogenated Amorphous Silicon Alloy Deposition Processes, *by Werner Luft and Y. Simon Tsuo*
2. Thallium-Based High-Temperature Superconductors, *edited by Allen M. Hermann and J. V. Yakhmi*
3. Composite Superconductors, *edited by Kozo Osamura*

Additional Volumes in Preparation

HYDROGENATED AMORPHOUS SILICON ALLOY DEPOSITION PROCESSES

WERNER LUFT · Y. SIMON TSUO

National Renewable Energy Laboratory
Golden, Colorado

Marcel Dekker, Inc. New York • Basel • Hong Kong

Library of Congress Cataloging-in-Publication Data

Luft, Werner.
Hydrogenated amorphous silicon alloy deposition processes / Werner Luft and Y. Simon Tsuo.
p. cm. -- (Applied physics series ; 1)
Includes bibliographical references and index.
ISBN 0-8247-9146-0 (acid-free paper)
1. Thin film devices--Design and construction. 2. Silicon alloys. 3. Plasma enhanced chemical deposition. 4. Thin films--Optical properties. 5. Thin films--Electric properties. I. Tsuo, Y. Simon. II. Title. III. Series: Applied physics series (Marcel Dekker, Inc.) ; 1.
TK7872.T55L84 1993
621.3815'2--dc20 93-18931
CIP

The publisher offers discounts on this book when ordered in bulk quantities. For more information, write to Special Sales/Professional Marketing at the address below.

This book is printed on acid-free paper.

Copyright © 1993 by MARCEL DEKKER, INC. All Rights Reserved.

Neither this book nor any part may be reproduced or transmitted in any form or by any means, electronic or mechanical, including photocopying, microfilming, and recording, or by any information storage and retrieval system, without permission in writing from the publisher.

MARCEL DEKKER, INC.
270 Madison Avenue, New York, New York 10016

Current printing (last digit):
10 9 8 7 6 5 4 3 2 1

PRINTED IN THE UNITED STATES OF AMERICA

SERIES INTRODUCTION

The Applied Physics Series represents a commitment by Marcel Dekker, Inc., to develop a book series that provides up-to-date information in the new and exciting areas of physics emanating from the explosion of discoveries in new materials and new processing techniques. The advances in amorphous materials, layered copper oxide high temperature superconductors, organic and inorganic cage structures, organic conductors, and a host of new thin film deposition and nanofabrication processes have lead to a host of new applications. It is the intent of the series editor to invite experts in the most important of these areas to assemble book-length manuscripts describing the latest developments. Many volumes will take the form of comprehensive multi-authored compendia that are linked logically by judicious choice of subject matter and editing.

The Applied Physics Series is designed to bring the non-specialist scientist/engineer to the forefront of each area. Researchers wishing to enter a particular area will find these volumes of significant benefit as a detailed introduction to the field. The books in the series are also expected to provide important summaries and to serve as reference guides to the experts in each field.

The scope of the series is broad. We will encourage volumes relating to new or improved sensors, memories, processing units, displays, energy conversion, storage and delivery, and to semiconducting and superconducting devices generating enabling technologies in areas ranging from communications and computers to medical physics.

We sincerely hope through this series that we can help today's discoveries rapidly become tomorrow's technology.

Allen M. Hermann

PREFACE

This book contains extensive practical information and references on the deposition processes and properties of hydrogenated amorphous silicon (a-Si:H) and related alloys (a-Si alloys). a-Si alloys are a new class of thin-film material with good semiconductor properties. While crystalline Si has a fixed set of well-defined properties, the electrical and optical properties of a-Si alloys can be adjusted over a wide range. For example, the optical bandgap of a-Si:H can be adjusted between about 1.5 and 2.0 eV. The optical bandgap can be further extended to between 1 and 5 eV by adding bandgap modifiers to form alloys of a-$Si_{1-x}Ge_x$:H, a-$Si_{1-x}C_x$:H, and a-$Si_{1-x}N_x$:H. Although there are many similarities between a-Si:H and crystalline Si, the processing and characterization methods discussed in this book are in many cases very different from those for crystalline Si, because of the thin-film nature of a-Si alloys.

Because the properties of a-Si alloys are very strongly dependent on the method and conditions of deposition, we believe every scientist or engineer who works with a-Si alloys should have a good knowledge of the film deposition processes. The objective of this book is to provide an easily understood review, with extensive bibliography, of the various a-Si:H deposition methods and to present, at the same time, concise reviews of film properties and common film and plasma diagnostic techniques. We hope this book will be useful to the students and process engineers who are new in the a-Si:H field in learning about and choosing film deposition and diagnostic methods. For the readers who are experienced in the field, we hope this book will be a useful resource for references on deposition processes, basic material properties, and characterization techniques for a-Si alloys. For this reason, a very extensive reference list is provided.

It has been less than 20 years since the discovery that thin films of hydrogenated amorphous silicon have good semiconductor properties and can be doped either p- or n-type. Efforts in developing a-Si:H photovoltaic solar cells began in 1974. Today, worldwide commercial a-Si:H solar cell shipments are about 14 MW, which accounts for about 25% of all photovoltaic solar cell shipments. The a-Si:H solar cell market is expected to expand rapidly in the coming years due to the introduction of new products, such as automobile sun roofs with integral amorphous silicon modules that generate electricity. In addition to photovoltaic solar cells, many other applications that require large-area, thin-film semiconductors are being developed using a-Si:H and related alloys. Examples include contact image sensors for fax machines, photoreceptors for electrophotography, charged particle and x-ray detectors, and thin-film transistors for controlling active-matrix liquid crystal displays.

The Department of Energy of the United States Government has spent over $110 million in research funds for the development of a-Si:H solar cells. Similar amounts of government funds have been spent by the European Community and by Japan. The private industries have spent even more on the development of a-Si:H solar cells. This book draws heavily on the knowledge gained from these studies of a-Si:H solar cells. However, it is by no means intended to be useful only for people working in the solar cell field. The deposition methods, film properties, and diagnostic methods discussed in this book should be applicable to all semiconductor applications of a-Si:H and related alloys.

We are greatly obligated to Prof. Cameron A. Moore of the Denver University and Dr. Changhua Qiu of the University of Colorado, Boulder, for their careful reading of the manuscript and their many helpful comments.

This work was supported by the National Renewable Energy Laboratory with funding from the U.S. Department of Energy under contract no. DE-AC02-83CH10093.

Werner Luft and Y. Simon Tsuo

CONTENTS

LIST OF SYMBOLS

α	Sub-bandgap optical absorption coefficient
η	Efficiency or quantum efficiency
μ	Mobility
σ_d	Dark conductivity
σ_l	Photoconductivity
τ	Lifetime
a-Si:H	Hydrogenated amorphous silicon
a-SiC:H	Hydrogenated amorphous silicon-carbon alloy
a-SiGe:H	Hydrogenated amorphous silicon-germanium alloy
C_H	Hydrogen content
E_a	Activation energy of dark conductivity
E_c	Conduction band energy (mobility edge)
E_f	Energy of Fermi level
E_g	Optical bandgap
E_u	Urbach energy
E_v	Valence band energy (mobility edge)
FF	Fill factor
$g(E_f)$	Defect density at Fermi level
$h\nu$	Photon energy
J_{sc}	Short-circuit current density
L	Liter
L_d	Ambipolar diffusion length
L_e	Diffusion length for electrons
L_h	Diffusion length for holes
n	Index of refraction
N_s	Density of states
$N_{s,0}$	Initial defect density
N_{sat}	Saturated defect density
s	Second
T_s	Substrate temperature
V_{bi}	Built-in voltage
V_{oc}	Open-circuit voltage

ACRONYMS

AC	Alternating current
AES	Auger electron spectroscopy
AFM	Atomic force microscope
AM1	Air mass one
AM1.5	Air mass one point five
AP	Atmospheric pressure
C-V	Capacitance vs. voltage
CARS	Coherent anti-Stokes-Raman spectroscopy
CMS	Columnar microstructure
CPM	Constant photocurrent measurement
CVD	Chemical vapor deposition
CW	Continuous wave
DC	Direct current
DLTS	Deep-level transient spectroscopy
DOS	Density of defect states
EBIC	Electron beam induced current
ECR	Electron cyclotron resonance
EELS	Electron energy loss spectroscopy
EPMA	Electron probe microanalysis
EPR	Electron paramagnetic resonance
ESD	Electron stimulated desorption
ESR	Electron spin resonance
FTIR	Fourier transform infrared spectroscopy
GC	Gas chromatography
HOMOCVD	Homogeneous chemical vapor deposition
HR-CVD	Hydrogen-radical-enhanced chemical vapor deposition
ICTS	Isothermal capacitance transient spectroscopy
IPA	Isopropyl alcohol
IR	Infrared
LICVD	Laser-induced chemical vapor deposition
LIF	Laser-induced fluorescence
MS	Mass spectroscopy
NMR	Nuclear magnetic resonance
OES	Optical emission spectroscopy
PAS	Photoacoustic spectroscopy
PDS	Photothermal deflection spectroscopy
PED-CVD	Periodic etching deposition chemical vapor deposition
PL	Photoluminescence

PVD	Physical vapor deposition
RF	Radio frequency
RGA	Residual gas analyzer
RTA	Rapid thermal annealing
SAXS	Small-angle x-ray scattering
SCCM	Standard cubic centimeter per minute
SCLC	Space-charge limited current
SEM	Scanning electron microscopy
SIMS	Secondary ion mass spectroscopy
SPC	Solid-phase crystallization
SPV	Surface photovoltage
SSPG	Steady State Photocarrier Grating Technique
STEM	Scanning tunneling electron microscopy
STM	Scanning tunneling microscope
TCO	Transparent conductive oxide
TEM	Transmission electron microscopy
TFT	Thin-film transistor
TOF	Time of flight
UHV	Ultra-high vacuum
UPS	Ultraviolet photoelectron spectroscopy
UV	Ultraviolet
XPS	X-ray photoelectron spectroscopy
YAG	Yttrium-aluminum-garnet ($Y_3Al_5O_{12}$)

1
INTRODUCTION

Initial studies of hydrogenated amorphous silicon (a-Si:H) were done by Sterling and Swann [1965] and by Chittick et al. [1969], who demonstrated the radio-frequency (RF) glow discharge deposition of a-Si:H from silane gas. This was followed by the work of Spear and LeComber [1976], who reported on successful n- and p-doping of a-Si:H. In 1974, the first photovoltaic solar cell using a-Si:H was fabricated [Carlson and Wronski 1976]. In 1975 and 1976, researchers confirmed that films deposited from silane by glow discharge contained hydrogen [Triska et al. 1975, Knights 1976]. The first commercial product based on a-Si:H—photovoltaic batteries for calculators—was introduced in 1980. Since then a-Si:H and related alloys (**a-Si:H alloys**) have found increased applications as large-area, thin-film semiconductors. This book discusses the various deposition techniques for a-Si:H alloys. It also introduces the reader to the basic properties and characterization techniques of a-Si:H alloys. In the first half of this chapter—section 1.1—we discuss some of the applications of a-Si:H alloys. In the second half of this chapter—section 1.2—we discuss the organization of this book.

1.1 POTENTIAL APPLICATIONS OF AMORPHOUS SILICON-BASED ALLOYS

Amorphous silicon-based alloy materials are increasingly being used in applications that require *large-area, thin-film semiconductors* [Madan and Shaw 1988, Wronski 1988, LeComber 1989, Kanicki 1991]. a-Si:H alloys have a short optical absorption length for visible light and good intrinsic and doped semiconductor properties, and they can be deposited easily, at low temperature and low cost, on inexpensive substrates of almost any size or shape by chemical vapor deposition methods. For example, a production line of monolithic photovoltaic solar modules with the size of 2.5 feet by 5 feet made by glow discharge deposition of a-Si:H on glass is in operation [Sabisky et al. 1989]. And there are commercial a-Si:H products deposited on curved substrates for solar electric automobile sunroofs and residential roof tiles [Hamakawa 1991]. Because of such advantages a-Si:H has become the most widely used thin-film material for photovoltaic solar energy modules [Stone 1990] and contact image sensors for facsimile machines [LeComber 1989, Rosan 1989, Weisfield 1989].

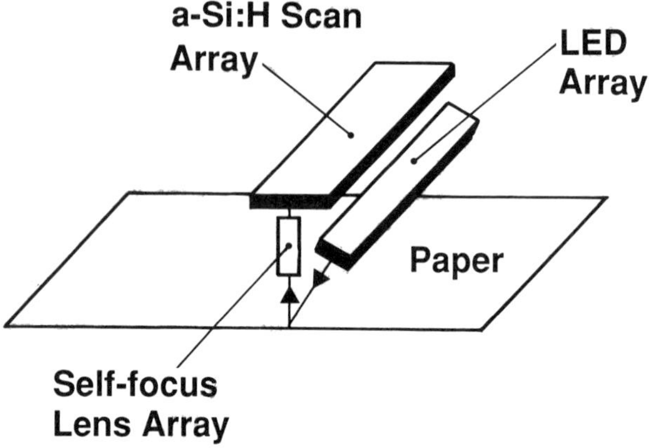

Figure 1.1-1 Schematic diagram of a-Si:H contact image sensor [Lecomber 1989]

A schematic diagram of a contact image sensor module is shown in **Fig. 1.1-1** [LeComber 1989]. Such page-width, high resolution a-Si:H sensor arrays coupled with a simple optical system for scanning documents have achieved widespread commercial use. Because of their high sensitivity to long-wavelength visible light, their high dark resistivity, and strong resistance to mechanical wear, a-Si:H alloys have been used as photoreceptors for electrophotography. The first commercial amorphous Si photosensitive drums were released by Canon, Inc. in 1984. The smaller bandgap of a-Si:H alloys compared to that of the traditional photoreceptor, amorphous selenium, allows for shorter discharge times which means higher copying speeds. In addition, a smaller bandgap makes it possible to write digital information directly to a copying drum using an inexpensive semiconductor laser. Due to the large electric field involved in electrophotographic processes, about 20 μm-thick a-Si:H alloys are required for such applications. This thickness requirement is an important consideration for the widespread use of a-Si:H alloys in electrophotography. Usual deposition rates of high quality a-Si:H alloys are below 0.4 nm/s. However, high deposition rates have been achieved, for example, the rate of 10 nm/s for device-quality material by a microwave plasma deposition method [Canon 1993]. High dark conductivity and high internal stress must be avoided in the use of a-Si:H alloys in electrophotography [Mort and Jansen 1986].

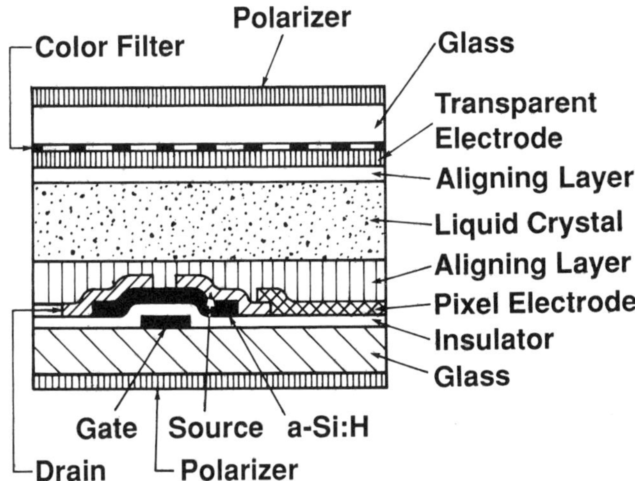

Figure 1.1-2 Cross section of an a-Si:H TFT/LCD color display [Yamano and Takasada, 1985]

The need for large-area charged particle and X-ray detectors for applications like medical imaging and calorimetry in high-energy physics experiments has stimulated significant investigations into using a-Si:H (50 to 70 μm thick) for such applications [Perez-Mendez et al. 1989, Pochet el al. 1989, Xi et al. 1991]. Other photodiode applications of a-Si:H alloys include ultraviolet light detectors [Huang et al. 1988], four-color discriminating sensors [Yang et al. 1988], edge detectors for application to neural network image sensors [Sah et al. 1990], and position sensors for telephone terminals [Takeda and Sano 1988].

There are many large-area electronics applications of a-Si:H that do not depend on the photoelectric properties, instead, they take advantage of the excellent area uniformity, high yield, and the relatively high field effect mobility of a-Si:H. One of these applications is the fabrication of thin-film transistors (TFT) that are most often used in controlling active-matrix liquid crystal displays (LCD) [Thompson 1986, Uchida et al. 1991, Crowley 1992]. A schematic of such a device is shown in **Fig. 1.1-2** [Yamano & Takesada 1985]. The current flowing through the a-Si:H thin-film (100 to 200 nm-thick) between the source and drain contacts is modulated by the gate electrode. Current modulation ratios of more than 10^7 can be achieved [Kanicki et al. 1991]. The field-effect mobility for a-Si:H TFT switching devices is up to 1.0 $cm^2\ V^{-1}\ s^{-1}$ [Powell 1989]. The existence of an electro-optic response in a-Si:H suggests potential use of a-Si:H as the core of a waveguide with dielectric claddings [Zelikson et al. 1992].

In recent years, the development of thin-film a-Si:H photovoltaic solar cells has been extensively pursued because such devices offer the potential of low-cost electricity, making them attractive as a source of utility and residential electric power. Single-junction a-Si:H p-i-n solar cells with a one-sun photovoltaic solar energy conversion efficiency of 12% have been achieved by several laboratories. The basic structure of a single-junction a-Si:H p-i-n cell consists of a very thin (less than 10-nm-thick, p-type layer), a low-defect, 200 to 600 nm-thick intrinsic layer, and a thin (about 30 nm-thick) n-type layer. To improve the efficiency and stability of a-Si:H solar cells, multiple-junction solar cell structures using a-SiC:H, a-Si:H, and a-SiGe:H alloys are being extensively studied [Yang et al. 1988]. A triple-junction a-Si-based tandem solar cell structure is shown in **Fig. 1.1-3**. This structure utilizes three different bandgap a-Si alloys to increase optical absorption efficiency by limiting heat loss of photon energies above bandgap. It also allows the use of thinner i-layers, which improves cell stability. Triple-junction a-Si:H solar cells have also been used in direct photoelectrolysis of water for hydrogen generation [Lin et al. 1989].

Sometimes microcrystalline Si doped layers are used instead of a-SiC:H and/or a-Si:H doped layers because of their high conductivity and low optical absorption. Graded-bandgap buffer layers are sometimes placed between p- and i-layers with different optical bandgaps to enhance the solar cell performance. The optical bandgap of a-SiC:H intrinsic layer is usually about 2.0 eV, a-Si:H is 1.75 eV, and a-SiGe:H is 1.5 eV. These three layers absorb light in different areas of the spectrum, as shown in **Fig. 1.1-4** [Hishikawa et al. 1989]. It is very likely a stabilized conversion efficiency of 10% or higher can be achieved with multijunction a-Si:H alloy solar modules with cells monolithically connected in series [Ichikawa 1990].

To achieve wide-spread usage in photovoltaic utility applications, flat-plate photovoltaic module conversion efficiencies of at least 10%, under terrestrial sun light conditions generally refferes to as air mass 1.5 (AM1.5) conditions [Green 1982], and module costs of $45-$80/m^2 (1986

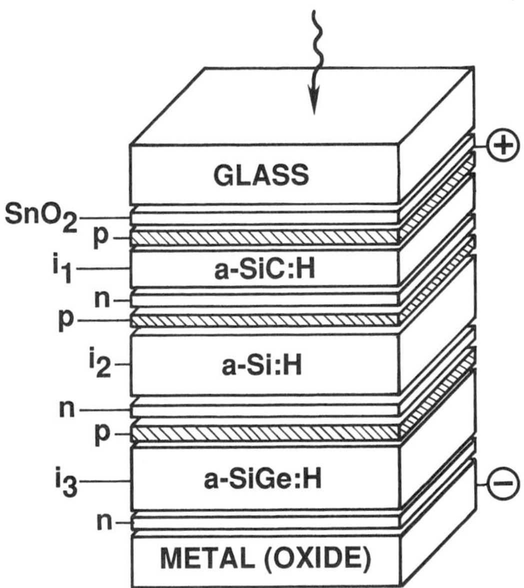

Figure 1.1-3 A triple-junction a-Si-based cell structure

dollars) are required [U.S. Department of Energy 1987]. To date, single-junction a-Si:H p-i-n solar cells with an efficiency of 12% and multijunction cells with an efficiency of 13.7% have been reported [Stone 1987]. The output characteristics from a typical a-Si:H module in shown in **Fig. 1.1-5**. The current at zero voltage is called the short-circuit current and the voltage at zero current is called the open-circuit voltage. The ratio of the product of current and voltage at the maximum power point to the product of short-circuit current and open-circuit voltage is called the fill factor. The efficiency is the ratio of the maximum power output to the solar energy input. First-order thermodynamic analyses predict a maximum efficiency of about 13% for single-junction a-Si:H solar cells with an open-circuit voltage (V_{oc}) of 0.99 V and a fill factor (FF) of 0.86 [Folkerts and Gordon 1988]. It is believed that AM1.5 efficiencies around 15% could be reached for single-junction a-Si:H cells if the V_{oc} could be increased to 1.2 V by reducing weak and strained Si-Si bonds during deposition. Currently, the band-tail states in the mobility gap limit the movement of quasi-Fermi levels and thereby limit the V_{oc} of a p-i-n a-Si:H solar cell. Using a comprehensive computer simulation model, Yamanaka et al. [1989] predicted that a 15.45%-efficient a-Si:H single-junction cell can be made if the density of dangling-bond states can be reduced to 1×10^{15} cm^{-3} and if a wide-gap

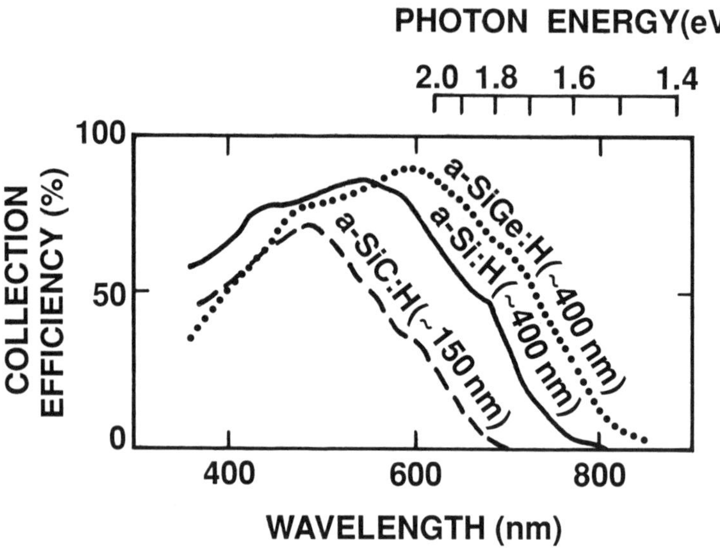

Figure 1.1-4 Collection efficiency spectra of the different active layers of a triple-junction a-Si-based solar cell

p-layer with a carrier concentration of 2 x 10^{18} cm^{-3} and an activation energy of 0.1 eV is used. Further improvement in a-Si:H cell efficiency would have to be obtained by using multijunction structures. The maximum conversion efficiency for triple-junction a-Si:H cells, using a-Si:H and SiGe-alloy materials, is believed to be about 24% [Kuwano 1982]. To realize high efficiencies in multijunction a-Si-based solar cells requires not only an improvement in the a-Si:H cells, but also an improvement in the low-bandgap hydrogenated amorphous silicon-germanium alloy (a-SiGe:H) and high-bandgap hydrogenated amorphous silicon-carbon alloy (a-SiC:H) cells. Some of the efforts to improve the quality of these alloys will be discussed in this book. **Table 1.1** shows a list of actual and potential applications of a-Si:H alloys.

1.2 DEPOSITION PROCESSES FOR AMORPHOUS SILICON-BASED ALLOYS

Glow discharge deposition (also known as plasma-enhanced or plasma-assisted chemical vapor deposition) has become the most common technique used in the deposition of a-Si:H alloy films and devices. The glow discharge is initiated by applying a direct current (DC) or

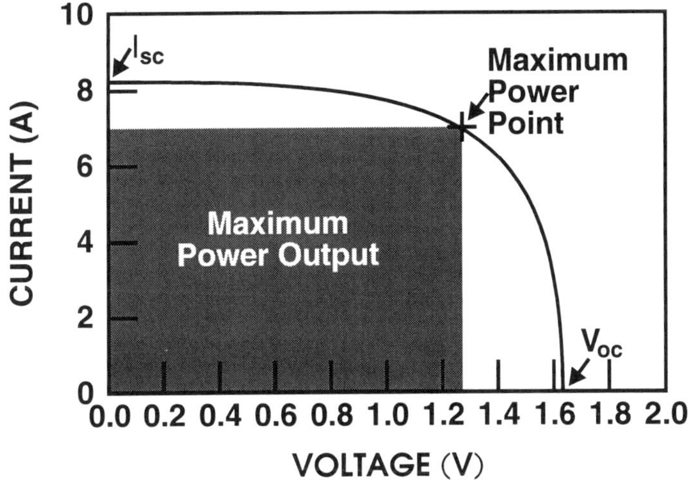

Figure 1.1-5 Current-voltage characteristic for an a-Si multijunction module

alternating current (radio-frequency or microwave) power between two electrodes to the deposition gas. It is a glowing electric discharge, similar to that present in a fluorescent light bulb. We devote a large portion of this book, chapters four to eight, to the discussions of glow discharge deposition processes. We summarize aspects of glow discharge processes relevant to the deposition of high quality a-Si:H alloys and elucidate the effects on these films and devices of various deposition parameters and other aspects of the growth process, to better understand how the films are formed. Such a basic understanding is important because the empirical optimization of the electronic properties of these devices has become increasingly more difficult as the device structures have become more complex. Moreover, the many strong interactions of plasma process parameters can cause widely varying properties in the resulting films.

One cannot discuss the deposition processes without some basic understanding of the properties of the deposited films and the characterization techniques used for measuring the properties. Chapter 2 provides material characteristics of a-Si:H alloys. Chapter 3 describes common diagnostic measurement techniques for determining plasma composition and film properties.

Table 1-1 Actual and Potential Applications of a-Si:H Alloys

Device	Product
Photovoltaic cells	Terrestrial solar power modules, Solar batteries for consumer electronic products, Battery chargers, Photoelectrolysis electrodes for hydrogen generation
Photosensors	Color sensors, Light sensors, Vidicons, Charge-coupled devices for large-area image sensors, Contact-type linear image sensors, Position-sensitive photodetectors, Optically addressed spatial light modulators, Electronic writing boards, High-speed detectors and modulators, Artificial Neural Networks
Photoreceptors	Electrophotographic printing, Laser printers
Charged particle & γ ray detectors	High energy radiation imaging
Optical recording media	Optical disks for information storage
Light emitting diodes	Large-area, light-emitting-diode, flat-panel displays
Thin-film transistors	Active-matrix flat-panel displays, Logic integrated circuits, Memory switching devices
Solar control layer	Heat-reflecting glass
Anti-reflecting/anti-static layer	Monitor and Television screens
Ambient sensors	Thermistors, Hydrogen sensor, Ion-selective and gas-sensitive field effect transistors
Semiconductor surface passivation	
Photolithographic masks	
Strain gauges	

Chapter 4 starts the discussions on the conventional glow discharge deposition process. Chapter 5 discusses aspects of glow discharge deposition-system design, especially the electrode geometry, bias voltage effects, and plasma confinement. Chapter 6 summarizes the effects of primary deposition parameters—such as power density, substrate temperature, feed-gas concentration, pressure, and magnetic field—on deposition rate, defect density, hydrogen content, dihydride-to-monohydride ratio, and photoconductivity. Chapter 7 describes the glow discharge deposition reaction chemistry.

Although widely used, the conventional glow discharge deposition method has its limitations, including low deposition rates, poor quality low-bandgap (<1.6 eV) and high-bandgap (>2.0 eV) alloy materials, poor doping efficiencies for both p- and n-type materials, and a limited selection of dopants. There has been very little real improvement in recent years in the defect and photostability properties of glow discharge-deposited a-Si:H films, indicating that some fundamental defect density limit has been reached in a-Si:H materials by these deposition methods. Thus it is important to investigate alternative deposition processes or even alternative materials, such as microcrystalline silicon. We devote chapters eight to twelve of this book to alternative deposition processes. Chapter 8 describes new deposition techniques that are only slight modifications of the conventional glow discharge technique. Chapter 9 reviews remote-plasma-assisted chemical vapor deposition methods. Chapter 10 reviews photochemical vapor deposition methods, Chapter 11 is on thermally induced CVD (pyrolysis), and Chapter 12 is on physical vapor deposition methods. Chapter 13 describes etching properties of amorphous silicon-based alloys, which are important to certain device fabrications and to the understanding of film deposition processes. Chapter 14 compares properties of materials deposited by the various methods. Chapter 15 discusses microcrystalline silicon and silicon-carbide films which are deposited using techniques similar to those for a-Si:H and are, in some cases, considered as alternatives to a-Si:H alloys. Chapter 16 is a brief discussion of some safety issues related to the use of hazardous gases for a-Si:H alloy depositions.

2
MATERIAL CHARACTERISTICS OF AMORPHOUS SILICON-BASED ALLOYS

In this chapter, we review some of the material characteristics of a-Si:H alloys that are important for the device applications mentioned in Section 1.1. There are several advantages of a-Si:H alloys that made the development of important device applications possible within a relatively short time of the discovery of the material, for example, the ability of using low-cost methods to deposit device-quality thin-film materials, at low substrate temperatures, on substrates of almost any shape, size, or composition. Another advantage is the ability to adjust the optical bandgap and electrical properties of the a-Si:H alloys to suit the needs of the particular device applications. The disadvantages of a-Si:H alloys include low charge carrier mobility and lifetime, low deposition rate, and unstable photoelectrical properties.

The most common a-Si:H material has a bandgap of 1.7 to 1.8 eV with a hydrogen content of about 10%. Bandgap modifiers, such as germanium, carbon, and nitrogen, can be added to the a-Si:H to change the bandgap. The properties of these alloys, a-SiGe:H, a-SiC:H, and a-SiN:H, will be discussed in this chapter. Although a-SiSn:H alloys with bandgap as low as 1 eV have also been studied, the charge transport properties and photosensitivities of a-SiSn:H alloys are much poorer compared to a-SiGe:H alloys of the same bandgap [Mahan et al. 1984, Vèrie 1984]. Important material characteristics of a-Si:H alloys and their significance are listed in **Table 2-1.**

Many of the material characteristics of a-Si:H alloys are interrelated. For example, the σ_l value is related to charge carrier lifetime (τ) and is directly proportional to the photocarrier generation quantum efficiency-mobility-lifetime product ($\eta\mu\tau$) of the majority carriers [see, for example, Fritzsche 1980]. The importance of film characteristics varies with the intended application. For thin-film transistor applications, high electron mobility is important. Low state density in the tail of the band energy distribution is needed for high-speed charge-coupled device applications. For solar cell applications, besides the bandgap, usually the minority carrier lifetime is the most important property. However, high performance a-Si:H solar cells operate with a built-in electric field across most of the cell thickness, and it is also important to have high charge carrier mobilities. Unfortunately, for a-Si:H alloys the value of $\mu\tau$ depends on the measurement method used [Crandall and Balberg 1991]. When comparing the quality of films with the same E_g but deposited by different methods, we will often use the values of one-sun photo-

Table 2-1 Important Material Characteristics for a-Si:H Alloys

Group	Characteristic	Significance
Electrical	Mobility (μ)	The quantity that describes how strongly the motion of a charge carrier is influenced by an applied electric field.
	Diffusivity (D)	The measure of how quickly carriers spread out by random motion.
	Lifetime (τ)	Average time before carrier recombination, the rate of which is limited by the capture of the minority carrier.
	Diffusion length (L)	Average distance an excess carrier will move before recombining, $L \propto \tau^{1/2}$.
	Drift length (l)	The average distance a carrier travels in an electric field E, $l = \mu\tau E$.
	Conductivity (σ_d)	$\sigma_d = nq\mu_n + pq\mu_p$, where q is electric charge, n and p are the carrier concentrations of electrons and holes, respectively.
	Activation energy (E_a)	At or above room temperature, the dark conductivity E_a is roughly equal to the energy difference between the conduction band edge, E_c, and the Fermi level, E_f.
	Density of defect states (N_s)	The density of defect-induced energy states per unit volume per eV within the energy gap.
Optical	Refractive index (n)	A quantity that is wavelength dependent and is often determined by $n = (1 + R^{1/2})/(1 - R^{1/2})$, where R is the reflectance.
	Optical bandgap (E_g)	An energy that is used to characterize the optical absorption edge.
	Photoconductivity (σ_l)	The ratio of photo-induced current density to electric field.
	Urbach energy (E_u)	At photon energies just below E_g, the absorption varies exponentially with energy (Urbach tail). The exponential slope of this sub-bandgap absorption region is E_u.
Physical	Hydrogen content (C_H)	The content of bonded hydrogen.
	Bonding configurations	The way atoms are bonded, such as SiH (silicon monohydride), SiH_2 (silicon dihydride), SiH_3, $(SiH_2)_n$, etc.
	Density and porosity	These quantities depend on alloy compositions and voids.

Table 2-1 Important Material Characteristics for a-Si:H Alloys (Continued)

Group	Characteristic	Significance
Physical	Surface morphology	Poor quality a-Si:H films has high density of bumps on the surface. May be influenced by columnar microstructure.
	Spacial distribution of elements	Bonding and compositional inhomogeneities.
	Intrinsic stress	Amorphous films are deposited either under compression or tension, depending on deposition techniques and conditions.

conductivity (σ_l), dark conductivity (σ_d), and one-sun photosensitivity (σ_l/σ_d). These values are the easiest properties of a-Si:H to measure and are available for almost all of the deposition methods discussed in this book.

2.1 DEFECTS IN AMORPHOUS SILICON-BASED ALLOYS

Defects in semiconductors, acting as traps and recombination centers for separated charges (electrons and holes), have strong influences on the electronic properties of the material. Amorphous semiconductors, unlike crystalline materials, have no clearly defined ideal structure and ideal properties. The only structural similarity between crystalline silicon and a-Si:H is that they both consist of groups of four tetrahedrally bonded atoms. In a-Si:H, these tetrahedras are randomly oriented and thus band structure theory based on Bloch waves is inapplicable [Smith 1978]. For a-Si:H and a-Ge:H, measurements suggest randomness in the dihedral angle or perhaps in coordinates corresponding to the fourth-bonded neighbors [Lannin 1988]. A *defect-free* continuous random network of a-Si:H may be possible in principle [Phillips 1987], but it has not been produced in practice. The composition, local bonding, and structure of an amorphous material vary with deposition condition and sample treatment. Defects common to crystalline solids, such as dislocations and grain boundaries, are not found in amorphous materials. However, because the semiconducting properties of amorphous materials depend on the short- (< 1 nm) and medium-range (1 to 10 nm) orders, deviations from such orders do create defects. a-Si needs monovalent atoms, such as hydrogen and fluorine, that form single-bonding alloys with silicon to relieve lattice strain [see, for example, Adler 1978 and Janai et al. 1985] and passivate dangling bonds [see, for example, Pankove et al. 1978 and Madan et al. 1979]. Although Si-F has a higher bonding energy (670 J/mole) than Si-H (377 J/mole) and F can also passivate dangling bonds, it is believed that a-Si:F or a-Si:H:F with more than 1 atomic percent (at.%) F do not have good electronic properties because fluorine cannot relieve strain in the a-Si network. It is not clear why the number, about 10^{22} cm^{-3}, of hydrogen atoms of a-Si:H alloys with good electronic quality is usually more than two orders of magnitudes higher than the number of dangling bonds, about 10^{19} cm^{-3}, measured by electron spin resonance in amorphous silicon without hydrogen. One explanation is that, instead of passivating dangling bonds, most of the singly-coordinated hydrogen atoms in a-Si:H are needed for reducing the intrinsic stress of the random network. The optical bandgap, E_g, of a-Si without hydrogen is about 1.5 eV. Alloying with hydrogen increases E_g because Si-H bonding energy (3.4 eV) is higher than the Si-Si bonding energy (2.2 eV). For a given H content, the more Si bonds are terminated by hydrogen, the higher is the E_g. Since hydrogen reduces the disorder in a-Si:H, we see that the less disorder, the higher the E_g for a given hydrogen content [Maley and Lannin 1987].

Defects in a-Si:H alloys include both positional and compositional disorders. The lack of a long-range order in a-Si:H inevitably causes distortions to the tetrahedral silicon network. This intrinsic disorder can be expressed in terms of bond angle and bond length deviations, and it is widely believed that the bond angle disorder is responsible for the band tail states in the optical gap. Specific defects or point defects, such as dangling bonds and impurities, are believed to cause deep (near midgap) defect states within the optical gap (E_g). The role of large bond angle distortions and their effects on the formation and stability of band edge and gap defect states are still not well understood [Lannin 1988]. In addition to electronic defects, large intrinsic mechanical stresses exist in most a-Si:H films [see, for example, Kurtz et al. 1986 and Stevens and Johnson, 1992]. Such stresses can cause adhesion problems for thick films and may have deleterious effects on certain electronic properties, such as dangling bond creation [Stutzmann 1985b].

2.1.1 Microstructural Defects

The structure on a scale of 10 nm down to the atomic level is generally referred to as the *microstructure* of the amorphous film. This concept of microstructure includes such aspects as the tetrahedral network, hydrogen bonding configurations [SiH, SiH_2, SiH_3, and polysilane ($(SiH_2)_n$) groups], multivacancies (up to three missing atoms), internal surfaces associated with microvoids, density fluctuations, bonded hydrogen distribution (clustered vs. dispersed), and unbonded hydrogen distribution (isolated and bulk molecular hydrogen). Features such as columnar structure, void volume fraction, void size distribution, void shape, and heterostructure ("islands" with low hydrogen concentration and "tissues" with high hydrogen concentration), although of micrometer scale, are also often included in the concept of microstructure. Certain aspects of the microstructure of a-Si:H films are desirable and others are undesirable. For example, we know that the following attributes of microstructure are associated with poor electronic material properties such as photoconductivity, dark conductivity, photosensitivity, ESR spin density, and Urbach energy: (1) columnar structure [Knights 1979, Knights et al. 1979a, and Drevillon et al. 1983], (2) heterostructure seen as islands and tissue of different density by transmission electron microscopy, especially for silicon-germanium alloys [Mackenzie et al. 1985a], (3) microvoids or density fluctuations [Mahan et al. 1987a], (4) polyhydride bonding [Knight and Lujan 1979, Hirose 1981, Ichimura et al. 1985, Meassen et al. 1987, Mahan et al. 1987a], and (5) clustered hydrogen [Shimizu et al. 1983]. Many of these attributes are interrelated, such as dihydride content and void fraction [Meassen et al. 1987a], clustered hydrogen and dihydride bonds [Schiff 1989], and clustered hydrogen and void fraction. Another aspect of poor structure is the presence of weak Si-Si bonds. For a-Si:H films deposited from silane, strained or weak Si-Si bonds may constitute 6-8% of all Si-Si bonds [Collins and Cavese 1988]. Breaking such bonds in films as a result of environmental factors, such as high temperature or light, deteriorates film quality because of an increase in dangling bond density; and such bond breaking may be the cause for the light- and thermally-induced instabilities [Stutzmann et al. 1985].

What affects the microstructure and what can be done to improve the microstructure? Many factors during deposition affect the microstructure. We only mention a few here: microparticles in the plasma, ion and electron bombardments of the growing film, presence of high-sticking-coefficient radicals, deposition temperature and pressure, feed-gas depletion, and doping species and the amount of such dopants.

It appears that microparticles as small as 2 nm in size result in film voids and poor structure and can have severly deleterious effects on the transport properties of a-Si:H films. Light-scattering experiments have shown microparticulates even when no dust can be seen by unaided visual inspection of the plasma deposition chamber. The microparticles in glow discharge are caused by gas-phase nucleations of ions and neutral radicals. Ion nucleation can be severe in low-field regions within the reactor if ions remain in such regions for a sufficient time to grow into microparticulates. Neutral nucleation is severe in high-field or high-gas-concentration regions of the plasma due to the higher densities of neutral radicals in these regions [Gallagher 1988a]. Plasma confinement can reduce high field regions, thereby reducing neutral nucleation. A short electrode distance (1-2 cm) minimizes ion nucleation. The amount of dihydride bonding in a-Si:H and a-SiGe:H has been correlated to the amount of plasma polymerization during growth [Slobodin et al. 1986]. The microstructure of a-Si:H may also be affected by the substrate. For example, Knights and Lujan [1979] reported that surface structures in aluminum-foil substrates and in some carbon films are found to act as either barrier to island growth or as preferred nucleation sites with the result that the interstitial regions are aligned parallel to these structures.

Whereas ion bombardment in glow discharge with ion energies below the atomic displacement energy (~50 eV for Si) can have positive effects on film quality, high-energy ion bombardment has detrimental effects. There have been many reports of the benefit of negative substrate bias that would increase ion bombardment, giving improved electronic properties [see, for example, Knights et al. 1979]. One possible explanation is that the low energy ion bombardment breaks up polymeric chains during growth and increases the percentage of monohydride bonds. On the other hand, the better properties (including higher σ_l and σ_l/σ_d, lower E_u, fewer SiH_2 bonds, and lower defect density) achieved by glow discharge with triode geometries as compared to diode geometry for otherwise equal conditions indicate the deleterious effects of high energy ion bombardment [Ichimura et al. 1985, Matsuda et al. 1985, and Tsai et al. 1987]. Deleterious ion bombardment is higher for inert-gas-diluted silane than for undiluted silane. The defect density and microstructural heterogeneity of a-SiH films increases with an increase in the atomic weight of the diluent gas for glow discharge from silane and inert gas mixtures [Knights et al. 1981]. The absence of ion bombardment in photochemical vapor deposition (photo-CVD) is believed by some to be one of the advantages of that method over glow discharge

[Konagai 1987]. On the other hand, photo-CVD does not have the benefit of low energy ion bombardment.

Radicals with high sticking coefficients (surface reaction probabilities) can acerbate both microscopic and macroscopic shadowing that results in void formation in a-Si:H films and the burial of bonded hydrogen before hydrogen elimination can occur [Kushner 1986]. SiH_3 radicals have a relatively low sticking coefficient of 0.2 and high surface migration; as a result, the SiH_3 deposition precursor produces good-quality films [Gallagher et al. 1989]. SiH_2 and SiH radicals have high sticking coefficients and produce poor quality material by physical-vapor-deposition-like growth [Tsai et al. 1987]. Surface diffusion of deposition precursors such as SiH_3 is believed to be the dominant physical phenomenon responsible for the bulk hydrogen content of a-Si:H films [Reimer et al. 1987]. Silane depletion during deposition reduces the availability of SiH_3 and increases the relative abundance of other radicals with higher sticking coefficients; it therefore results in poorer film quality [Gallagher 1988]. The addition of dopants affects the microstructure of a-Si:H films, with phosphorous causing larger H clusters to form and boron reducing H clustering [Gleason et al. 1987a].

Increasing the substrate temperature for glow discharge deposition decreases the amount of clustered hydrogen in a-Si:H films and increases the amount of dispersed hydrogen [Shimizu et al. 1983]. Hydrogen clusters of 5-7 atoms are found in films prepared at 270-324°C, and lower-temperature films have even larger hydrogen clusters [Gleason et al. 1987a]. Increasing the deposition temperature decreases the amount of SiH_2 bonding and decreases the amount of microvoids. For a-Si:H films produced by radio frequency (RF) glow discharge from silane, the optimum hydrogen content seems to be between 8 and 12 at.%. For a-Si:H films with hydrogen contents more than about 12 at.%, most of the additional hydrogen is incorporated as SiH_2 bonds that increase the void fraction and do not increase the optical bandgap [Meassen et al. 1987]. Hydrogenation affects the optical bandgap only indirectly by reducing disorder and thus removing states from the top of the valance band [Adler 1984].

2.1.2 Electronic Defect States

Distorted or deformed bonds in a-Si:H, where the bond angle deviates from the normal cubic tetrahedral 109.5° angle, can cause localized band-tail states. Threefold-coordinated silicon with dangling (broken) bonds give rise to localized defect states in the energy gap of the semiconductor. These mid-gap defect states can serve as recombination centers that degrade the electronic properties of the semiconductor. Fivefold-coordinated silicon with an floating bond has also been proposed as a possible defect center for a-Si:H [Pantelides 1986, Fedders and Carlson 1988]. However, there seems to be a lack of evidence supporting such a theory. The dangling bonds in a-Si:H can be positively charged when unoccupied (sp^2), neutral when singly occupied (sp^3), and negatively charged when doubly occupied (s^2p^3). Neutral dangling bonds (D^o) can be measured by electron spin resonance (ESR). Although there have been discussions of positive and negative dangling bonds (D^+ and D^-) in the as-deposited or annealed state of bulk undoped a-Si:H [Branz 1990, Schumm et al. 1992], so far, there is little experimental evidence of the existence of them. The minimum value of dangling bond density in a-Si:H can be as low as 7×10^{14} $eV^{-1}cm^{-3}$ [McMahon 1991]. This density depends strongly on the preparation and the history of the sample. The low densities of midgap defect states in a-Si:H alloys makes it possible to control the Fermi level position by doping, which is necessary for semiconductor device fabrication.

A commonly used band model of a-Si:H is shown in **Fig. 2.1.2-1.** Unlike crystalline semiconductors, amorphous semiconductors have a continuous distribution of defect states in the energy gap. For a-Si:H, these defects states are reduced by hydrogenation. The density of states of a-Si:H at the Fermi level can be lower than 10^{15} $eV^{-1}cm^{-3}$, that is less than one dangling bond defect per 10^7 atoms. Because the localized states and extended states cannot coexist at a given energy, there are sharp divisions between them that form the valence and conduction band edges, E_v and E_c. Because these band edges separate the high-mobility regions in the conduction and valence bands from the low-mobility localized states where the only allowed conduction mechanism is thermally activated hopping in tail states [Monroe 1985, Fritzsche and Pollak 1990], the energy gap in a-Si:H is actually a mobility gap. Hopping is defined as tunneling of electrons from localized states into adjacent localized states. The energy difference between these states is compensated by the emission or absorption of phonons. This mode of conduction is not important for good quality a-Si:H at room temperature or above. The Fermi level position, E_f, for undoped a-Si:H is usually slightly closer to E_c than to E_v.

The exponentially decreasing tails of states near the valence and conduction band edges are usually described by a characteristic energy, E_u. E_u is often called the Urbach energy because the exponential

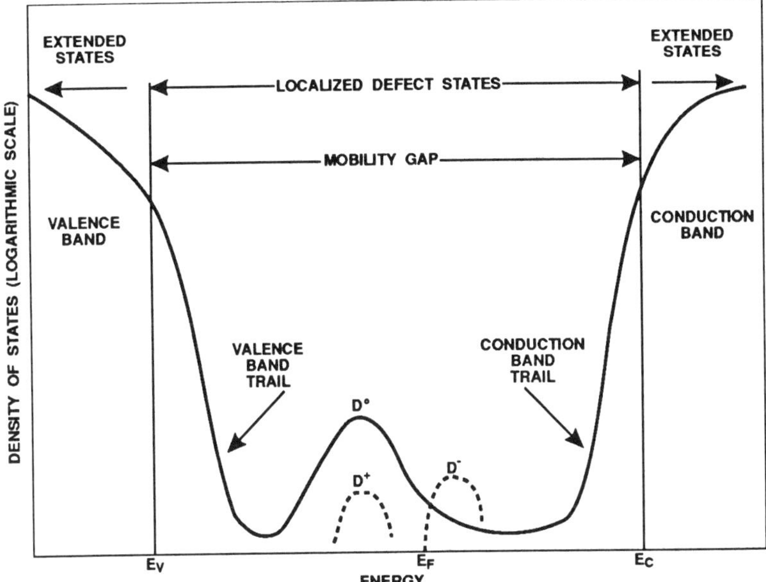

Figure 2.1.2-1 Electronic density of states in the mobility gap of a-Si:H

absorption edges were first observed by Urbach in the absorption spectrum of AgBr [Urbach 1953, Ley 1984]. The valence band tail characteristic energy of a-Si:H is usually determined by $d(h\nu)/d(\ln\alpha)$ from the straight line region of the sub-bandgap optical absorption coefficient α (in logarithmic scale) vs. photon energy ($h\nu$) plot. The smaller the value of E_u, the better the film quality. For a-Si:H, E_u for the valence band tail is larger than 42 meV, and E_u for the conduction band tail is usually smaller than 35 meV [Winer et al. 1988, Aljishi et al. 1990]. This asymmetry in E_u values of the band tails is because the valence band tail is more susceptible to disorder than the conduction band tail [Allan and Joannopoulos 1984].

The density and distribution of dangling-bond defects in a given sample of hydrogenated amorphous silicon depends on thermal history and on electron and hole densities. As a consequence, the defect density of a-Si:H changes during their use in devices. Hata and Wagner [1992] have developed a defect model that accounts for the effects of the temperature of film growth; the rate of film growth; the film thickness; light-soaking intensity, time, and temperature; and the temperature and duration of thermal annealing. This model is based an the assumption of a limited pool of defects with a Gaussian distribution of thermal annealing energies. They demonstrate the applicability of the model by many examples drawn from measurements of dark and photo-conductivities, transmission spectroscopy, and subgap optical absorption.

2.2 CHARACTERISTICS OF HYDROGENATED AMORPHOUS SILICON

The properties of a-Si:H have been well characterized and extensively reviewed [Pankove 1984, Joannopoulos and Lucovsky 1984, Brodsky 1985, Madan and Shaw 1988, Street 1991]. Selected room-temperature characteristics of device-quality, undoped a-Si:H, deposited by glow discharge decomposition of silane, with an E_g value of 1.7 to 1.8 eV and a hydrogen content of about 10 at.%, are listed in **Table 2.2-1.**

2.2.1 Material Properties and Depositions

Many of the parameters listed in **Table 2.2-1** are interrelated, optimizing one parameter often improves several other parameters also. For example, when optimizing processing conditions in a new a-Si:H deposition system, one usually begins by minimizing the SiH_2/SiH ratio detected by infrared transmission spectroscopy (§ 3.1.5). This ratio does not degrade with light soaking, and it is a necessary requirement for good quality a-Si:H to have a SiH_2/SiH ratio that is almost undetectable by IR absorption. Further optimizations of a-Si:H properties are complicated by the instability problem of the electrical properties (§ 2.6) and the fact that the values of some a-Si:H characteristics depend on the film thickness (§ 2.2.2) and the measurement methods used. Traditionally, as-deposited or annealed values of σ_l, σ_d, E_a (§ 3.3.1), L_h (either by surface photovoltage or photocarrier grating, § 3.1) and E_u (by sub-bandgap optical absorption, § 3.1.4) are used for optimization of a-Si:H films with desired E_g and C_H (§ 3.1.5). Although optimizing the stabilized (degraded, metastable) properties of a-Si:H is more relevant to device applications, it seems to be usually the case that a-Si:H films with optimized properties in the as-deposited state also have good properties in the metastable state.

The σ_d value of a-Si without hydrogen is usually higher than 10^{-7} S/cm. Adding hydrogen, either during deposition or after deposition [Tsuo et al. 1987a], reduces σ_d by passivating dangling-bond defects and reducing stress in the a-Si network. Only monohydride silicon-hydrogen

Table 2.2-1 Typical Room Temperature Characteristics of Device-Quality, Undoped a-Si:H

Property	Typical Values	Units
optical bandgap, E_g	1.7 to 1.8	eV
hydrogen content, C_H	8 to 15	at.%
refraction index, n, at 600 nm	~ 4.3	
electron drift mobility, μ_e	≥ 1	$cm^2V^{-1}s^{-1}$
hole drift mobility, μ_h	≥ 0.008	$cm^2V^{-1}s^{-1}$
electron lifetime, τ_e	$\geq 2 \times 10^{-7}$	s
hole lifetime, τ_h	$\geq 10^{-6}$	s
$\mu\tau$ (electron)	$\geq 2 \times 10^{-7}$	cm^2V^{-1}
$\mu\tau$ (hole)	$\geq 10^{-8}$	cm^2V^{-1}
hole diffusion length, L_h	0.3	μm
AM1 photoconductivity, σ_l	5×10^{-5} to 10^{-4}	S/cm
dark conductivity, σ_d	10^{-11} to 10^{-10}	S/cm
σ_d activation energy, E_a	0.7 to 0.9	eV
AM1 photosensitivity, σ_l/σ_d	$\geq 1 \times 10^6$	
valence band tail slope, E_u	42 to 50	meV
conduction band tail slope	~ 25	meV
ESR spin density	7×10^{14} to 10^{16}	$eV^{-1}cm^{-3}$
density of states at E_f	5×10^{14}	$eV^{-1}cm^{-3}$
SiH_2/SiH	~ 0	
microvoids (d > a few Å)	~ 0	
intrinsic stresses	≤ 400	MPa

(SiH) bonds are observed in the infrared absorption spectra of device-quality a-Si:H films, whereas dihydride, trihydride, and polymeric bonds are seen in poor quality low-temperature-deposited films with a high hydrogen content. Hydrogenation with predominantly SiH bonds improves the photosensitivity of a-Si:H from negligible to over six orders of magnitude. A σ_d value higher than 10^{-10} S/cm for a-Si:H with a 1.75 eV E_g usually indicates insufficient dangling bond passivation or impurity contamination. It is important to note that reported conductivity values in thin films are not always reliable because they can be affected by band bending caused by surface defects and ambient conditions [see, for example, Fritzsche 1980 and Ye et al. 1988]. For some a-Si:H materials, σ_l and σ_l/σ_d values can also be enhanced by impurities [Hishikawa et al. 1990].

High quality a-Si:H alloys are usually deposited by processes, such as glow discharge, that allow simultaneous etching of the deposited film during the deposition process. Etching provides a mechanism for selecting low energy configurations of the microstructure at low temperatures by preferentially eliminating energetically unfavorable configurations. In this non-equilibrium process of a-Si:H growth, etching lowers the temperature required for annealing out weak or strained bonds to yield device-quality material [Tsai et al. 1990]. Scanning tunneling electron microscopy (§ 3.6.2) of a growing a-Si:H film surface has shown that some atomically flat regions occur, even at 400 nm film thickness [Gallagher et al. 1992]. Deuteron magnetic resonance measurements (§ 3.2.2) have shown that high quality a-Si:H films have negligible content of microvoids with a diameter larger than a few Angstroms.

Good quality a-Si:H films typically contain less than 5 x 10^{18} cm^{-3} oxygen, 5 x 10^{18} cm^{-3} carbon, and 5 x 10^{17} cm^{-3} nitrogen. Normally, for depositing high quality a-Si:H, the deposition system leak rate should be less than 1 x 10^{-5} torr·L/s. For research systems, leak rates on the order of 10^{-7} torr·L/s are commonly used. For example, the impurity levels of 6 x 10^{18} cm^{-3} oxygen, 6 x 10^{17} cm^{-3} carbon, and 1 x 10^{17} cm^{-3} nitrogen were reported by Miyachi et al. [1987] for a reactor system using a 10^{-8}-torr background pressure with an outgassing rate of 2 x 10^{-7} torr·L/s. At a deposition rate of 0.3 nm/s from disilane they made a-Si:H films with values of σ_l = 1.2 x 10^{-5} S/cm, σ_d = 1.7 x 10^{-12} S/cm, σ_l/σ_d = nearly 1 x 10^7, space charge density = 1 x 10^{15} cm^{-3}, and minority carrier (hole) diffusion length = 1 μm. The properties of a-Si:H are strongly dependent on deposition parameters, such as deposition rate (**Fig. 2.2.1-1**). This figure is for ~1.83 eV a-Si:H films deposited by DC glow discharge at 85°C with 4:1 H_2 dilution. The absorption and Urbach energy data are from PDS measurements.

The relationship between E_g and the total bonded-hydrogen content, C_H, the hydrogen bonding configuration, and the molecular hydrogen content of a-Si:H depends on the deposition method and condition used. For example, a-Si:H deposited by a remote hydrogen plasma [§ 9.1] has 1 at.% molecular hydrogen content which is a factor of 10 higher than that for glow discharge material. This high molecular hydrogen content is probably caused by a number of large voids in the material [Ready et al. 1990]. **Fig. 2.2.1-2** shows the relationship between C_H as derived by infrared absorption (§ 3.1.5), and E_g as derived by Tauc plot (§ 3.1.5), of glow discharge-deposited a-Si:H films by conventional RF-diode glow discharge-decomposition of undiluted SiH_4 using low RF power at 250°C and films that have been dehydrogenated and rehydrogenated. The solid dots are for films with all or part of their hydrogen content removed by heating. The open dots are for films with hydrogen content reintroduced by low-energy ion implantation [Tsuo et

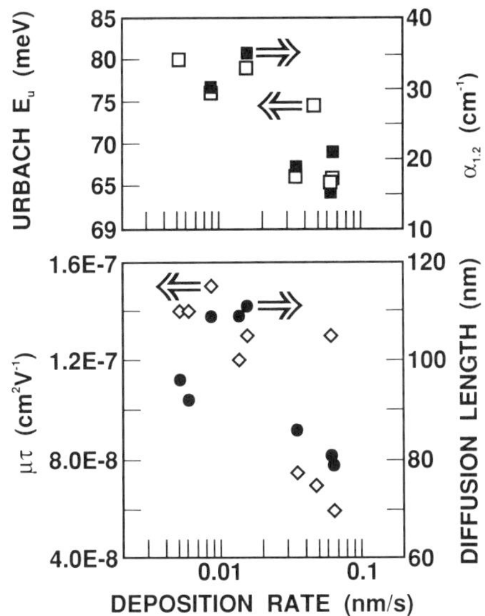

Figure 2.2.1-1 The dependence of ambipolar diffusion length, μτ-product, Urbach energy, and sub-bandgap absorption at 1.2 eV on the deposition rate [Catalano et al. 1991b].

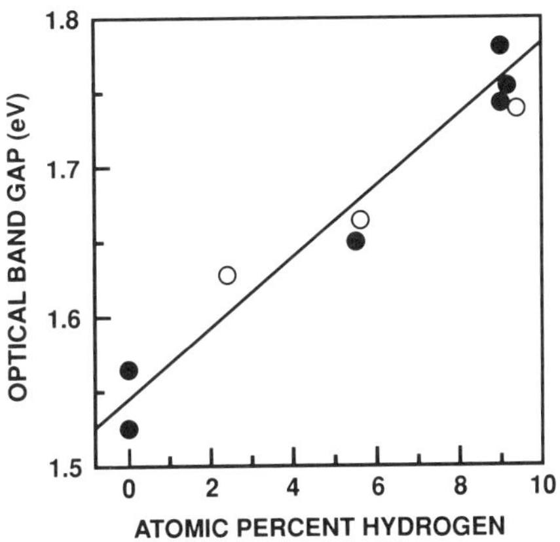

Figure 2.2.1-2. Effect of hydrogen content on the optical bandgap [Tsuo et al. 1987a].

al. 1987a].

The room temperature conductivity of a-Si:H can be changed by over 10 orders of magnitude by doping. The doping atoms, mostly boron for p-type doping and phosphorous for n-type doping, are electrically active (generate excess carriers) when they substitute for the fourfold coordinated Si atoms in the tetrahedral structure. When the doping atoms

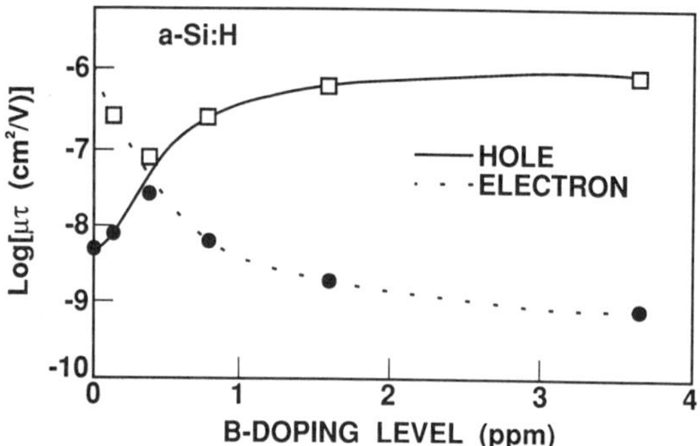

Figure 2.2.1-3 $\mu\tau$ products calculated from σ_l as a function of doping in a-Si:H. Solid and dashed lines represent the behavior of holes and electrons, respectively [Catalano 1989d].

reside interstitially or threefold coordinated in the Si network, they are not electrically active. The doping efficiency, the fraction of substitutional dopants, is low for a-Si:H alloys [Street 1991]. Doping changes the transport properties of electrons and holes, as shown in **Fig. 2.2.1-3**. The conductivity of doped a-Si:H remains thermally activated because the large band tail states and the low doping efficiency prevents the Fermi level from reaching the extended states. The E_a for n-type a-Si:H, about 0.15 eV for high levels of phosphorus doping, is smaller than the E_a for p-type a-Si:H, about 0.3 eV for high levels of boron doping, because of the narrower conduction bandtail.

Doping induces high levels of defect density in a-Si:H. The defect density increases with the square root of the dopant concentration in the gas phase for P, B, and As doping [Street 1991]. In phosphorous-doped, n-type a-Si:H, E_g is not affected by the doping amount, but in boron-doped, p-type a-Si:H, E_g decreases with increased doping level. In p-i-n solar cells with a p-type window layer, boron-doped a-$Si_{1-x}C_x$:H is usually used as the window layer to enhance the optical transmission.

2.2.2 Effects of Film Thickness on Measured Properties of a-Si:H

Many of the electrical and photoelectrical properties of a-Si:H depend on the film thickness [Ye et al. 1988, Catalano et al. 1990a & 1991a]. Usually, when the a-Si:H film thickness is less than 1 µm, the thicker the film the better the measured film properties. It is not clear whether this is due to surface band bending or due to film properties

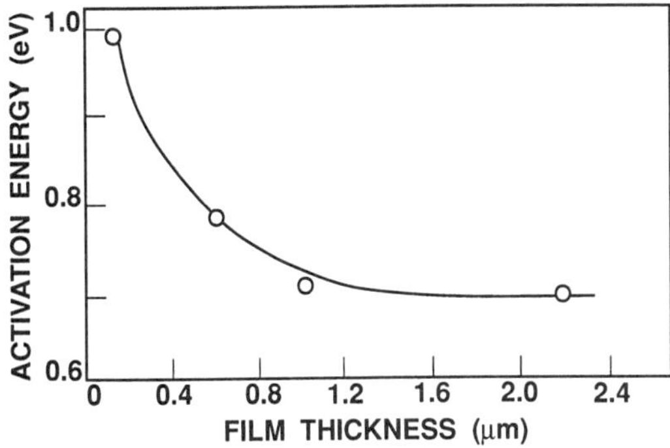

Figure 2.2.2-1 Dependance of the dark conductivity activation energy on thickness of undoped a-Si:H [Ye et al. 1988].

improving with deposition thickness. Ye et al. [1988] showed that the surface layer of undoped a-Si:H can be more photosensitive than the bulk and that the E_a decreases with increasing film thickness (**Fig. 2.2.2-1**). This thickness dependence of measured electrical and photoelectrical properties of a-Si:H alloys is different for different deposition conditions [Hishikawa et al. 1990].

Yang et al. [1990] found that σ_l and ambipolar diffusion length of a-Si:H deteriorate rapidly with decreasing film thickness, as shown in **Fig. 2.2.2-2**. In this figure, solid and dashed curves are calculated for surface recombination velocities of S = infinite, 100, and 1 cm/s. They attribute this to the effect of surface recombination. They recommend to use 1 µm or thicker samples and red light to obtain the photoelectronic bulk properties of a-Si:H and related materials. However, the need for thick specimens sometimes constitutes problems because of the long deposition times (deposition rates of 0.3 - 1 nm/s), and poor adhesion to substrates of glass, quartz, crystalline silicon at low substrate temperatures.

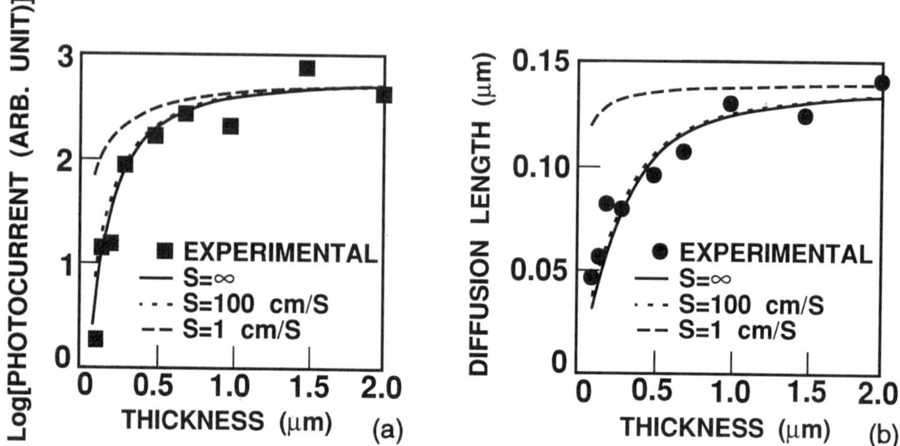

Figure 2.2.2-2 Thickness dependence of photocurrent (a) and ambipolar diffusion length (b) of a-Si:H measured with a He-Ne laser at about 600 W/m^2 [Catalano et al. 1991a].

Catalano et al. [1990a] have shown that the Urbach energy as measured by photothermal deflection spectroscopy (section 3.1.4) is a strong function of film thickness. He gives an example of films having an Urbach energy of 53 meV for a thickness of 4 μm that increases to 68

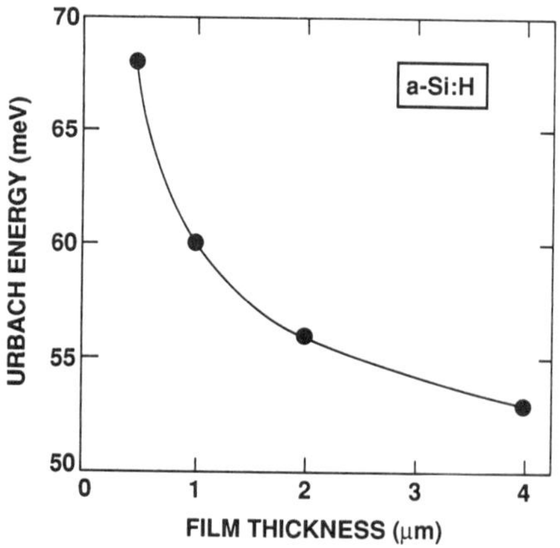

Figure 2.2.2-3 Urbach energy for a-Si:H measured by photothermal deflection spectroscopy as a function of film thickness [Catalano et al. 1990a].

meV for a thickness of 0.5 μm (See Figure **2.2.2-3**). There is a positive correlation between Urbach energy and defect density in the annealed state condition ($N_{s,o}$) of a-Si:H. $N_{s,o}$ increases from ~1 x 10^{15} cm^{-3} at 42 meV Urbach energy to 2 x 10^{16} cm^{-3} at 60 meV [Park et al. 1990].

In summary, it is better to use a-Si:H alloys with film thickness of 1 μm of more for film characterizations. When film thickness is less than 1 μm, it is important to compare films with similar thickness. Measurement techniques that are sensitive to surface properties should be avoided. For example, photothermal deflection spectroscopy is sensitive to surface effects even for film thickness of 4 μm, as shown in Fig. 2.2.2-3. Thus, the E_u measured by PDS is not a good film quality indicator.

2.3 CHARACTERISTICS OF HYDROGENATED AMORPHOUS SILICON GERMANIUM

Hydrogenated amorphous silicon germanium alloys (a-$Si_{1-x}Ge_x$:H or simply a-SiGe:H) are, so far, the only type of low-bandgap materials used in multiple-junction amorphous silicon solar cells. The reason for the interest in multijunction devices is their potentially higher conversion efficiencies than single-junction devices. For example, to achieve high efficiency in a-Si:H tandem (dual-junction) cell structures with the high-bandgap material being the standard 1.75 eV a-Si:H, the low-bandgap material should have an optical bandgap of less than 1.5 eV, preferably as low as 1.2 eV. This bandgap value is calculated on the basis of good photoelectronic properties for the low-bandgap material. However, in reality, the electronic properties of amorphous alloys deteriorate significantly as the optical bandgap is reduced from about 1.7 eV. Part of the reason for this is the dramatic increase of Ge dangling bonds when the Ge content is increased.

Although a-SiGe:H is currently the best low-bandgap material, it has much poorer electrical and photoelectric properties than a-Si:H (for example, reduced σ_l and photoluminescence, and increased N_s). The photoelectronic properties deteriorate with increasing Ge content, and the deterioration proceeds rapidly as the Ge content of the alloy exceeds 40% (corresponding to an E_g of approximately 1.5 eV) [Mackenzie et al. 1985a&b].

Some of the best reported characteristics for glow discharge-deposited a-SiGe:H or a-SiGe:H:F materials with E_g values of 1.5 eV are: σ_l = 3.0 x 10^{-4} S/cm [Tsuda et al. 1985], $\eta\mu\tau$ = 1 x 10^{-6} cm^2/V [Nakamura et al. 1982], σ_d = 1.4 x 10^{-10} S/cm [Shing 1987b], σ_l/σ_d = 1.8

x 10^4 [Shimizu 1985], Urbach parameter = 46 meV [Tsuda et al. 1987a&b] (43 meV for a 1.4 eV material reported by Guha [1988a]), ESR spin density = 6.5 x 10^{15} cm^{-3} [Tsuda et al. 1987a&b], and density of states at the Fermi level obtained by measuring space-charge-limited currents using n^+-i-n^+ structures = 2 x 10^{16} $eV^{-1}cm^{-3}$ (3 x 10^{16} $eV^{-1}cm^{-3}$ for a 1.4 eV a-SiGe:H:F material) [Guha 1988a].

The best a-Ge:H films with 1.01 eV E_g were made by DC glow discharge. They have E_a = 0.45 - 0.49 eV, σ_d = 10^{-4}-10^{-5} S/cm, E_u = 55 - 60 meV, and relatively strong IR absorption peaks at 1870 cm^{-1} (Ge-H bond stretching mode) and 560 cm^{-1} (H-bond bending mode) [Newton and Kritikson 1990].

Improvement of a-SiGe:H prepared from gas mixtures of silane, germane, and hydrogen by RF glow discharge has been attempted using ultrahigh-vacuum (UHV) (10^{-9}-torr base vacuum) deposition systems [Tsuda et al. 1987a,b]. Impurities in the resulting films were reduced: oxygen from 4 x 10^{19} to 1 x 10^{19}, nitrogen from 8 x 10^{18} to 4 x 10^{17}, and carbon from 4 x 10^{19} to 2 x 10^{18} cm^{-3} as compared to films deposited by the same researchers in a conventional glow discharge system. The photoluminescence was more than 5 times higher than that for conventional glow discharge films, suggesting that non-radiative recombinations and, possibly, the tail state density in the films were reduced by the reduction in impurities. The refractive indices were higher than that of the conventional glow discharge films over the bandgap range of 1.42 - 1.55 eV (e.g. 3.77 vs. 3.72 at 1.5 eV), indicating higher structural density. Other parameter comparisons between the 1.52 eV E_g UHV-deposited films and conventional films were: σ_l = 2.7 x 10^{-4} vs. 1.6 x 10^{-4} S/cm, E_u = 46 vs. 49 meV, ESR spin density = 6.5 x 10^{15} vs. 1.1 x 10^{16} cm^{-3}. Besides the ESR spin density, however, results as good or better had previously been achieved by other researchers without using a UHV system.

The question, then, is whether anything can be done to avoid this deterioration of photoelectronic properties with increasing germanium content. This section describes some of the attempts that researchers have made to accomplish this and their results.

2.3.1 Effects of Germanium Content

The optical bandgap is directly related to the Ge content of the alloy; as the Ge content, x, of a-$Si_{1-x}Ge_x$:H increases from zero to 100% ($0 \leq x \leq 1$), the optical band gap decreases monotonically from about 1.75 eV to 1.1 eV (see **Figure 2.3.1-1**). The E_g in eV is mathematically

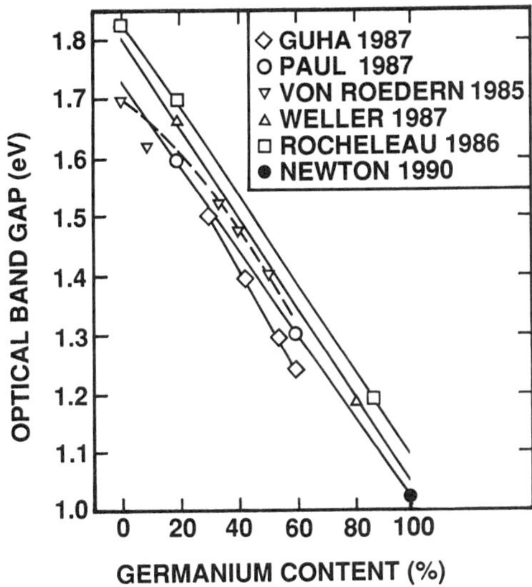

Figure 2.3.1-1. The effect of germanium content on the optical bandgap for a-SiGe:H(F) for various deposition methods and conditions as shown in Table 2.3.1-1

described as roughly 1.76 - 0.78x [Mackenzie et al. 1988, Chen et al. 1989]. The deposition parameters for the data in Figure 2.3.1-1 are listed in **Table 2.3.1-1**. E_g also depends on certain deposition conditions, such as temperature [Hegedus et al. 1988a]. The substrate temperature during deposition in turn affects the hydrogen content in the film; increased temperature results in lower hydrogen content and lower E_g. However, the H and Ge contents of a-SiGe:H films are not independent parameters [von Roedern et al. 1982, Mitchell et al. 1985]. Fortmann and Tu [1988] showed the interrelation between germanium content, temperature, hydrogen content, and optical bandgap. The bandgap is essentially independent of the deposition method, the feed-gas type, and the diluent [von Roedern et al. 1985, Rocheleau et al. 1986, Guha et al. 1987, Paul & Mackenzie 1987, Weller et al.1987, Hegedus et al, 1988a]. Thus, the differences in E_g that various researchers have observed for a given amount of Ge are attributable to variations in hydrogen content and in the determination of the optical bandgap (normally established from Tauc plots), rather than to the deposition method or feed-gas.

Most frequently, the various film characteristics are reported as a function of E_g rather than as a function of the germanium content, most likely because E_g is easily determined from the Tauc relationship, whereas accurate determination of the germanium content requires microprobe measurements (with ±1 at.% accuracy) [von Roedern et al. 1982, Mackenzie et al. 1985] or atomic emission spectroscopy (±3% accuracy) [Muramatsu et al. 1988].

The Urbach parameter, E_u, increases for films deposited by glow

Table 2.3.1-1. Deposition Parameters for Data in Figure 2.3.1-1.

Method	Electrode Configuration	Feed-gas	Diluent	Temp. (°C)	Reference
RFGD	Diode	$Si_2H_6/GeH_4/SiF_4$	H_2	225	Guha 1987
RFGD	Diode	SiF_4/GeF_4	H_2	300	Paul 1987
RFGD	Diode	SiH_4/GeH_4	None	320	von Roedern 1985
DCGD	Triode	SiH_4/GeH_4	H_2	NA	Weller 1987
Photo	---	Si_2H_6/GeH_4	H_2/He	240	Rocheleau 1986
DCGD	Diode	GeH_4	H_2/Ar	240/340	Newton 1990

RFGD=RF glow discharge, DCGD=DC glow discharge, Photo=Photo-CVD
Temp.= Substrate temperature, NA=Not available

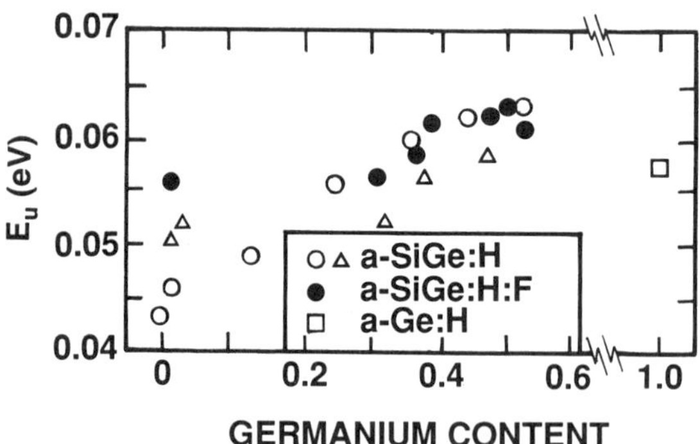

Figure 2.3.1-2 Effect of Germanium content on the Urbach parameter E_u for a-SiGe:H and a-SiGe:H:F [Mackenzie et al. 1985a (○,●), von Roedern et al. 1985 (Δ), Newton and Kritikson 1990 (□)].

discharge from about 43 meV for a-Si:H to 56 meV for a-SiGe:H with a 40% Ge content. It is about the same for a-SiGe:H and a-SiGe:H:F [von Roedern et al. 1985, Mackenzie et al. 1985a] (see Figure **2.3.1-2**). However, Guha et al. [1987] have reported a constant E_u of 50 meV for band gaps from 1.72 eV to 1.25 eV for a-SiGe:H:F films produced by RF glow discharge from a disilane/germane/ silicon tetrafluoride/hydrogen mixture. Rocheleau et al. [1988] obtained E_u of 44 and 46 meV for band gaps of 1.65 and 1.53 eV, respectively using Hg-sensitized photo-CVD of hydrogen diluted silane and germane. Different Urbach energy measurements cannot be directly compared unless the film thickness is reported.

Most researchers have found that σ_l decreases and σ_d increases as E_g decreases (with increasing Ge content), resulting in decreases in the photosensitivity with decreasing E_g (Figure **2.3.1-3**). Photosensitivity is the ratio of AM1.5 σ_l to σ_d. The logarithm of the photosensitivity decreases nearly linearly from 7 x 10^6 at 1.75 eV to 10^3 at 1.35 eV.

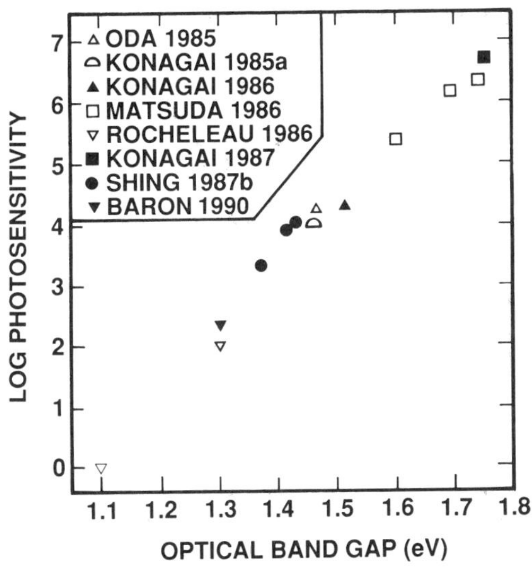

Figure 2.3.1-3. The effect of the optical bandgap on the best photosensitivities recorded for a-SiGe:H films produced by glow discharge and photo-CVD

Examples of the change in AM1.5 σ_l with E_g are from 2 x 10^{-4} S/cm for a-Si:H to 2 x 10^{-5} S/cm for 1.5 eV a-SiGe:H; the corresponding change in σ_d is from 5 x 10^{-11} S/cm to 2 x 10^{-9} S/cm [Watanabe et al. 1987]. However, there are a few exceptions. Dalal [1985] and Konagai et al. [1985] showed an essentially constant σ_d versus E_g for a-SiGe:H films produced by glow discharge and photochemical vapor deposition. The best photosensitivities obtained to date are 1 x 10^7 for a-Si:H [Miyashi et al., 1987] and 1 x 10^4 [Shing et al. 1987b] for 1.42 eV a-SiGe:H.

The photoconductivity of a-SiGe:H is also a strong function of the hydrogen content. Fortmann [1990] showed a decrease of σ_l from 1.1 x 10^{-6} S/cm at 3 at.% H to 3 x 10^{-7} S/cm at 9 at.% H for a film with 60 at.% Ge. In the best films produced by glow discharge, the $\eta\mu\tau$ product for electrons decreases from about 1 x 10^{-5} cm^2 V^{-1} for a-Si:H to

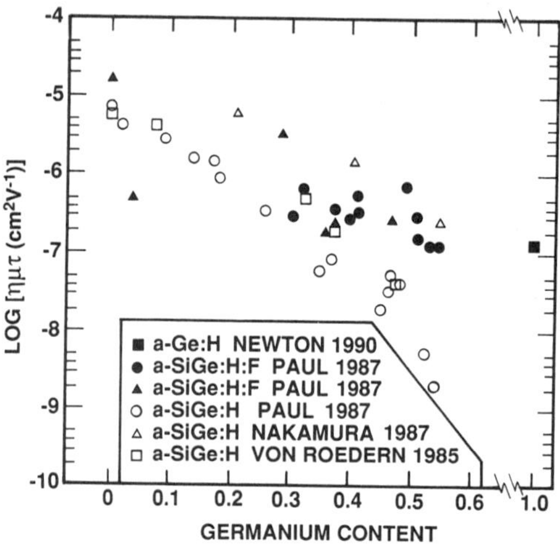

Figure 2.3.1-4. The effect of germanium content on electron $\eta\mu\tau$ product for a-SiGe:H and a-SiGe:H:F produced by glow discharge.

1 x 10^{-6} cm^2 V^{-1} for a-SiGe:H and a-SiGe:H:F with a 40% Ge content (see Figure **2.3.1-4**). In Figure 2.3.1-4, the data from Paul and Mackenzie and Nakamura et al. are for a photon energy of 1.96 eV (630 nm) and a flux of 10^{15} $cm^{-2}s^{-1}$ [Paul and Mackenzie 1987, Nakamura et al. 1982]. Data from von Roedern et al. [1985] were measured at 600 nm and at a flux of 3.5 x 10^{15} $cm^{-2}s^{-1}$. The $\mu\tau$ product for holes is about 1/3 that for electrons. Films deposited at higher temperatures (e.g., 270°C

instead of 200°C) have superior electron transport and show a relatively small change in $\mu\tau$ in the E_g range of 1.3 to 1.5 eV compared to that in the 1.5 to 1.65 eV range [Fortmann & Tu 1988].

For a-SiGe:H,F, the diffusion length is about 50 nm for material having 1.1 - 1.5 eV E_g. The diffusion length is correlated with E_u: the log(diffusion length) decreases linearly with increasing E_u, from 0.1 μm at 50 meV to 0.05 μm at about 70 meV and then remains constant. The diffusion length is a weak function of the defect density, decreasing from

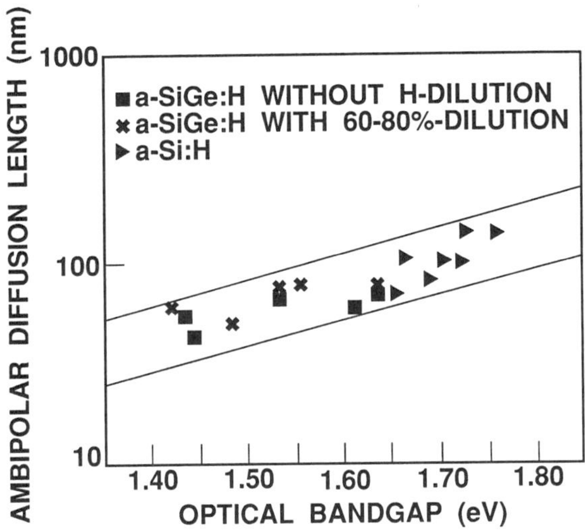

Figure 2.3.1-5 L_d vs. E_g for a-SiGe:H deposited with and without hydrogen dilution and for a-Si:H deposited with different T_s.

0.1 μm at 10^{16} cm^{-3} to 0.03 μm at 10^{18} cm^{-3} [Liu et al. 1990]. Tsuo et al. [1991a] showed that the ambipolar diffusion length, measured by SSPG (§ 3.3), increases with increasing E_g for a-SiGe:H deposited with and without hydrogen dilution. They also showed that high-temperature-deposited a-Si:H films, with E_g down to 1.65 eV, have L_d comparable to a-SiGe:H with similar E_g **(Fig. 2.3.1-5).**

Mitchell et al. [1985] found for a-SiGe:H material fabricated by RF glow discharge from a silane-germane-hydrogen mixture that the total H content decreases from 10 at.% for films with no Ge to 4 at.% for films with 100% Ge content; the dihydride content rapidly increases in

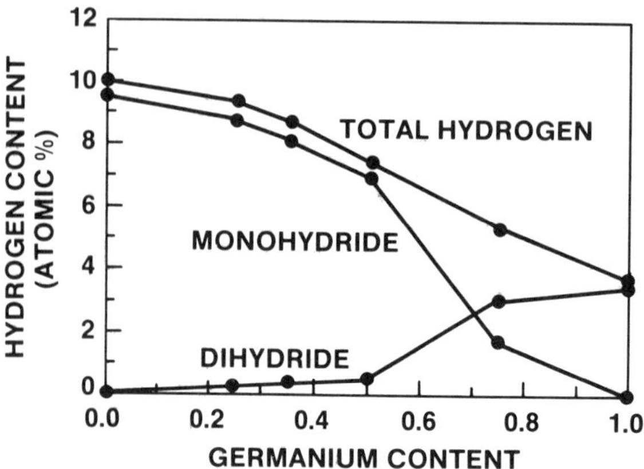

Figure 2.3.1-6. Hydrogen content as a function of germanium content for a-SiGe:H prepared by glow discharge from silane-germane mixtures [Mitchell et al. 1985].

films having more than 50% Ge content, and the monohydride content decreases even more rapidly above that limit, falling to zero at 100% Ge content (see **Figure 2.3.1-6**). Likewise, Shimizu, T., et al. [1986] showed for a-SiGe:H films produced by glow discharge an increase in the silicon-dihydride content together with a monotonic reduction in the total H content, clustered as well as dispersed, with increasing Ge content, i.e., from 11 at.% total hydrogen for films with no Ge to 4 at.% total hydrogen for films with 100% Ge content (Fig. **2.3.1-7**). In contrast, Hegedus et al. [1988a] observed for films by photo-CVD a maximum in the dihydride/monohydride ratio at 60% germanium. They also found the amount of dihydride to decrease with increasing deposition temperature (160-250°C) and being higher for films from disilane-germane than from silane-germane feed-gases, and independent of hydrogen dilution up to 90%. Rocheleau et al. [1988] showed 20% silicon dihydride bonding at E_g = 1.40 eV and 10% at E_g = 1.20 eV. They believe that microstructure indicated by dihydride bonding is not the only factor affecting film properties at low bandgaps, but that deep-level defect states from Ge dangling bonds play a role.

One reason for this reduction in the total H content with increasing Ge content in the alloy is that the hydrogen attaches preferentially to silicon rather than to germanium [Paul, D., et al. 1980, von Roedern et

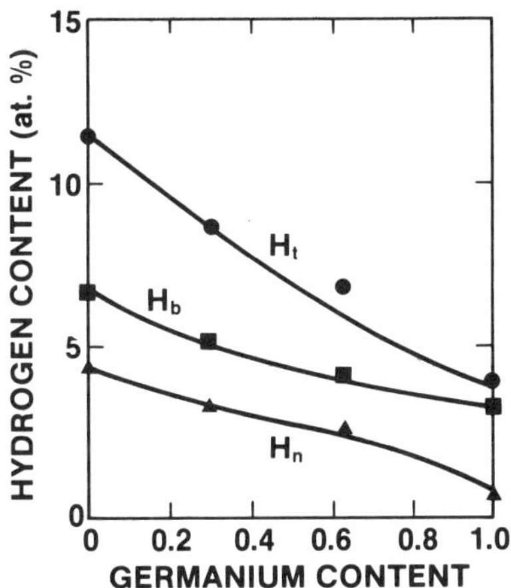

Figure 2.3.1-7. The effect of germanium content on the hydrogen content, total (H_t), clustered (H_b), and dispersed (H_n), for a-SiGe:H by glow discharge [Shimizu, T., et al. 1986].

al. 1982, Mackenzie et al. 1985, Ichimura et al. 1985, Guha 1985]. Paul, D. K., et al. [1980] reported preferential H attachment to Si of about 10 (in a range of 8.2 to 20) for a-SiGe:H at various Ge contents. Guha [1985] found a value of 1.5 for a-SiGe:H:F with an E_g of 1.5 eV, and Guha et al. [1987] and Guha [1988] reported constant values in a range from 2.8 to 4 for a-SiGe:H:F with Ge contents from 30% to 57%. Hegedus et al. [1988] showed the preferential attachment of hydrogen to silicon to increase with increasing germanium content in a film from 1 at 20% Ge to 4 at 70% Ge. But both Si-H and Ge-H bondings decrease with increasing germanium content [Sato et al. 1988].

As a result of the preferential attachment of hydrogen to silicon rather than to germanium and the weakness of the Ge-H bonds, relatively more defects in a-SiGe:H alloys are Ge dangling bonds than Si dangling bonds [Paul et al. 1981]. The ESR spin density for a-SiGe:H films fabricated by glow discharge increases with increasing Ge content, from about 5 x 10^{-16} cm^{-3} in films with zero Ge to about 2 x 10^{-18} cm^{-3} in films with 100% Ge content [Shimizu et al. 1986], which is consistent with the results by Hegedus et al. [1988a] (see **Fig. 2.3.1-8**). Stutzmann et al. [1989] observed that, when the Ge content in a-SiGe:H is more than 50%, the Ge dangling bond density in high-quality films deposited by RF glow discharge decomposition of SiH_4 and GeH_4 is more than 5 x 10^{17}

cm^{-3}, whereas the Si dangling bond density in the same films is less than 10^{15} cm^{-3}.

Slobodin et al. [1986] believe fluorine to be a dangling-bond passivator because of its greater bond strength in comparison to H. However, Okada et al. [1986] could not see any Ge-F bonds in a-SiGe:D:F, although they did observe Si-F and Si-F_2 bonds. It must be noted that most a-SiGe:H:F films contain less than 1 at.% F. Thus, any improvement in film properties achieved with F-containing gases is more likely attributed to F etching of weak Si-Si bonds and subsequent changes in the growth kinetics, rather than to F acting as a significant dangling-bond passivator.

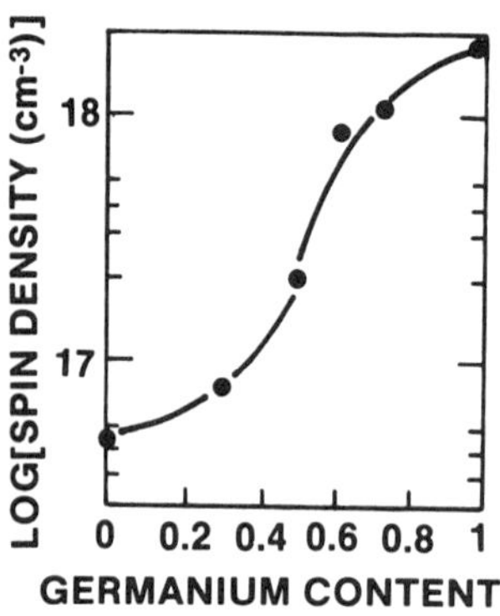

Figure 2.3.1-8. The effect of the germanium content on the ESR spin density for glow discharge [Shimizu, T., et al. 1986].

During the glow discharge deposition of a-SiGe:H, there is a strong tendency for Ge radicals to preferential incorporate into the film [Street & Winer 1989]. Part of the reason is because GeH_4 gas has a lower bond dissociation energy and thus decomposes in a plasma by electron impact dissociation more readily than SiH_4. Tsuo et al. [1991a] found that the ratio of film Ge concentration ([Ge] / ([Ge] + [Si]) and feed-gas GeH_4 concentration (GeH_4 / (SiH_4 + GeH_4)) is 2.5 for a feed-gas mixture of pure GeH_4 and SiH_4 with 20% GeH_4 concentration. This ratio decreases with increasing GeH_4 concentration: it is 3.2 and 1.6 at 10% and 50% GeH_4 concentrations, respectively.

2.3.2 Material Properties and Microstructure

Amorphous silicon-germanium alloys have a significant degree of preferential Ge-Ge bonding in excess of that predicted for an ideal alloy in alloys with relatively low (0.1-0.5) Ge concentration. Ge-Ge clustering reduces transport properties such as the $\mu\tau$ product and diffusion length. Even Ge-Ge clustering expected for a totally random alloy would lead to a significant deterioration in transport properties of a-SiGe:H relative to those for a-Si:H [Catalano et al. 1989].

Germanium-rich a-SiGe:H alloys have a lower density network structure than that for a-Si:H. This is attributed to the highly reactive surface of the a-SiGe:H alloy and the subsequently suppressed surface diffusion of adsorbed radicals [Matsuda et al. 1986a,b and Yang 1991]. Even small fraction of germane added to silane in glow discharge deposition increases the Si-H_2 bonding and increases the microvoid fraction [Yang, L., et al. 1991]. The question is whether the microstructure can be directly related to the photoelectronic properties.

Columnar microstructure, for which transmission and scanning electron microscopy (TEM and SEM) and small-angle neutron scattering have provided direct evidence [Knights 1985], has long been associated with poor electronic material properties in a-Si:H (see, e.g., [Knights and Lujan 1979 and Drevillon et al. 1983]). Although, on a 5 to 10 nm scale, the microstructure probably is different for a-Si:H and a-SiGe:H, columnar microstructure has also been associated with poor electronic properties in a-SiGe:H. For example, in comparing a-SiGe:H films made by two glow-discharge methods, Ichimura et al. [1985] found that the poorer film consisted of void-rich material having a columnar structure surrounded by low-atom-density tissue, whereas the better film had a homogeneous structure with a higher σ_l, a lower ESR spin density (4.7×10^{-16} cm^{-3}), a lower E_u (50 meV), and lower Si and Ge dihydride-to-monohydride ratios, which Ichimura et al. attributed to lower compositional disorder.

Similarly, Mackenzie et al. [1985] found that poorer photoelectronic properties, including an increased dihydride-to-monohydride ratio, in Ge-rich alloys are the result of increased heterostructure, that is, more low-density tissue relative to denser islands. Differences in material properties between a-SiGe:H and a-SiGe:H:F are also attributed to differences in microstructure, specifically to the difference in heterostructure [Paul and Mackenzie 1987]. One of the difficulties one encounters in trying to prove that assumption is our inability to establish the composition of the islands and the tissue in the heterostructure.

Scanning tunneling microscopy, applied to thin films, may be of use in analyzing at the 0.5 nm level the various heterostructures of a-SiGe films that heretofore have been observed only at a grosser level by transmission electron microscopy. An interesting model of a fractal structure to explain the different composition of islands and tissue has been proposed by Muramatsu et al. [1988] based on extensive structural investigation.

The effect of dispersed and clustered hydrogen on the ESR spin density (and thus on dangling bonds) has been shown for a-Si:H. The spin density decreases with an increase in dispersed H (and a concomitant decrease in clustered H) in a-Si:H prepared by glow discharge both from silane [Shimizu et al. 1983] and from disilane [Kumeda et al. 1985]. The same trend can be expected for a-SiGe:H films. The ESR spin density increases as the bandgap is decreased (more Ge or lower H content) and values of 1.6 x 10^{-16} cm^{-3} at 1.55 eV have been measured for a-SiGe:H films prepared by the microwave excited plasma CVD method [Perry et al. 1988].

Mahan et al. [1987a] have defined a quantitative measure of the amount of microstructure in terms of a dihydride-to-monohydride ratio; they have shown that the microstructure fraction so defined (called R^*) is related to the Ge content, the material density, the Urbach parameter, and

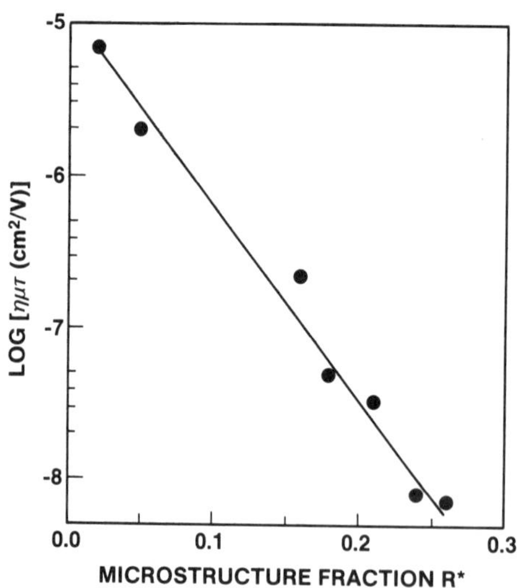

Figure 2.3.2-1. Photoconductivity as a function of microstructure fraction R^* for a-SiGe:H [Mahan et al. 1987a].

the photoconductivity. Thus, they have established a quantitative measure of material quality in terms of one aspect of microstructure. Figure **2.3.2-1** shows how $\eta\mu\tau$, which is proportional to the photoconductivity, decreases with increasing R^*.

E_u increases with increasing R^* and with decreasing film density. Mahan et al. reason that the increase in E_u is not due to greater compositional or structural disorder but due to local microstructural effects, namely, local disturbances in the vicinity of the microstructure. Mahan et al. [1987a] attribute the greater effect of microstructure on the quality of a-SiGe:H than on that of a-Si:H to dangling bonds associated with Ge.

2.3.3 Effects of Deposition Methods, Feed-Gases, and Diluents

In an effort to avoid or minimize the deterioration of photoelectronic properties with increasing germanium content, researchers have investigated different deposition methods and used a variety of feed-gas mixtures. These deposition methods include RF glow discharge with both a diode and a triode geometry, DC proximity glow discharge, and photo-CVD. The gas mixtures evaluated are hydrogen-diluted and undiluted mixtures of silane/ germane, disilane/germane, silane/germanium-tetrafluoride, and silicon-tetrafluoride/germanium-tetrafluoride.

If a comparison of various researchers' results is to be meaningful, the data to be compared must be for materials with the same bandgap. In this section we generally report on properties at a 1.5 eV E_g (as determined from the Tauc relationship). Since the most common data in the literature for a-SiGe:H alloy materials, either as a function of the Ge content or as a function of E_g, are for σ_l and σ_d, we concentrate on these parameters. To achieve high quality material, *both* σ_l and the photosensitivity must be high. There is a positive correlation between the photosensitivity and the $\mu\tau$ product for holes and electrons [Sato et al. 1988].

When the deposition method is DC proximity glow discharge, there is no essential difference in the relationship of the Ge content to E_g among films deposited from SiH_4-GeH_4, Si_2H_6-GeH_4, or SiH_4-GeH_4-H_2 gas mixtures [Catalano et al. 1987]. There is also no significant difference in E_u whether DC proximity glow discharge or RF glow discharge is used to prepare films from SiF_4-GeF_4-H_2 or SiH_4-GeF_4-H_2 [Aljishi et al. 1986]. DC proximity glow discharge of a hydrogen-diluted SiH_4-GeH_4 mixture yields about the same σ_l as do diode and triode RF glow discharges.

Guha et al. [1987] reported to have grown films by RF glow discharge from a mixture of disilane, germane, silicontetrafluoride, and hydrogen, in which H is bonded to Si and Ge without any preference, that show a constant E_u (about 50 meV) down to 1.25 eV E_g. Morin et al. [1992] showed that low-gap (<1.5 eV) a-SiGe:H alloys of device quality can be deposited only from gas mixture including SiF_4, either $SiF_4/GeF_4/H_2$ or $SiF_4/SiH_4/GeF_4/H_2$, and that the latter mixture is required to reach higher growth rates (0.25-0.55 nm/s).

Several researchers have reported producing a-SiGe:H films with a higher photoconductivity from undiluted silane and germane using RF glow discharge with a *triode* rather than a diode geometry [Matsuda et al. 1985, Ichimura et al. 1985]. However, the best films, for both electrode geometries, produced from *hydrogen-diluted* silane and germane mixtures have about the same σ_l (2×10^{-5} S/cm at 1.5 eV E_g). The triode material (from undiluted silane and germane with an E_g of 1.52 eV) had a more homogeneous structure, a lower Si dihydride-to-monohydride ratio, and no Ge dihydride, in comparison to the diode geometry material, which had a Si dihydride-to-monohydride ratio of about 1 and a Ge dihydride-to-monohydride ratio of about 0.25 [Ichimura et al. 1985]. When a mixture of $SiF_4/GeF_4/H_2$ was used, a σ_l value of 3×10^{-4} S/cm was achieved with a triode geometry [Tsuda et al. 1985]; however, the σ_l of that film was about the same (within a factor of two) as that of the best diode films. The reason that better performance is achieved with a triode rather than with a diode geometry, under otherwise similar conditions, is that high energy ion bombardment and impingement by neutral radicals such as SiH_2 on the growing film is eliminated under triode conditions [Matsuda et al. 1985]. When good properties are achieved with a diode geometry, high energy ion bombardment may have been eliminated by other means, such as biasing of the substrate, that are often not reported [Luft and Tsuo 1988].

The best reported films with a 1.5 eV E_g of a-SiGe:H prepared by photo-CVD of $SiH_4/GeH_4/H_2$ have σ_l values five times higher than those of films made by glow discharge: 1×10^{-4} [Itozaki et al. 1985] versus 1.9×10^{-5} S/cm [Matsuda et al. 1986] (see **Table 2.3.3-1**). However, the photosensitivity is not improved by using photo-CVD. The highest photosensitivity for material with an E_g of 1.5 eV produced by glow discharge is 1.2×10^4 [Matsuda et al. 1986]; using photo-CVD, it is 1×10^4 (both deposition methods used a silane-germane-hydrogen mixture) [Sichanugrist et al. 1985]. Note, however, that by glow discharge of a hydrogen-diluted silane/germane mixture, Shing et al. [1987b] achieved a photosensitivity of 1×10^4 for a 1.42 eV E_g material, a value that corresponds to 8 x 10^4 for a 1.5 eV E_g material. Baron et al. [1990] reported a photosensitivity of 2 x 10^2 for 1.3 eV E_g material deposited by photo-CVD from hydrogen diluted silane/germane mixture (σ_l = 4 x 10^{-6} and σ_d = 2 x 10^{-8} S/cm).

Table 2.3.3-1. The Best Photoconductivities for a-SiGe:H and a-SiGe:H:F at E_g = 1.5 eV [Luft 1989]

Feed Gas	RFGD (diode)	RFGD (triode)	MWEP	DCPGD (triode)	Photo-CVD
SiH_4+GeH_4	2.0E-5	5.0E-5	-	-	1.0E-7
Si_2H_6+GeH_4	2.0E-5	-	-	-	6.0E-6
SiH_4+GeF_4	-	-	-	2.0E-6	-
SiH_4+GeH_4+H_2	1.9E-5	1.5E-5	1.5E-5	1.5E-5(a)	1.0E-4
Si_2H_6+GeH_4+H_2	1.0E-5	-	-	-	1.6E-4(b)
SiF_4+GeF_4+H_2	-	3.0E-4	-	1.0E-4	-
SiH_4+GeH_4+Ar	-	-	3E-5	-	-

RFGD = radio-frequency glow discharge; MWEP = microwave-excited plasma;
DCPGD = direct-current proximity glow discharge.
(a) E_g = 1.57 eV, (b) E_g = 1.55 eV.

When RF glow discharge was used with a SiF_4-GeF_4-H_2 gas mixture instead of a SiH_4-GeH_4-H_2 mixture, a marginally higher photosensitivity (1.8 x 10^4 versus 1.0 x 10^4 for 1.5-eV-E_g material) was obtained (see **Table 2.3.3-2**).

Local passivation of dangling bonds by F rather than H is not regarded as an important issue, since only about 1 at.% of F is incorporated in a-SiGe:H:F films; instead, the difference in material properties is attributed to differences in the heterostructure (island and tissue distribution, as seen in TEM) [Paul and Mackenzie 1987]. The best photosensitivity for photo-CVD material from disilane in a hydrogen-diluted germane mixture is not better than that achieved with a silane-germane-hydrogen mixture. The lack of improvement in the photosensitivity observed when disilane is used rather than monosilane may be because less work has been done with high-purity disilane than with high-purity silane. Disilane increases the hydrogen content especially as silicon dihydride.

Table 2.3.3-2. The Best Photosensitivities for a-SiGe:H and a-SiGe:H:F at E_g = 1.5 eV [Luft, 1988a]

Feed-Gas	RFGD (diode)	RFGD (triode)	MWEP	DCPGD (triode)	Photo-CVD
SiH_4+GeH_4	5.0E3	4.4E3	-	-	5.0E2
Si_2H_6+GeH_4	1.0E3	-	-	-	6.0E1
SiH_4+GeF_4	-	-	-	2.0E3	-
SiH_4+GeH_4+H_2	1.2E4	1.0E4	3.1E3(a)	3.7E4(b)	1.0E4
Si_2H_6+GeH_4+H_2	2.0E3	-	-	-	2.0E4(c)
SiF_4+GeF_4+H_2	-	1.8E4	-	2.5E3	-
SiH_4+GeH_4+Ar	-	-	1.0E4	-	-

RFGD = radio-frequency glow discharge; MWEP = microwave-excited plasma; DCPGD = direct-current proximity glow discharge.
(a) E_g = 1.46 eV, (b) E_g = 1.57 eV, (c) E_g = 1.55 eV.

Hydrogen dilution reduces the tendency of Si-H_2 bond formation and minimizes the light-induced degradation [Yang, L., et al 1991]. Hydrogen (or inert gas) dilution of germane-silane mixtures appears to be essential if high photosensitivity is to be obtained in a-SiGe:H alloys with any of the deposition methods covered in this section. Hydrogen dilution results in higher photosensitivity by a factor of four (or greater) than do undiluted mixtures (2×10^4 versus 5×10^3) (see **Table 2.3.3-2**). Raman spectroscopy studies show that Ge-Ge clustering in a-SiGe:H films is consistently lower for films made with hydrogen dilution than for those without such dilution [Catalano 1990a]. An increase in Ge-Ge clustering as indicated by the Raman ratio I_{Ge-Ge}/I_{Si-Ge} leads to a decrease in $\mu\tau$ (from 4×10^{-8} cm^2/V at a ratio of 0.75 to 7×10^{-9} cm^2/V at a ratio of 1.4) and a decrease in diffusion length (from 80 nm at a ratio of 0.75 to 40 nm at a ratio of 1.4) [Yang, L., et al. 1989]. Hydrogen dilution does not always increase $\eta\mu\tau$ and L_d for a-SiGe:H [Tsuo et al. 1991a]. It is possible that the geometry of the deposition apparatus may have an effect on whether hydrogen dilution produces better material.

Increasing the germane fraction in the feed-gas mix of silane and germane, and thus the germanium content of the resulting film, increases the silicon dihydride bonding as indicated by the absorptance at 2080 cm^{-1}. Diluting the gas mixture with hydrogen reduces the dihydride bonding at all germanium fractions as seen in **Figure 2.3.3-1.** The ratio

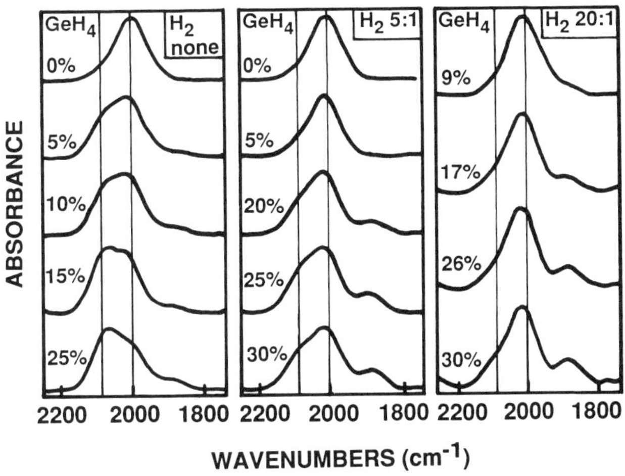

Figure 2.3.3-1 Infrared spectra of a-SiGe:H made with (a) no H_2 dilution, (b) 5:1 H_2 dilution, and (c) 20:1 H_2 dilution [Catalano et al. 1991b]

of dihydride to monohydride bonding is much reduced at germane fractions of the silane/germane mixture of up to 0.3 as evident from **Figure 2.3.3-2.** For instant at 30% germane, the ratio is reduced from 1.7 to 0.8. The germane gas fraction gives essentially the same germanium fraction in the resultant film independent of the hydrogen dilution. Ishihara, T., et al. [1987] showed that the dihydride-to-monohydride ratio for a-SiGe:H from silane and germane by glow discharge decreases from 0.6 to zero with increasing H dilution (from 80% to 98%). The Si dihydride bonds decrease and the Ge monohydride bonds increase with increasing hydrogen dilution over the same range. Slobodin et al. [1986] showed that H dilution of SiH_4/GeF_4 in DC proximity glow discharge reduces Si-H_2 bonds, which is attributed to a lesser plasma-phase polymerization. For a-SiGe:H prepared by RF glow discharge, Perry et al. [1988] attribute the better material properties resulting from H dilution to a reduction in the gas-phase germane polymerization and associated particle formation. For a-SiGe:H produced from disilane by photo-CVD, Sichanugrist et al. [1985] showed a slight increase in the total bonded hydrogen (from 4 x 10^{21} to 5 x 10^{21} cm^{-3} as the H dilution is increased from 5% to 10%.

The $\mu\tau$ product and L_d (as measured with SSPG, § 3.3) are also improved by hydrogen dilution as shown in **Figures 2.3.3-3a and b.** Thus, at 1.5 eV E_g, $\mu\tau$ increases from about 2 x 10^{-9} cm^2/V to 5 x 10^{-8} cm^2/V and L_d from 40 nm to 75 nm as a result of 5:1 hydrogen dilution as compared to no dilution. The photodegradation of cells with a 1.5 eV

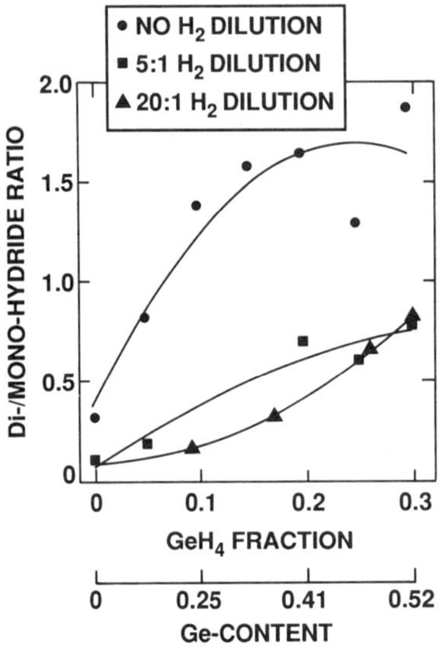

Figure 2.3.3-2. Ratio of SiH_2/SiH bonding of a-SiGe:H films prepared without hydrogen dilution, with 5:1 and 20:1 H_2 dilution vs. GeH_4 fraction [Catalano et al. 1991a].

a-SiGe:H i-layer after 850 hours exposure to 1000 W/m^2 was found to be less for a-SiGe:H made with hydrogen dilution than without hydrogen dilution (18% versus 30% degradation) [Catalano et al. 1990]. Higher hydrogen dilution (20:1) of silane in glow discharge gave better a-SiGe:H material (lower SiH_2/SiH ratio) than 5:1 dilution for 0% to 20% germane fraction and higher $\mu\tau$-product over the optical gap range of 1.46 to 1.62 eV. For instance, at 1.50 eV, the $\mu\tau$-product was $3x10^{-7}$ versus $5x10^{-8}$ $cm^2 V^{-1}$. At higher optical gap, the difference is even greater [Catalano et al. 1990b].

Tsuo et al. [1991a] found helium dilution of silane/germane gas mixtures more effective than hydrogen dilution in improving a-SiGe:H. Comparing a-SiGe:H films deposited using 65% helium dilution with films deposited using 65% hydrogen dilution, they found that films with He dilution have longer L_d and higher $\eta\mu\tau$ values. They also found that

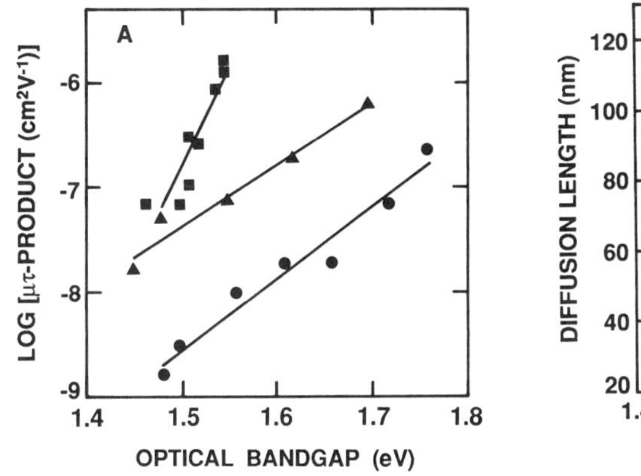

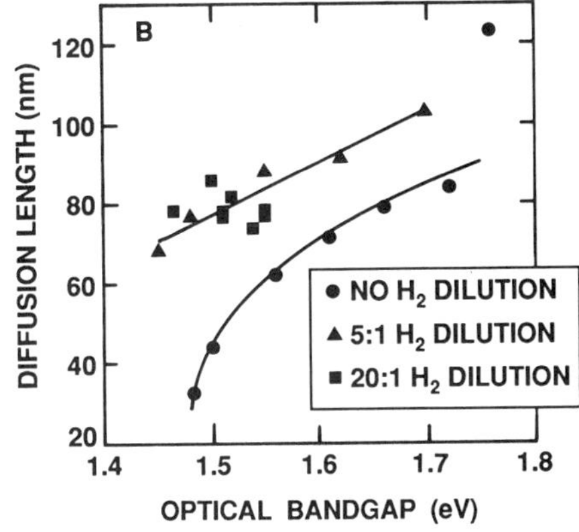

Figure 2.3.3-3 (a) mobility-lifetime product and (b) diffusion length (measured by SSPG) for a-SiGe:H prepared with various hydrogen dilution ratios vs. optical bandgap [Catalano et al. 1991b].

the incorporation of Ge atoms in the a-SiGe:H films is more efficient with He dilution than with H_2 dilution. This means less germane gas concentration is needed to achieve a certain E_g of a-SiGe:H with He dilution than without dilution or with H_2 dilution.

2.4 CHARACTERISTICS OF HYDROGENATED AMORPHOUS SILICON CARBIDE

Hydrogenated amorphous silicon carbide alloys ($a\text{-}Si_{1-x}C_x$:H or a-SiC:H) is used in p-i-n solar cell devices as: 1) a wide bandgap (about 2.0 eV) intrinsic layer in multijunction solar cells, 2) a wide bandgap p-type window layer for a-Si:H or a-SiGe:H cells, and as 3) a compositionally graded (1.9 - 1.7 eV) interface buffer layer (10 - 20 nm thick) between the p-type a-SiC:H:B layer and the intrinsic a-Si:H layer or as a compositional graded interface buffer layer between the a-SiC:H:B and the intrinsic a-SiGe:H layer in multijunction cells. a-SiC:H is also used in thin-film visible light emitting diodes [Kanicki 1991].

2.4.1. Intrinsic a-SiC:H Wide Bandgap Films

Intrinsic wide bandgap a-SiC:H films are used in the top cell of multijunction stacked cells. Successful incorporation of a wide-bandgap material in the front junction can significantly increase the V_{oc} of the device. The greater thickness of an a-SiC:H intrinsic layer as compared to that for a-Si:H reduces shunt-related problems in multijunction modules.

The optical bandgap of intrinsic a-SiC:H increases monotonically as the carbon content increases from that of a-Si:H of about 1.75 eV to 2.0 - 2.15 eV at a 15 at.% carbon content as shown in Figure **2.4.1-1**.

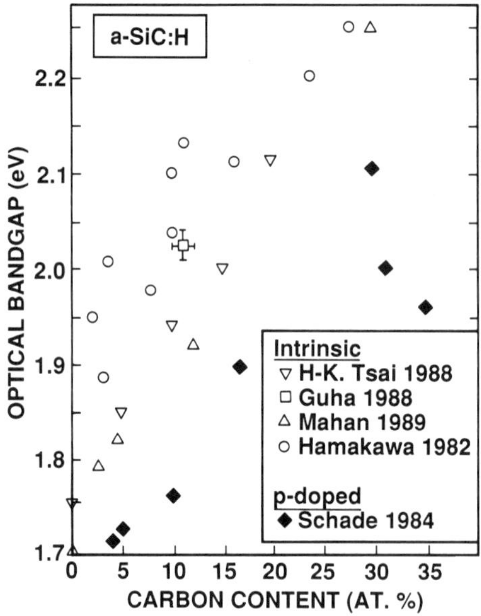

Figure 2.4.1-1 Optical bandgap for intrinsic and p-type a-SiC:H vs. carbon content. The data by Mahan [1989] are for films from undiluted SiH_4 and CH_4.

Most a-SiC:H films are deposited using a mixture of silane and methane (CH_4) gases, either with or without hydrogen dilution. Nevin et al. [1991] reported that, using mixtures of xylene and silane, they have produced a-SiC:H films with 2.2 to 3.5 eV E_g with carbon contents from 0.4 to 0.9 atomic fraction. A typical E_g for a-SiC:H films is 2.0 eV for 10 at.% carbon content. In combination with a-Si:H in dual-junction cells, the desired E_g for a-SiC:H is 2.25 eV giving a maximum theoretical efficiency of 22.8% (V_{oc} = 2.64 V, J_{sc} = 10.8 mA/cm^2, and FF = 0.80) [Hollingsworth et al. 1987a]. For triple-junction cells, Kuwano [1986] calculated a maximum efficiency of 24% for a E_g combination of 2.0/1.7/1.4 eV.

Various feed-gas mixtures have been used to deposit a-SiC:H films as shown in **Table 2.4.1-1.** a-SiC:H films grown from silane and methane have better quality than those grown from silane and ethylene (C_2H_4) [Tawada et al. 1982]. Another feed gas combination that has been used is SiH_4 and C_2H_2 [Nakayama et al. 1988]. The rationale of experimenting with feed-gases such as dimethylsilane (DMS) and trimethylsilane (TMS) is that in such carbon sources there are no methyl groups (CH_3 and CH_2, respectively) which, once incorporated in the film, are suspected to cause structural distortion of the lattice and lead to poor electronic properties. It is expected that the built-in SiC bonds in DMS or TMS will be preserved in the deposited film to help minimize random bonding of the carbon species [Catalano et al. 1991b]. Films from silylmethane (CH_3-SiH_3) incorporate carbon more efficiently than films from a silane-methane mixture [Catalano et al. 1989]. An even lower optical absorption coefficient (at 1.2 eV) as measured by PDS, lower Urbach energy, and higher ambipolar diffusion length than for silylmethane has been observed for a-SiC:H films from trisilylmethane $(SiH_3)_3$-CH [Li and Fieselmann 1991].

For a-SiC:H films deposited using an undiluted silane-methane mixture, small-angle X-ray scattering measurements indicate that such films have voids, the radii of which increase from 0.43 to 0.7 nm as the carbon content increases from 0 to 30%; the number density of voids increases also with increasing carbon content. Mohring et al. [1991] achieved Urbach energies (by PDS) of 46 meV at 1.75 eV E_g and 53 meV at 1.95 eV E_g using undiluted silane/methane gas mixture in DC glow discharge in an ultra-high vacuum deposition system; photosensitivity of 10^6 over the E_g range of 1.75 to 1.95 eV, activation energy 0.9 eV at 1.95 eV E_g, $\mu\tau$ product for electrons 6 x 10^{-7} cm^2/V at 1.75 eV E_g dropping to 1.5 x 10^{-8} cm^2/V at 1.95 eV E_g, and 10^{-8} cm^2/V for holes at 1.80 eV E_g at a generation rate of 10^{18} cm^{-3}s^{-1}, dropping to 2 x 10^{-9} cm^2/V at 1.95 eV.

Table 2.4.1-1 Feed-Gases and Diluents for Deposition of a-SiC:H

Feed-Gas		Diluent	Reference
SiH_4+CH_4	Silane/Methane	H_2, He, Ar	Catalano et al. 1991b
SiH_4+C_2H_2	Silane/Ethylene	H_2, He, Ar	Nakayama et al. 1988
SiH_4+C_2H_4		H_2, He, Ar	Tawada et al. 1982
SiH_4+C_2H_6		H_2, He, Ar	Catalano et al. 1991b
$(SiH_3)CH_3$	Silylmethane	H_2	Catalano et al. 1989 Matsuda et al. 1989
$(SiH_3)_2CH_2$	Disilylmethane		Catalano et al. 1991b Beyer et al. 1989
$(SiH_3)_3CH$	Trisilylmethane		Catalano et al. 1991b
$(SiH_3)_4C$	Tetrasilylcarbon		
SiH_4/CCl_3D	Carbontrichlorodeuterium		Catalano et al. 1992
SiH_4/CH_3- C_6H_4-CH_3	Silane/Xylene		Nevin et al. 1991

a-SiC:H films deposited from silane-methane diluted in either H_2 or Ar are of superior quality to those deposited from silane/methane with no dilution. This assertion is based on IR absorption peaks around 2000 cm^{-1} (indicative of monohydride bonding in a dense network) and around 2070 cm^{-1} (indicative of monohydride or dihydride bonding in films having larger microvoids) [Catalano et al. 1989], an increase in σ_1, photosensitivity, and L_d, lower defect density as determined from space-charge-limited current measurements and from photothermal deflection spectroscopy measurements, lower E_u (65 meV at 2 eV E_g), and lower absorption coefficient at 1.2 eV.

The electronic properties vary with the dilution ratio. For hydrogen dilution of CH_4 + SiH_4 of material with E_g ~2 eV, the optimum ratio $[H_2]$ /($[SiH_4]$ + $[CH_4]$) of 25 gives the lowest E_u, sub-bandgap absorption at 1.2 eV (a measure of deep-state density), and the highest σ_1. Comparison of H_2 evolution spectra suggests that H_2 dilution during deposition produces a denser network structure with fewer voids because of the increased surface mobility of absorbed radicals during film growth.

When comparing parameters of a-SiC:H alloys deposited at different substrate temperatures, T_s, it must be considered that higher T_s reduces the hydrogen content in the film and thus leads to lower E_g. Thus, it appears that the $\mu\tau$ product increases and the Urbach energy and optical absorption coefficient decrease as the T_s is increased. However,

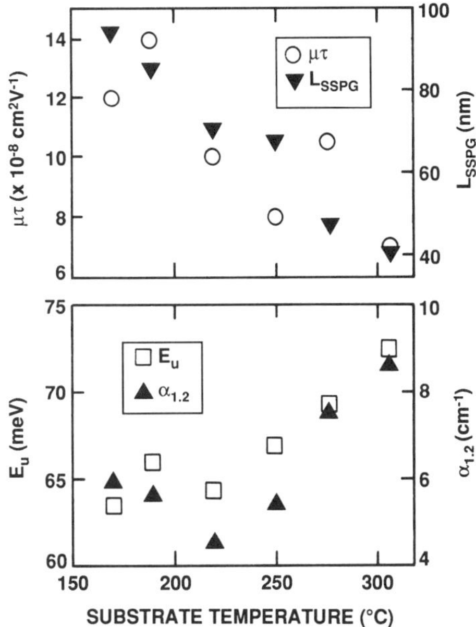

Figure 2.4.1-2 Variations of transport and optical properties with substrate temperature for 2 μm-thick a-SiC:H films [Catalano et al. 1992a]

when E_g is held constant, then one finds that the $\mu\tau$ product and the ambipolar diffusion length decrease over the T_s range of 150 - 300°C and the Urbach energy and the optical absorption coefficient increase over the range of 250 - 300°C (**Figure 2.4.1-2**). Fig. 2.4.1-2 is for films with the same E_g (~ 1.96 eV) deposited from hydrogen diluted SiH_4/CH_4 [Catalano et al. 1992a]. When measuring transport and optical properties for films from hydrogen-diluted silane/methane, it must be remembered that $\mu\tau$, L_d, E_u, optical absorption coefficient, σ_d, and E_a are all thickness dependent up to thicknesses of 2 to 5 μm [Catalano et al. 1992a].

Lu and Petrich [1992] reported achieving less microstructure, wider optical bandgaps, and higher deposition rates when depositing

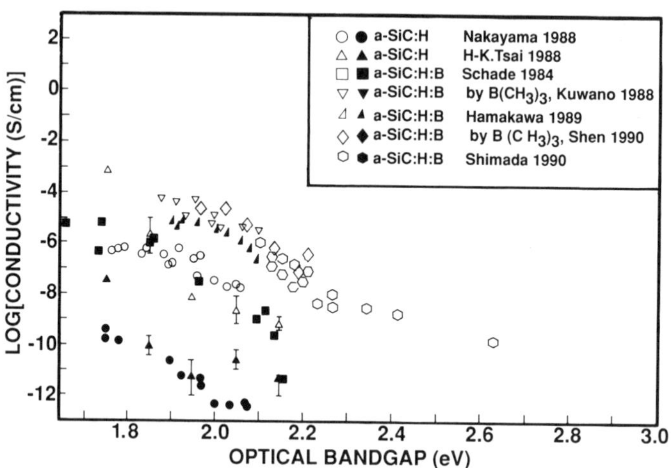

Figure 2.4.1-3 Dark and Photoconductivities vs. Optical Bandgap for Intrinsic and P-Type a-SiC:H from various studies.

a-SiC:H at lower substrate temperatures and with an external dc bias applied to the rf-exited powered electrode.

The photoconductivity decreases with increasing carbon content because of more and larger voids in the film. The σ_l and σ_d values for intrinsic a-SiC:H films vs. E_g in the range 1.65 to 2.2 eV and the comparison to those of doped a-SiC:H:B are shown in Figure **2.4.1-3**. In Figure 2.4.1-3, open symbols are for σ_l, and filled symbols are for σ_d. The dark conductivity for intrinsic a-SiC:H ranges from about 5×10^{-8} to 5×10^{-10} S/cm at an E_g of 1.75 eV to 1×10^{-11} to 2×10^{-13} S/cm at an E_g of 2.1 eV. The corresponding values for boron-doped a-SiC:H:B are 1×10^{-5} and 1×10^{-9} S/cm.

The activation energy for undoped and n- and p-type a-SiC:H is shown in Figure **2.4.1-4.** For intrinsic a-SiC:H, E_a increases from about 0.83 eV at E_g = 1.75 eV to 1.1 eV at E_g = 2.05 eV and then levels off. Likewise, doped a-SiC:H has a lower E_a, but increases with increasing E_g.

As the carbon content increases from 0 to 15%, E_u increases from 50 to 80 meV and the midgap density of states increases from 5×10^{15} to 8×10^{16} cm^{-3}. The general trend for E_u (derived from PDS) is to increase with increasing E_g from about 58 meV at 1.75 eV to 98 meV at 2.1 eV for a-SiC:H films prepared from silane/methane without dilution or with hydrogen or argon dilution, and from SiH_4/C_2H_6 with and without hydrogen dilution [Catalano et al. 1989]. The deposition temperature is

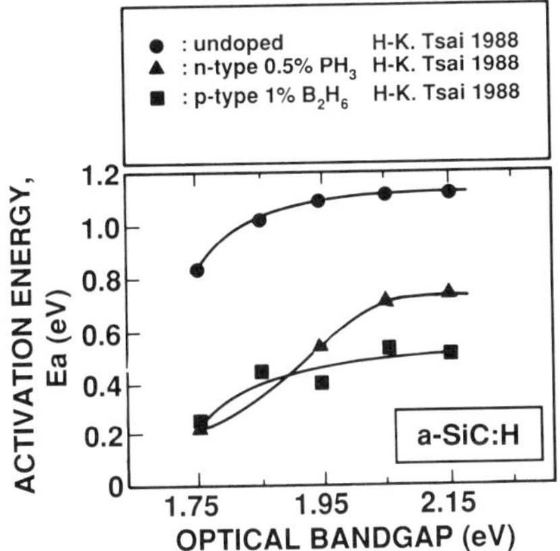

Figure 2.4.1-4 Activation Energy vs. E_g for doped and undoped a-SiC:H [Tsai, H.-K. 1988].

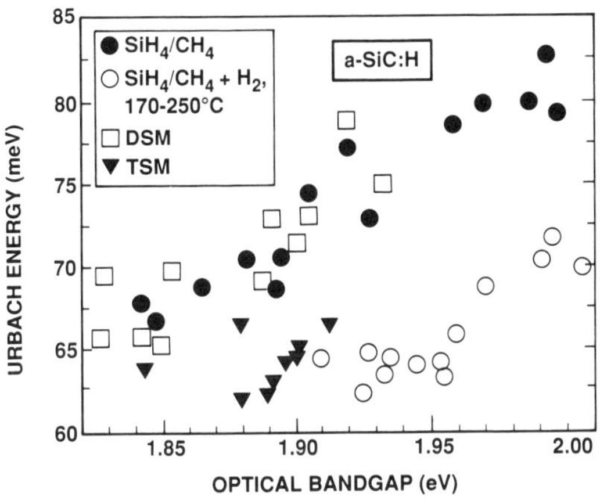

Figure 2.4.1.-5 Urbach energy from PDS vs. E_g for a-SiC:H prepared from various feed-gases.

also a factor. Thus, E_u for films from DMS or diluted silane/methane mixture deposited at 300 to 320°C is about the same for films from undiluted silane/methane (80 meV at 2 eV), whereas that for films from TMS or from diluted silane/methane deposited at 170 to 250°C is about 65 meV at 1.9 eV and 70 meV at 2 eV [Catalano et al. 1991a] **(Figure 2.4.1-5)**. The Urbach energy increases also with increasing applied power density (above 50 mW/cm^2) and with increasing dihydride and monohydride bonding ratio from about 60 meV at a microstructure fraction of 0.45 to 0.80 meV at a microstructure fraction of 0.9 [Mahan et al. 1988].

The ambipolar diffusion length for a-SiC:H as measured by the steady state photograting technique decreases with increasing E_g from 125 nm at 1.75 eV to 35 nm at 1.95 eV for films deposited using undiluted SiH_4/CH_4 and from about 180 nm at 1.8 eV to 80 nm at 1.95 eV for films deposited using diluted SiH_4/CH_4 or from TMS as shown in **Figure 2.4.1-6.** Films deposited from TMS and hydrogen diluted SiH_4/CH_4

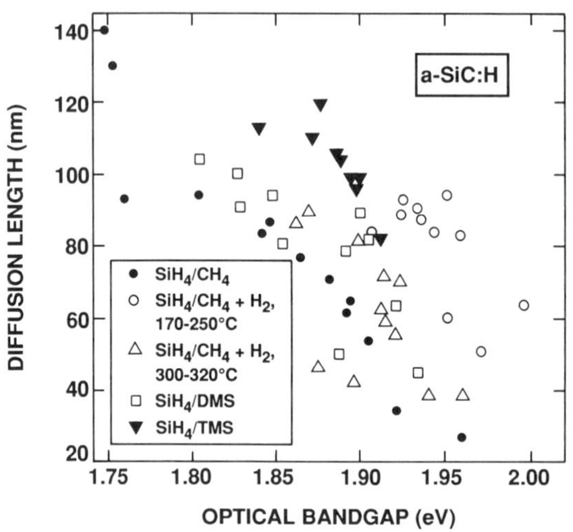

Figure 2.4.1-6 Ambipolar diffusion length (from SSPG, § 3.3) vs. E_g for a-SiC:H prepared from various feed-gases [Catalano et al. 1991b and 1992].

deposited at 170 to 250°C substrate temperatures have a greater diffusion length than the other films.

The $\mu\tau$ product for intrinsic a-SiC:H decreases with increasing E_g, because of decreasing mobility. The $\mu\tau$ product is much higher for hydrogen diluted SiH_4/CH_4 than for undiluted SiH_4/CH_4, DSM, or TSM (**Figure 2.4.1-7**). There is no difference in $\mu\tau$ for films made from hydrogen diluted SiH_4/CH_4 deposited at 170 to 250°C and at 300 to 320°C. Films from CCl_3D had even lower $\mu\tau$ and σ_d than those from undiluted SiH_4/CH_4.

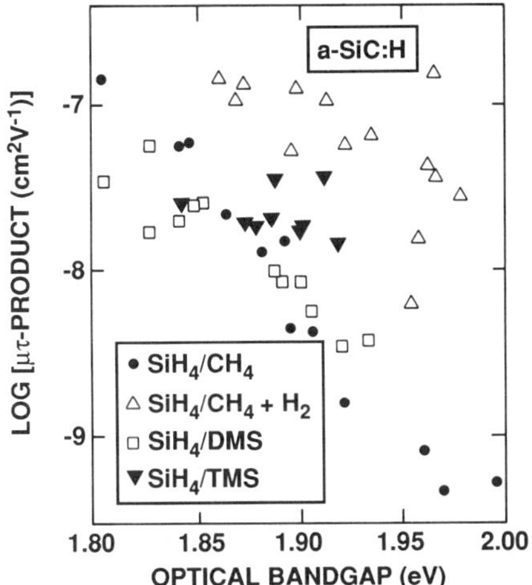

Figure 2.4.1-7 Mobility-lifetime product vs. E_g for a-SiC:H films prepared from different feed-gases [Catalano et al. 1991b].

The optical absorption coefficient at 1.2 eV for a-SiC:H as measured by photothermal deflection spectroscopy is lower for films deposited from TMS and hydrogen diluted SiH_4/CH_4 mixture at 170 to 250°C T_s than for films from undiluted SiH_4/CH_4 or from DSM. Films from diluted SiH_4/CH_4 at 300 to 320°C have intermediate values of optical absorption coefficients (**Figure 2.4.1-8**). The best a-Si:H materials at 1.75 eV E_g have even lower E_u than those shown in Figure **2.4.1-8.** This also applies to the conductivity. Thus, the various quality indicators, σ_l, $\eta\mu\tau$, E_a, and E_u all show a deterioration in a-SiC:H film quality as E_g increases above that for a-Si:H.

2.4.2 Doped a-SiC:H Layers

a-Si:H homojunction p-i-n solar cells suffer from poor short wavelength response because of high optical absorption in the a-Si:H:B p-layer and recombination at the p/i interface. The short-wavelength response is improved by the use of a heterojunction front contact such as p-type a-SiC:H:B (see Figure **2.4.2-1**). The high-E_g p-layer reduces optical losses in the p-layer and also reduces back diffusion and recombination of carriers [Catalano & Wood 1988].

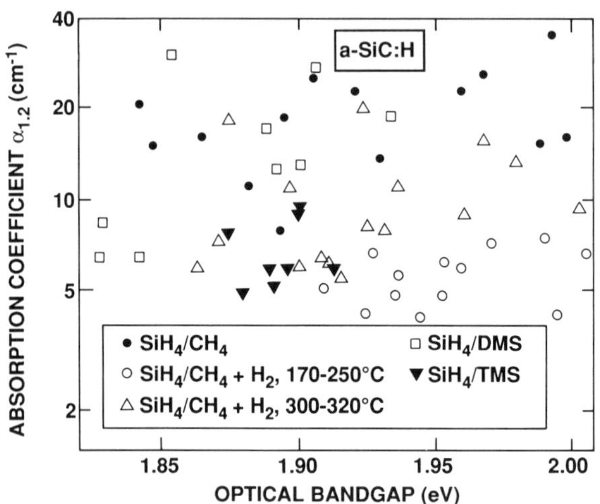

Figure 2.4.1-8 Optical Absorption Coefficient at 1.2 eV (from PDS) vs. E_g for a-SiC:H from various feed-gases [Catalano et al. 1991b and 1992].

The higher optical bandgap for the p-layer of SiC compared to that of Si will increase the built-in potential; thus heterojunction p(SiC)/i(Si) devices have a higher built-in field than homojunction p(Si)/i(Si) devices. However, the relationship is not simple since the activation energy of a-SiC:H is a function of the optical bandgap; it increases as the bandgap increases (as was shown in Figure 2.4.1-4).

a-SiC:H/a-Si:H heterojunctions show not only a higher V_{oc}, but also higher J_{sc} and efficiency. These improvements are caused by a wide optical bandgap of the p-layer which reduces the optical absorption loss in the p-layer, and the large internal field which increases the collection probability of free carriers [Tawada et al. 1982].

The optical bandgap for a-SiC:H:B increases linearly with increasing carbon content was as shown in Figure 2.4.1-1. The dark conductivity for a-SiC:H:B decreases from 10^{-5} to 10^{-10} S/cm with E_g increasing from 1.75 eV to 2.1 eV and σ_l decreases from 10^{-4} to 10^{-7} S/cm as E_g increases from 1.75 eV to 2.2 eV (as was shown in Figure 2.4.1-3).

The narrowing of the optical bandgap by boron doping is mainly caused by a decrease in the content of hydrogen attached to carbon [Hamakawa 1982]. By doping a-SiC:H:B with trimethylboron, $B(CH_3)_3$ or TMB) instead of the traditional doping gas B_2H_6 and using the

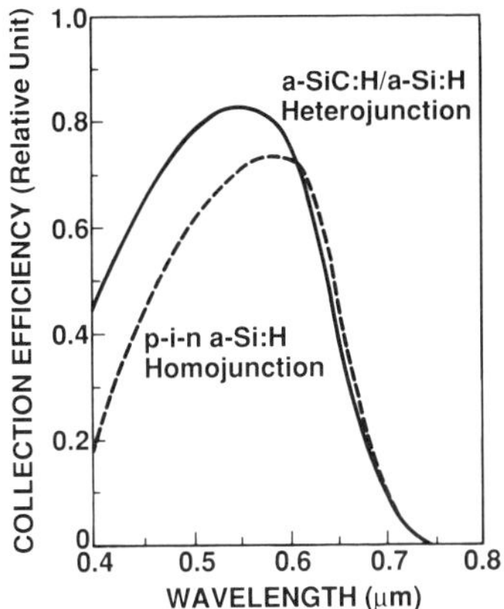

Figure 2.4.2-1 Collection efficiency of p-i-n a-Si:H homojunction compared to a-SiC:H/a-Si:H heterojunction solar cell [Tawada et al. 1982].

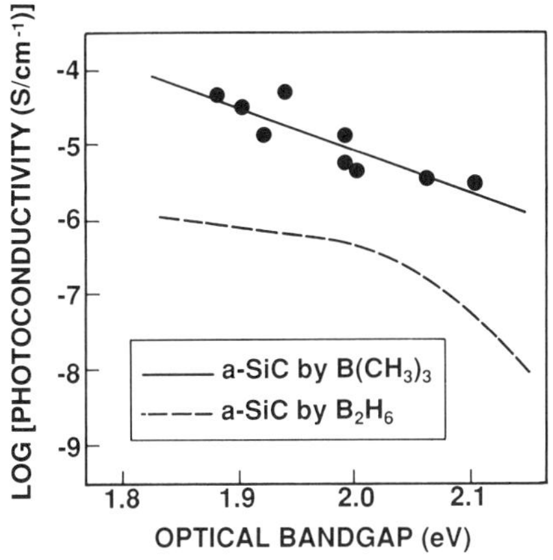

Figure 2.4.2-2 Photoconductivity of p-type a-SiC:H,B films using two different doping gases vs. optical bandgap [Kuwano & Tsuda 1988].

controlled plasma magnetron CVD method, p-layer films of high σ_l (10^{-4} to 5 x 10^{-6} S/cm) over the E_g range of 1.9 to 2.1 eV has been achieved as shown in **Figure 2.4.2-2**. a-SiC:H films prepared by glow discharge from a silane-methane mixture doped with TMB exhibit at high boron concentrations (>5 x 10^{19} cm^{-3}) higher hydrogen content (2 to 9 at.%), higher E_g (50 to 100 meV), and lower σ_d than films doped with diborane having the same boron concentration in the film [Lechner et al. 1990]. The use of TMB is also attractive due to its higher thermal stability than diborane [Wu et al. 1989b, Roca i Cabarrocas et al. 1989]. The higher conductivity also improves the quantum efficiency in the 350 to 500 nm wavelength range [Kuwano & Tsuda 1988]. Improved doping efficiency and high photoconductivity of boron-doped hydrogenated amorphous silicon carbide films have also been observed when BF_3 was used [Asano and Sakai 1989].

High quality a-SiC:H:B p-layers have been produced using silylmethanes [$(SiH_3)CH_3$ and $(SiH_3)_2C_2$] instead of methane [Fieselmann et al. 1987]. The structure of such films is strongly dependent of the pressure during deposition [Ryners et al. 1991].

The refractive index for a-SiC:H:B as a function of E_g is shown in Figure **2.4.2-3**; it decreases with an increase in E_g and is somewhat lower than that for undoped a-Si:H. For the higher E_g films, the index of refraction is closer to that for conductive transparent oxides (n = 1.9 at 500 nm for SnO_2:F or ZnO:F) on glass and thereby reduces front reflection losses for devices with a structure of glass/CTO/a-SiC:H:B [Schade et al. 1984].

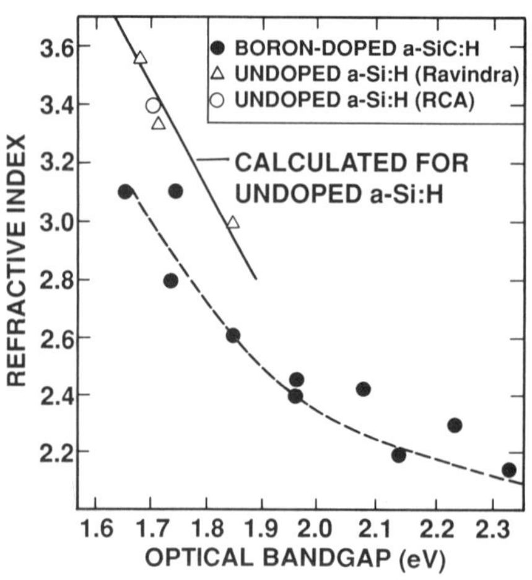

Figure 2.4.2-3 Refractive index for a-SiC:H:B vs. optical bandgap [Schade et al. 1984].

Willeke & Martins [1988] describe structural properties of a-SiC:H:P n-layers deposited by remote means using a 2-chamber system. The films consisted mostly of silicon microcrystals (5 to 10 nm) embedded in an a-SiC:O:H matrix.

2.5 CHARACTERISTICS OF HYDROGENATED AMORPHOUS SILICON NITRIDE AND HYDROGENATED AMORPHOUS SILICON TIN

2.5.1 Hydrogenated Amorphous Silicon Nitrogen

Plasma-deposited a-$Si_{1-x}N_x$:H (a-SiN:H) is mostly used as the gate dielectric in thin-film transistors and as a passivation layer coating over integrated circuits made on crystalline silicon [Adams 1986]. It protects active areas of the integrated circuits from contamination and scratching. Favorable properties are obtained when films are deposited under conditions which produce nitrogen rich compositions ($x>1.3$). Because of the large optical bandgap and high defect density, a-SiN:H has not been used much as an active device material. (However, films with a dark resistivity of $\geq 1 \times 10^{14}$ ohm-cm could be used as photoreceptors for electrophotography.) Possible exceptions are the use of a-SiN:H as a wide-bandgap semiconductor or window material in a multiple-junction a-Si:H alloy solar cell structure, the use of a-SiN:H in thin-film transistor and light-emitting diode structures [Kanicki 1991], and as the storage medium in metal-nitride-oxide-semiconductor memory devices [Robertson 1991]. Alternating layers of a-Si:H and a-SiN:H have been studied for possible superlattice properties [Ibaraki and Fritzsche 1984, Street and Thompson 1984].

2.5.1.1 *Effects of Nitrogen content*

The optical bandgap of a-SiN:H increases smoothly from 1.75 to 5.5 eV with increasing nitrogen content from 1% to 30% without any appreciable deterioration in semiconducting properties [Hirose 1983]. The gap of a-$Si_{1-x}N_x$:H is calculated to widen roughly symmetrically about midgap as $x \to 4/3$ [Robertson 1991]. **Figure 2.5.2-1** shows the optical bandgap and the N/Si compositional ratio vs. the molar fraction of NH_3 to SiH_4 [Kurata et al. 1981].

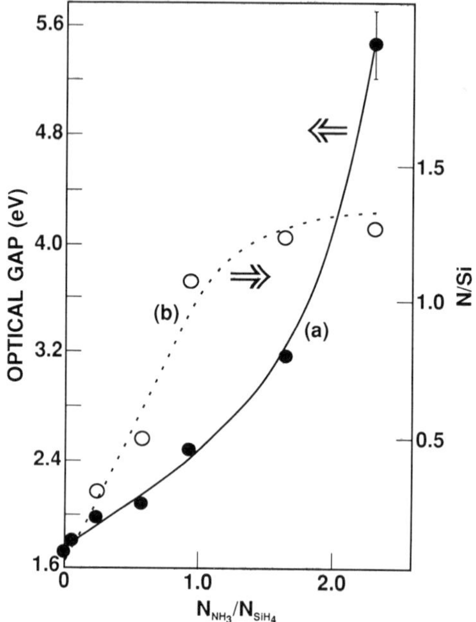

Figure 2.5.1-1 Optical bandgap and N/Si composition for a-SiN:H films vs. molar fraction NH_3 to SiH_4. [Kurata et al. 1981]

Both the dark and photoconductivities increase with increasing ammonia molar fraction until it reaches 0.26 (see **Figure 2.5.1-2**) Above a fraction of 0.26, both dark and photoconductivities decrease rapidly. The activation energy decreased with increasing molar fraction up to 0.26 and then increases rapidly. Watanabe et al. [1982] reached dark conductivities below 1×10^{-14} S/cm suitable for photoreceptors for electrophotography by using N_2 in a diluted silane mixture with a N_2 mole fraction above 80%. The increase in defect density with increasing nitrogen content in PECVD alloys is attributed to the breaking of Si-Si bonds in the valence-band tail.

Thermal stability of a-SiN:H is enhanced over a-Si:H. The SiH bond in a-SiN:H is stable up to 400°C and 50% of the incorporated hydrogen remains after annealing at 550°C. In contrast, hydrogen evolution in a-Si:H generally occurs at 350°C and more than 80% of bonded hydrogen is lost by annealing at 550°C [Hirose 1983].

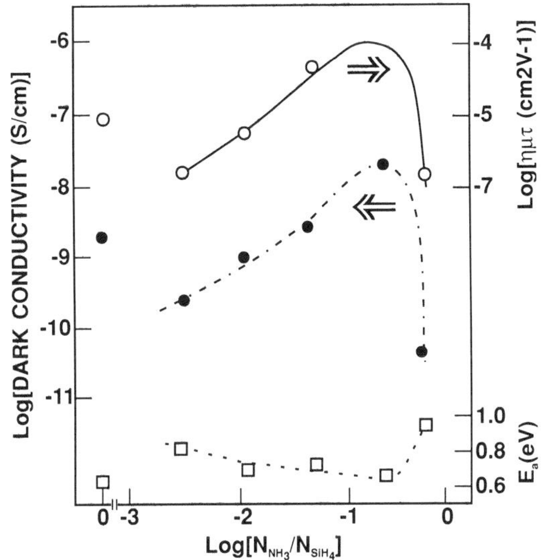

Figure 2.5.1-2 Dark conductivity, normalized photoconductivity (ημτ), and dark conductivity activation energy vs. molar fraction NH_3 to SiH_4 [Kurata et al. 1981].

2.5.1.2 *Materials properties*

Substitutional doping with B or P of a-SiN:H is possible in the same manner as for a-Si:H, a-SiC:H or a-SiGe:H. **Figure 2.5.1-3** shows the conductivity and its activation energy as a function of the doping ratio for a molar fraction NH_3/SiH_4 of 0.26. The conductivity of boron-doped a-SiN:H is at a minimum at a doping ratio of 10^{-3}, where the conversion of n-type conduction to p-type occurs and the activation energy correspondingly exhibits a maximum [Hirose 1983].

The interface between hydrogenated amorphous silicon and silicon nitride is of particular interest because this combination is used in thin-film transistor devices and in superlattice studies. The electronic states at the a-Si:H/silicon nitride interface are observed to depend strongly on the order of deposition. When the nitride is on the top of a-Si:H, there is an interface charge of ~$5x10^{11}$ electrons cm^{-2} and the interface Fermi level is 0.25 eV from the conduction band. The bottom nitride has also an electron accumulation but with a much smaller space charge [Street and Thompson 1984]. By controlling deposition conditions, very smooth films have been obtained. Thin-film transistors

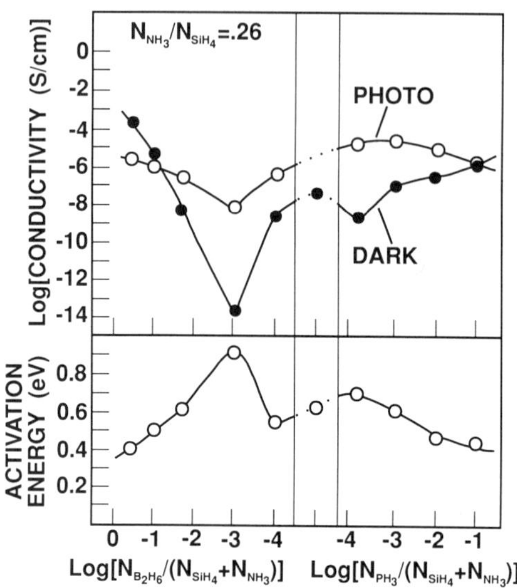

Figure 2.5.1-3 Photoconductivity, dark conductivity, and its activation energy vs. doping ratio [Kurata et al. 1981].

with smooth a-Si:H on smooth SiN exhibit both moderate mobility (1 $cm^2V^{-1}s^{-1}$) and high stability important to high-pixel-density liquid crystal displays [Uchida, H., et al. 1991].

2.5.1.3 *Effects of deposition methods, feed-gases, and diluents*

a-SiN:H is deposited by glow discharge or sputtering. a-SiN:H is prepared by glow discharge from silane or hydrogen diluted silane and N_2, NH_3, NH_4, C_2H_2, or CH_4.

The deposition rate increases with increasing nitrogen mole fraction from 0.01 μm/minute at zero N_2 to over 0.02 μm/minute at 80% N_2 in the gas mixture [Watanabe et al. 1982]. The optical bandgap for a-SiN:H increases with increasing RF power levels and the refractive index decreases slightly [Alvarez and Chambouleyron 1984].

The photosensitivity ranges from 10^3 to 10^5 depending on the power level [Alvarez and Chambouleyron 1984]. Guha & Ovshinsky [1988] have achieved 2 S/cm conductivity, E_a = 0.05 eV, and absorption coefficient of 3 x 10^4 cm^{-1} for 2 eV E_g a-SiN:H films 50 nm thick, having 70% volume fraction of microcrystallites with 6-10 nm size.

2.5.2 Hydrogenated Amorphous Silicon Tin

2.5.2.1 *Effects of Tin content*

Tin is another alloy element for a-Si:H to reduce the bandgap. a-$Si_{x-1}Sn_x$:H (a-SiSn:H) has been produced by glow discharge from a mixture of SiH_4, H_2, and either $SnCl_4$ or $Sn(CH_3)_4$ [Mahan et al. 1984b]. Doping with $SnCl_4$ leads to high contamination of Cl (~15% at 1.5 eV), and doping with $Sn(CH_3)_4$ to C incorporation (20% at 1.2 eV).

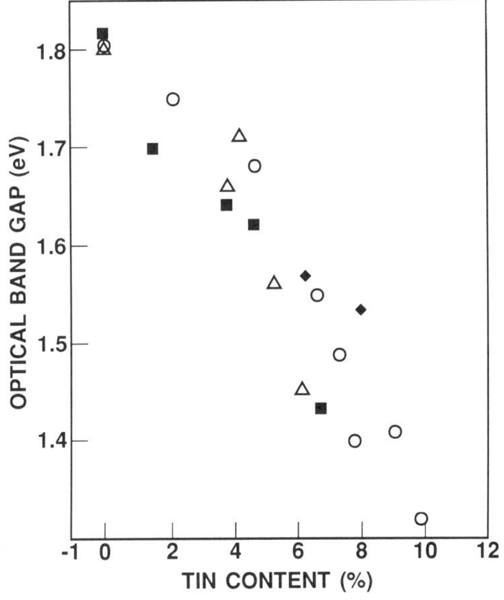

Figure 2.5.2-1 Optical Bandgap vs. tin content for RF CVD of a-SiN:H at three power levels [Mahan et al. 1984b]

The reduction of the optical bandgap with increasing tin content is shown in **Figure 2.5.2-1.** 10% tin reduces the bandgap to about 1.3 eV and about 17% reduces the bandgap to 1.0 eV. The different symbols in Figure 2.5.2-1 indicate different power levels during deposition.

Reducing the bandgap leads to a rapid decrease in the dark conductivity reaching a minimum at about 1.73 eV as seen in **Figure 2.5.2-2a.** The dark conductivity then increases from that minimum with further reduction in the bandgap. Typical dark conductivity data for a-SiGe:H [Nakamura et al. 1982] are shown for reference. The increasing

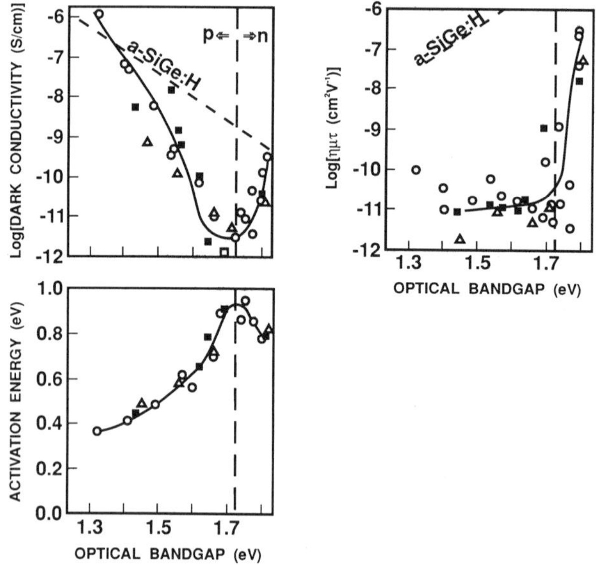

Figure 2.5.2-2 Dark Conductivity, activation energy, and ημτ-product vs. optical bandgap for a-SiSn:H [Mahan et al. 1984b].

doping leads to a change from n-type material to p-type material with a change in the conductivity mechanism from extended-state conduction to hopping. The corresponding change in the dark conductivity activation energy is shown in **Figure 2.5.2-2b.** The photoresponse (ημτ-product) measured at 600 nm with an incident photon flux of 3.4 x 10^{15} photons cm^{-2} s^{-1} decreases by 5 orders of magnitude from 1.8 eV to 1.72 eV and then remains relatively constant with further reduction in bandgap (**Figure 2.5.2-2c**).

The reduction in the ημτ-product may be partly due to the change in conduction mechanism from electron majority carrier controlled to hole carrier controlled [Mahan et al. 1984b].

2.5.2.2 *Materials properties*

Charge collection experiments indicate that the carrier range for both electrons and holes is strongly reduced with incorporation of Sn. **Figure 2.5.2-3** shows the μτ-product vs. the density of midgap states as obtained by space charge limited current measurements for a variety of undoped a-Si:H and a-SiSn:H samples. For the Sn-alloy material, the μτ-

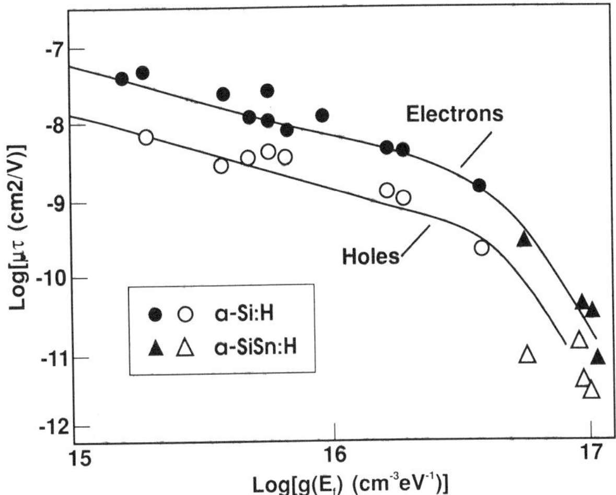

Figure 2.5.2-3 μτ-product for electrons and holes vs midgap defect density for a-Si:H and a-SiSn:H [Mahan et al. 1984a].

product decreases rapidly above 5 x 10^{16} cm^{-3} eV^{-1} [Mahan et al. 1984a]. The stability of hydrogen in coevaporated a-$Si_{1-x}Sn_x$:H (x = 0 to 0.2) alloys was found to decrease with increasing tin concentration [Houssaini et al. 1993].

2.6 METASTABLE PROPERTIES OF AMORPHOUS SILICON-BASED ALLOYS

Most of the properties of a-Si:H alloys discussed in this book so far are initial, as-deposited or annealed properties. In reality, many of these properties of a-Si:H alloys, such as σ_l, σ_d, E_a, N_s, E_u, and $\eta\mu\tau$, are unstable. All a-Si:H alloys exhibit reversible degradations in many material characteristics as a result of light exposure, charge injection, or thermal quenching. The light-induced metastability is often called the Staebler-Wronski effect after the initial observers of the phenomenon [Staebler & Wronski 1977]. Many of the electrical and optical properties listed in Table 2-1 are affected. However, there is no conclusive evidence that E_g and any of the physical properties listed in Table 2-1 are affected by light soaking. For a-Si:H solar cell devices, the characteristics most affected are V_{oc} and FF, due mostly to the degradation of the i-layer.

Research on the causes of this metastability has been ongoing since the discovery of the Staebler-Wronski effect, and numerous papers on the subject have been published [See, for example, the three American Institute of Physics conference proceedings on a-Si:H - Taylor & Bishop 1984, Stafford & Sabisky 1987, Stafford 1991]. Although there is no universally accepted model for the basic mechanisms that cause metastability in a-Si:H alloys, the following experimentally observed effects are widely accepted:

- The light-induced degradation has been observed in all high-quality a-Si:H alloys regardless of the method of preparation.
- The metastability is a bulk effect not a surface effect.
- Light soaking with photon energy larger than 1.1 eV, bombardments by high energy particles (e.g., X-rays or electrons), quenching from high temperatures, and charge injection can all induce metastable defects.
- Photons generate metastable defects only indirectly, via photo-generated charge carrier recombinations. At room temperature, the light-induced defect generation rate is low, one defect is produced per 10^8 recombination events and this rate decreases until saturation is reached [Adler 1983].
- The magnitude of light-induced change decreases as the sample temperature increases.
- The metastable defects can be removed by thermal annealing at 100°C to 200°C for about 1 hour. Infrared irradiation has no effect on annealing. Visible light irradiation increases the thermal annealing rate.
- The density of neutral silicon dangling bonds are increased after degradation. The saturated light-induced defect density, N_{sat}, is between 5×10^{16} and 2×10^{17} cm^{-3} and drops with decreasing E_g and with decreasing C_H from 2×10^{17} cm^{-3} at 1.85 eV E_g to ~5 x 10^{16} cm^{-3} at 1.6 eV E_g. N_{sat} is not correlated with the initial defect density, $N_{s,o}$, nor with the Urbach energy [Park et al. 1990]. **Figure 2.6-1** shows $N_{s,o}$ and N_{sat} of a-Si:H, a-Ge:H, and a-SiGe:H from E_g of 1.15 eV to 1.83 eV The films were deposited by either conventional glow discharge or hot-wire (HW) CVD [§ 11.3] [Wagner 1992]. All the data are by CPM except as indicated.
- The neutral Si dangling bond states formed in the metastable state of a-Si:H are located about 0.8 eV above the valence band mobility edge.
- After light soaking, σ_l and σ_d decrease, E_a and sub-bandgap

absorption increase, the main 1.2 eV luminescence peak intensity decreases while the 0.7 to 0.9 eV defect luminescence increases.

- Although light soaking increases the deep-level defect density, it has little effect on the bandtail states [Wang et al. 1992]. Thus, the decreases in electron and hole $\mu\tau$ products are mainly due to the decreases in lifetimes not drift mobilities.
- The slow relaxation towards the lowest energy configurations of light-induced defect density, band tail carrier density, or conductivity at different annealing temperatures follows a stretched exponential time dependence, $\Delta = \Delta_0 \exp[-(t/\tau)^\beta]$, with Δ being the normalized change, Δ_0 and τ are constants, and the value of β is between 0 and 1 depending on the temperature. This time dependence is observed in a wide variety of disordered materials [Street 1991].
- The metastability is likely an intrinsic property of a-Si:H alloys. Impurities below 1×10^{18} cm^{-3} have no observable effects on the metastability [Tsai et al. 1984]. The concentration of most impurities, such as P, B, C, and N, can be reduced to levels lower than the concentration of light-induced defects without improving the stability of the material.
- The metastability is not directly related to intrinsic mechanical stress [Kurtz et al. 1986].
- No change in hydrogen bonding structure (SiH_n and CH_n) for a-Si:H and a-SiC:H (up to 2.0 eV bandgap) has been observed after light soaking using infrared absorption. There is also no evidence that light soaking releases hydrogen in a-Si:H alloys. However, most researchers believe hydrogen is involved in the creation and annealing of metastable defects either directly, by migrating between alternative bonding sites, or indirectly, by stabilizing defects or determining the defect structure.

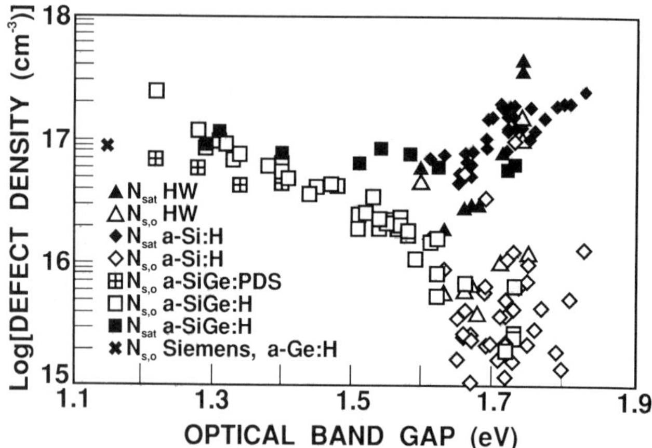

Figure 2.6-1 Initial, $N_{s,o}$, and saturated, N_{sat}, defect densities of a-Si:H, a-Ge:H, and a-SiGe:H from E_g of 1.15 eV to 1.83 eV. [Wagner 1992]

2.6.1 Light- and Charge Carrier-Induced Metastability

The most obvious metastable property of a-Si:H alloys is the reductions of σ_l and σ_d after light soaking. However, the following precautions have to be taken when doing studies of the metastability using conductivity measurements:

(1) The sample temperature has to be well controlled, especially when doing accelerated degradation using concentrated light, because both the conductivity values and the saturated defect density depend strongly on the sample temperature during light soaking. Samples with different E_g may absorb light differently and reach a different stable temperature during light soaking.

(2) Make sure the material under study is fully annealed and then fully degraded. For doing AM1 light soaking, this means at least 250 hours of light soaking is needed. Otherwise, one is only studying the degradation rate of the material, not the total degradation.

(3) Keep in mind that materials with poorer initial quality degrade less, and the degradation rate depends on the history of the sample. The defect densities in poor quality a-Si:H films are sometimes so large that they mask the effects produced by

metastable defects. It is also possible that the defect density and position in the bandgap are such that they short out the charge recombination processes that induce the metastable defects. For example, a-Si:H alloys with lower initial σ_l show less σ_l degradations after light soaking than materials with higher initial σ_l. The relative changes in σ_l as a function of the initial σ_l for undoped a-Si:H and a-Si:H:F samples deposited by different methods are shown in **Fig. 2.6.1-1** [Smith, E., et al. 1989]. For films with initial σ_l less than 1 x 10^{-5} S/cm, the relative change, $\Delta\sigma_l/\sigma_l$, is so small that σ_l appears to be stable. It is often misleading to draw conclusions of metastability based on materials with quality inferior than the best glow discharge-deposited a-Si:H. Historically, many a-Si:H films deposited by new deposition techniques were mistakenly believed to be more stable than conventional a-Si:H.

In Fig. 2.6.1-1 the relative changes in photoconductivity, $\Delta\sigma_l/\sigma_l$, after light soaking are plotted against the initial σ_l values for a-Si:H and a-Si:H:F deposited by different methods (where GD = glow discharge deposition, GD(XeF_2) = GD with the addition of a XeF_2 etch gas, GD/implant = GD followed by rehydrogenation by a hydrogen ion beam, GD/(RF-H) = GD followed by RF posthydrogenation, PED(RF-H) = periodic-etching-deposition by GD using an RF-generated hydrogen plasma for the etching cycle, PED(XeF_2) = PED using XeF_2 vapors for the etching cycle).

The simplest model for the metastability is a two-level system of defect configurations. A configuration diagram often used for a-Si:H is shown in **Fig. 2.6.1-2**. There is a high potential barrier separating the annealed state at q_0 and the metastable state at x_1. The activation energy for annealing is smaller, by ΔE, than the activation energy for defect generation. The energy barrier, E_0, and the separation, x_1-x_0, are small enough to allow tunneling transitions between the configurational states. For a-Si:H, E_0 is of the order of 1 eV. At low temperatures, there is not enough thermal energy to overcome the barrier and external sources of energy, such as photons, are needed for the defect generation. At high temperatures, thermal activations over the potential barrier are the dominant generation and annealing mechanisms. While this model appears capable of phenomenologically describing the kinetic behavior of the degradation, it lacks the capability to predict the microscopic structure of the convertible sites.

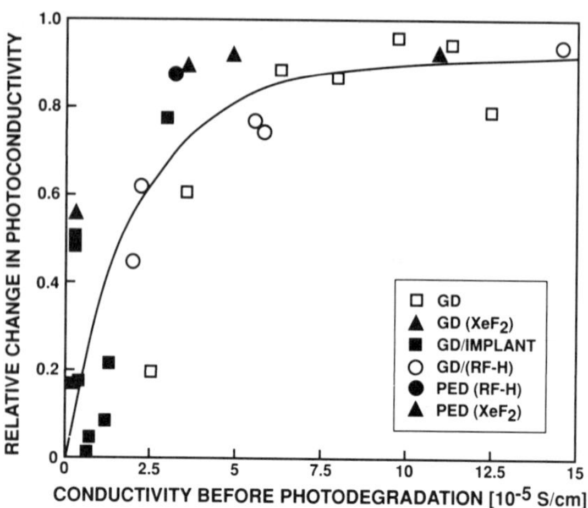

Figure 2.6.1-1 Relative changes in photoconductivity, $\Delta\sigma_l/\sigma_l$, after light soaking vs. the initial σ_l values for a-Si:H and a-Si:H:F deposited by different methods.

Several models, although not necessarily mutually exclusive, of the microscopic mechanism for the metastability in a-Si:H alloys have been proposed. Although the weak-bond breaking model is the most popular, total validity of any of the following models has not been demonstrated. Some of the well known models are the following:

1) Weak-bond-breaking Model: In this model, weak or highly strained Si-Si bonds are broken (Fig. **2.6.1-3**) by the energy released by nonradiative recombination of photo-generated charge carriers and resulting in two neighboring neutral threefold coordinated Si dangling bonds [Eliot 1979, Stutzmann et al. 1985]. To prevent the two neighboring bonds from recombining, a neighboring hydrogen atom moves in to separate the two neighboring dangling bonds and leaves a dangling bond in its original position [Dersch et al. 1980]. Two Si dangling bonds are formed by this process. In the annealing process, the hydrogen atom moves back to its original position and the two dangling bonds face each other again and recombine. Hydrogen is directly involved in both the defect generation and annealing of this model.

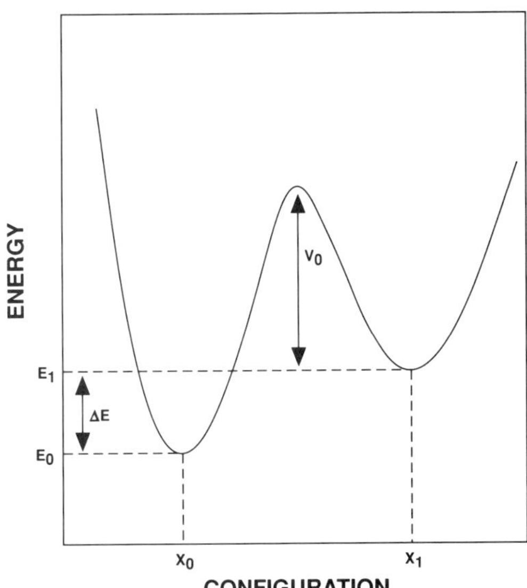

Figure 2.6.1-2. Energy vs. configuration diagram used for a phenomenological description of the two-level systems and metastable defects in a-Si:H [Stutzmann et al. 1985].

2) Charge-defect Model: Charged states of dangling bonds, which are not detected by ESR, are converted to neutral dangling bonds during degradation [Adler 1982]. Theoretically, these charged dangling bond defects may be found in inhomogeneous local environments, such as sites affected by C, O, or N contaminants or sites having H-clustering or excessive local stresses [Bar-Yam et al. 1986]. This model differs from the weak-bond breaking model in that it requires the existence of about 10^{17} cm^{-3} of charged defects in the annealed state. However, currently, there is no experimental evidence for the existence of charged defects in the annealed state. Branz and Silver [1990] suggested that positively correlated charged dangling bond defects can exist in a-Si:H due to long range potential fluctuations.

3) Rehybridized Two-site Model: This model by Redfield and Bube [1990] is based on the involvement of impurity atoms. It attempt to describe the degradation process using simple differential rate

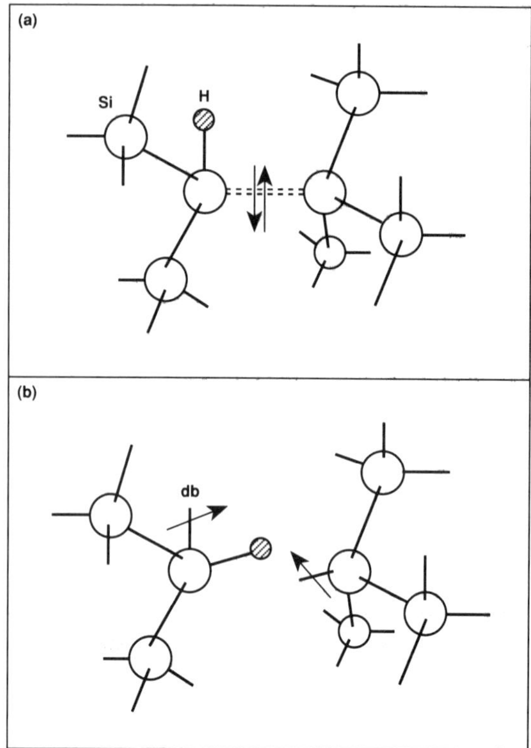

Figure 2.6.1-3 A possible microscopic process leading to the creation of metastable dangling bonds [Stutzmann et al. 1985].

equations, based on the following considerations: (1) the number of sites at which light-induced defects can be generated is fixed for a given material, and (2) the number of convertible sites increases because of contamination and doping. The fixed number of convertible sites determines the saturated defect level at low temperatures, which appears to be temperature independent below 70°C and, thus, not determined by a competitive annealing process.

4) It has also been suggested that the doped a-Si:H behaves differently from undoped a-Si:H. In doped a-Si:H, light-induced changes have been explained by changes in dopant coordination [Branz 1988], by the creation of midgap states, or by a combination of both.

Most of the models involve the presence of hydrogen in the material. Experimental correlations have been established between the kinetics of annealing Staebler-Wronski defects and hydrogen motion (H diffusion coefficient). This has led some researchers to attempt to find a clue for the light-induced effect by manipulating the hydrogen content or bonding. However, it presently is not clear whether hydrogen is the driving force behind the effect, or whether electronic defects present may determine the mobility of hydrogen.

Microvoids have been suggested as metastability center in a-Si:H [Carlson 1986], and there is some direct experimental correlation for this model. Guha et al. [1993] established correlation between increased microvoid density resulting from higher deposition rate as measured by SAXS and decreasing initial and light-degraded solar cell performance. McMahon and Crandall [1989] proposed a concept of "safe hole traps" that do not trap electrons in the annealed state and become efficient recombination centers after illumination. The safe hole traps model can explain some features of the temperature dependence of σ_l, but it does not provide a microscopic explanation of the Staebler-Wronski effect.

High impurity levels may affect the metastability of a-Si:H alloy. Nakamura et al. [1989] show that light-induced degradation of a-Si:H deposited by glow discharge increases with increasing SiH_2 bonding and increasing oxygen, nitrogen and carbon impurities above 10^{18} cm^{-3} and that after light soaking the ESR spin density increases from $3x10^{16}$ to $6x10^{16}$ cm^{-3} and the fill factor for cells decreases with increasing number of SiH_2 bonds in the 10^{21} to 10^{22} cm^{-3} range. Heavily compensated materials [Stutzmann, 1990] show very few light-induced neutral dangling bond defects after 20 hours of light soaking while the initial defect density remains low.

a-SiC:H films of 1.90 eV deposited from hydrogen diluted silane/methane, disilylmethane (DSM), and trisilylmethane (TSM) have considerable increased absorption coefficient at 1.2 eV (50-60 cm^{-1}) after light soaking, but the absorption coefficient is nevertheless still lower than for films from undiluted silane/methane (97 cm^{-1}). For comparison, a-Si:H films have an absorption coefficient of 29 cm^{-1} after the same light soaking. The Urbach energies increase only slightly as a result of the light soaking. For a-SiC:H alloys, comparison of photoconductivity degradation data with device degradation data indicates that photoconductivity alone is not an appropriate indicator for stability for bipolar devices such as solar cells [Catalano et al. 1992]. Similarly, for a-SiGe:H there is a lack of correlation between the change in defect density calculated from the integrated sub-bandgap absorption below the

Urbach energy as measured by CPM and device degradation due to light soaking [Guha et al. 1992].

2.6.2 Thermally Induced Metastability

Recent studies have found that defect state densities in a-Si:H alloys are not entirely defined at the time of film deposition [Street 1991] but reach thermal equilibrium in the temperature range 100° to 300°C. This thermodynamic equilibration determines the defect density in both doped and intrinsic a-Si:H [McMahon and Tsu 1987]. The thermal equilibrium defect density increases with sample temperature [Street and Winer 1989]. The thermal equilibrium defect densities of a-Si:H from 200 to 250°C is between 1 and 3 x 10^{15} cm^{-3} [McMahon 1991]. At temperatures below the thermal equilibration temperature, metastable defects are frozen in because of the long equilibration time, and the electronic properties of a-Si:H are not unique but dependent on the thermal history of the sample. For example, the conductivity of doped a-Si:H films was found to depend on the cooling rate [Ast and Brodsky 1979, Street et al. 1986]. Compared to light-soaking defects, frozen-in defects anneal more slowly and saturate at a lower level (10^{15} vs. 10^{17} cm^{-3}).

The defect structure thermal equilibration of a-Si:H alloys can be understood by analogy to the glass transition temperature in glassy solids. Glasses have very similar structures in both the liquid and solid states. One can think of a glass as a liquid whose atoms have been frozen in place at the glass transition temperature. Thermal equilibrium in a-Si:H alloys occurs rapidly above the glass-transition (equilibration) temperature (200°C for undoped, 130°C for n-type, and 80°C for p-type a-Si:H) [Street 1991], but at lower temperatures the relaxation time is very long, so that the material is in a frozen-in state. The glass transition temperature depends on the rate of cooling and defect density. Rapid quenching from high temperatures create quenched-in defects that anneal themselves back slowly.

Because the thermal equilibrium defect density increases with temperature, there seems to be two possible ways of reducing the defects in a-Si:H: (1) decreasing the deposition temperature, T_s, and thus reducing the defects present at film growth, and (2) increasing the speed of defect equilibration at low temperature. Smith and Wagner [1987] have studied the effects of reducing T_s on defect density. They found that the neutral dangling-bond density measured at room temperature decreases with T_s

from 600°C to about 250°C. When T_s is reduced to below 250°C, the low substrate temperature reduces the surface mobility of film growth constituents and causes a strongly inhomogeneous film structure and poor electrical properties. One possible method of overcoming this is to use high hydrogen dilution, because the enhanced etching by atomic hydrogen creates the effects of a higher substrate temperature without raising the actual deposition temperature. It would also be interesting to investigate whether the defect equilibration speed can be increased by some of the postdeposition treatment techniques, such as posthydrogenation.

2.6.3 Photostability of Photovoltaic Solar Cells

Unstable material properties are a problem in most device applications of a-Si:H, particularly photoelectric applications such as solar cells. Although, the research community has not reached a consensus on the basic mechanisms that cause the metastability, the device research efforts have been somewhat successful in reducing the solar cell performance degradations by using mostly empirical approaches. The proven techniques yielding increased stabilized solar cell output are the following [Luft et al. 1991]: (1) thinner i-layers [Bennett and Rajan, 1988], (2) maximizing optical absorption, (3) better p+ and transition layers, (4) optimizing transparent conductive oxide layers, and (5) compositional grading [Guha et al. 1988]. Implementation of these techniques (1 through 4) has resulted in 10% stable cell efficiency in 1990 [Ichikawa et al. 1990b].

So far, the optimized stabilized performance has been obtained with materials that also yield the best initial performance. It appears that when a material is produced under "non-optimum" conditions, any potential gain in stability is overwhelmed by a loss in initial performance, and that degradation and the initial performance cannot be addressed separately. The majority of research to date has concentrated on optimizing the properties of the initial (as-grown) state. In many instances the initial material parameters found to be important for the optimization process were used to model solar cell performance. There has been a lack of correlation between the stabilized materials parameters and the stabilized device efficiency.

In order to further improve the photostability of a-Si:H solar cells, a number of questions must be answered: (1) For device-quality materials, are there important cause-and-effect relationships between material quality indicators, such as ESR spin density, photoconductivity, and electron or hole life time, and device quality indicators, such as open-circuit voltage, short-circuit current density, and fill factor? (2) Provided there is a correlation, what are the limits of these main material quality indicators at which they cease to be of significance for device performance? (3) Is the degradation inherent and intrinsic to a-Si:H alloys, and if so, can the degree of instability be reduced? (4) Is the change in material quality indicators resulting from light soaking reflected in a corresponding change in device quality indicators? Are there correlations of importance that have been overlooked? (5) What are the preferred instability indicators? Is there only one or are there several? Should they be device quality indicators? If so, should they be relative changes or absolute changes. (6) How do light-induced changes in doped layers or at interfaces affect the stabilized device performance? How can the influence of doped layers and interfaces in a device be separated from changes in the intrinsic layers? (7) Are the currently used optimization procedures for best initial material properties necessary or sufficient for optimized stabilized device performance?

3
FILM DIAGNOSTIC MEASUREMENTS

A variety of diagnostic measurements of the deposited films are employed to determine the film quality. Diagnostic measurements of films include *in-situ* and *ex-situ* ellipsometry, Raman scattering, photoluminescence (PL), nuclear magnetic resonance (NMR), transmittance spectroscopy, etc. Fritzsche [1980], Knights [1984], and Pankove [1984] describe a number of measurement techniques for film diagnostics. An extensive review of the properties of amorphous silicon films reported before 1985 has been published by the Institute of Electrical Engineers [Properties of Amorphous Silicon 1985]. Some of the relevant diagnostic techniques for films are discussed in this chapter. **Table 3-1** lists important film characteristics and the techniques to measure them.

Before we discuss specific measurement methods, we need to realize that the measured properties of a-Si:H often depends strongly on film thickness [§ 2.2] and on the measurement method used. For example, $\mu\tau$ products measured by the steadystate photoconductivity methods are up to three orders of magnitude larger than the $\mu\tau$ products measured by time-of-flight charge-collection method [Crandall and Balberg 1991]. In another example, the charge carrier diffusion lengths measured by the surface photovoltage method are more than an order of magnitude larger than those measured by the steadystate photocarrier grating technique [Ritter et al. 1987 and Balberg et al. 1988]. When comparing properties of a-Si:H, it is important to compare only properties measured by similar techniques on films of similar thickness.

3.1 OPTICAL MEASUREMENTS

3.1.1 Ellipsometry

Ellipsometry is extremely sensitive to near-surface (the first few Angstroms) structural changes and provides a measure of the Si-Si bond packing density. It can be used for *in-situ* growth analysis of amorphous and microcrystalline film depositions [Collins and Pawlowski 1986, Antoine et al. 1987]. The ellipsometry can be either single-wavelength or spectroscopic [Collins et al. 1985a]. Spectroscopic ellipsometry is effective in detecting heterogeneity and the onset of crystallinity in film deposition [Collins et al. 1985b]. Jackson et al. [1985] measured the complex dielectric functions of a-Si:H from 1.5 to 5.9 eV using a rotating-polarizer ellipsometer. Lin et al. [1988] showed that a growing a-Si:H film has a very thin (1 to 2 atomic layers) hydrogen-rich surface.

Table 3-1 Important Film Characteristics and How They are Being Measured

Characteristic	Measurement Technique	Section
Absorption Coefficient	Absorption, transmission	3.1.4, 3.1.5
Activation Energy	Dark conductivity vs. temperature	3.3.1
Bond-angle Disorder	Raman scattering	3.1.3
Bonded Hydrogen Content	Infrared transmittance spectroscopy, NMR	3.1.5, 3.2.2
Columnar Structure	TEM, SEM	3.6.1
Crystallinity	Raman scattering, ellipsometry, X-ray	3.1.1, 3.1.3, 3.6.3
Dark Conductivity	Current vs. voltage	3.3.1
Defect Density	Field effect method	3.2.1
(Density of Defect States)	MOS capacitance-voltage	3.2.1
	Deep level transient spectroscopy	3.2.1
	Isothermal capacitance transient spec.	3.2.1
	Electron spin resonance	3.2.2
	Space-charge-limited current	3.2.3
	Luminescence	3.1.2
	Sub-bandgap absorption	3.1.4
	Transmittance	3.1.5
	Constant photocurrent measurement	3.1.4
	Photothermal deflection spectroscopy	3.1.4
Diffusion Length	Photocarrier gratings, surface photovoltage	3.1.6, 3.3.2
Fermi Energy	Dark conductivity vs. temperature	3.3.1
Heterostructure (Islands and Tissue)	TEM, ellipsometry	3.1.1, 3.6.1, 3.6.2
Hydrogen Bonding	Infrared transmittance spectroscopy, NMR	3.1.5, 3.2.2
Impurity Depth Profile	SIMS, AES, XPS, EPMA	3.5
Microstructure	TEM, SAXS	3.6
Minority Carrier Diffusion Length	Surface photovoltage, SSPG	3.1.6, 3.3.2
Minority Carrier Lifetime	Transient grating (τ), photoconductivity ($\mu\tau$)	3.1.6, 3.3.1
Mobility	Hall measurement, time-of-flight	3.2.4, 3.3.3
Mobility-Lifetime Product	Quantum efficiency, photoconductivity	3.2.1, 3.3.1
Neutral Dangling Bonds	Electron spin resonance	3.2.2
Optical Bandgap	Optical absorptance vs. energy (Tauc plot)	3.1.5
Photoconductivity	Conductivity under illumination	3.3.1
Surface Morphology	SEM, STEM	3.6.1, 3.6.2
Total Hydrogen Content	Gas evolution	3.5
Urbach Energy	Sub-bandgap absorption	3.1.4
Void Fraction, Size, and Number	Small angle x-ray scattering	3.6.3

3.1.2 Photoluminescence and Electroluminescence

Luminescence is a technique widely used in studying carrier recombination mechanisms and the localized states within the bandgap of a semiconductor. It is particularly applicable to amorphous semiconductors with high densities of localized states [Street et al. 1978, Pankove 1984, Taylor and Bishop 1984]. It has been shown that there is an inverse correlation between luminescence intensity and electron spin resonance (ESR) spin density in a-Si:H [Street et al. 1982]. In a luminescence process, electron-hole pairs are first excited, usually by externally injected photons (photoluminescence) or charge carriers (electroluminescence). Electrons and holes then relax downward in energy (called thermalization) and eventually (in about 10^{-12} s) end up in band-edge localized states. Finally, they recombine either radiatively (luminescence) or nonradiatively (energy converted to heat). Dangling bonds and defects are predominant nonradiative recombination centers. By studying the relaxation process and the intensity and spectral distributions of the radiated signal, one can obtain detailed information about the nature of defect states [Bauer and Bilger 1981, Pankove 1984].

Photoluminescence (PL) measurements of a-Si:H usually start at the liquid helium temperature of about 4 K, the PL signal increases with temperature due to reductions in Auger recombinations of neighboring electron-hole pairs. The PL intensity reaches a peak at 50 to 100 K (depending on the sample) and then decreases with temperature. There is no detectable PL signal in a-Si:H at room temperature because most of the photogenerated charge carriers recombine through the defects rather than from band tail to band tail. Usually there are two clearly observable PL peaks in a-Si:H - a dominant peak near 1.4 eV and a smaller peak near 0.9 eV. The 1.4 eV peak is identified as a radiative tunneling transition between band-tail states. The 0.9 eV peak seems to originate from radiative recombinations through defect states [Street in Pankove 1984].

Electroluminescence (EL) can be observed in a-Si:H by forward biasing a p-i-n device. The EL signal comes from the i-layer near the p/i interface [Carius 1990]. Care must be taken to avoid sample heating due to large current flowing through the sample. Sample heating can cause the EL peak positions to shift to higher energies via bandgap widening. The nature and energy positions of EL and PL peaks are similar. Wang et al. [1993] measured the EL lifetime distribution of a-Si:H and found two peaks: one occurs at 10^{-6} s and is independent of temperature, another one changes from 10^{-3} to 10^{-6} s with increasing temperature.

3.1.3 Raman Scattering

Raman scattering is being used extensively as a characterization tool for crystalline semiconductors [Pollak and Tsu 1983, Lannin 1988]. In the Raman effect, a photon is scattered inelastic by a solid material with the subsequent creation (Stokes effect) or annihilation (anti-Stokes effect) of one or two phonons (equivalent to 20 to 40 meV of energy gain or loss). These electron-phonon interactions, which are responsible for the scattering, are very sensitive to local environments in the material. Raman scattering provides a fast, convenient method for determining whether a semiconductor such as silicon is crystalline or amorphous. Crystalline Si exhibits a sharp Raman peak near 520 cm^{-1} due to the zone-center optical phonon; a-Si:H has a much broader peak centered at 480 cm^{-1}. Integrated widths of these 480 cm^{-1} and 520 cm^{-1} peaks can be qualitatively used to judge relative crystalline content. Raman scattering may also be used to estimate the microcrystalline particle size [Pollak and Tsu 1983]. In addition, detailed studies of the Raman spectra can provide dynamic and structural information about amorphous solids, such as bond angle disorder [Pankove 1984 Part B, Tsu et al. 1984, Hishikawa et al. 1985]. For example, Sakata et al. [1985] studied by Raman scattering a-Si:H films deposited from undiluted silane and disilane at rates of 1 to 10.5 nm/s and found that the structural disorder increases proportionally to the deposition rate.

3.1.4 Sub-Bandgap Optical Absorption Spectra Measurements

Direct optical transmission and reflection measurements can provide information about defect states in a-Si:H [Bauer and Bilger 1981]. However, when the defect density is small and the absorption coefficient is below 10^3 cm^{-1} as is often the case with a-Si:H, more than 10 µm-thick films are required for accurate measurements. This is often impractical. In such cases, indirect measurements that determine the spectral absorption - such as photoconductivity [Moddel et al. 1980, Abeles et al. 1980, Chahed et al. 1987], electroabsorption [Eggert and Paul 1987], photoacoustic spectroscopy (PAS) [Jackson and Amer 1980], photothermal deflection spectroscopy (PDS) [Jackson et al. 1981, Pankove 1984 Part B], and constant photocurrent method (CPM) [Moddel et al. 1980, Vanecek et al. 1984] - have greater sensitivities to determine the defect-state densities. Among these techniques, PDS is the most commonly used. In this technique, the sub-bandgap absorption is measured by heating the material with an intensity-modulated monochromatic light and modulating the index of refraction near the material's surface with a laser probe. PDS is very surface-sensitive, which is a severe disadvantage if the behavior of the surface is different from that of the bulk. Winer et al. [1988], using total-yield photoelectron spectroscopy in combination with the Kelvin probe, found the existence of an excess density of deep defect states within about 10 nm of the surface of an 1 µm-thick, undoped a-Si:H sample corresponding to an equivalent surface-state density of 3×10^{11} cm^{-2}. From the PDS measurement one can obtain information about the density of defect states and the slope of the exponential sub-bandgap absorption region known as the Urbach tail region [Joannopoulos and Lucovsky 1984]. This slope (between the log of absorption and the wavelength) is related to the width of the density of gap states near the band edges. This width in the unit of eV is called Urbach energy, E_u. Because the valence band tail is usually much wider than the conduction band tail (§ 3.5), it dominates the optical absorption process and mostly determines E_u. E_u gives a direct measure of structural and compositional disorder, because it relates to the localized defect states in the valence band tail. A small E_u is desirable for good charge transport properties. For high quality intrinsic a-Si:H, the Urbach energy, E_u, is usually less than 50 meV. However, for thin samples or samples with large surface defects, the PDS-measured E_u can be significantly larger than the bulk value.

CPM was found to be relatively insensitive to surface defects [Xu et al. 1987, Shimizu et al. 1988] and it is also a simpler method than PDS. In CPM, the light intensity of a monochromatic light is adjusted to keep the photocurrent constant while the wavelength of the monochromatic light is being adjusted. By matching CPM measured absorption with direct optical absorption at short wavelengths (with photon energies between 1.75 and 1.85 eV), we obtain a plot of the sub-bandgap absorption coefficient as a function of wavelength, which gives us information about the defect states.

Photocurrent or photoconductivity due to sub-bandgap photon absorption can be used to determine both band-tail and mid-gap defect states in a device or film [Crandall 1980, Okamoto et al. 1985, Hegedus et al. 1986, Smith et al. 1987]. Primary photocurrent measured on a reverse-biased junction is dominated by the mobile hole density generated by the sub-bandgap photon absorption by electrons moving from the valence band to defect states. In a primary photocurrent measurement, a charge is counted only once per photon absorbed and, ideally, the current is independent of voltage. Estimations of $\mu\tau$ can be obtained from photocurrent vs. reverse bias [Crandall 1983, 1984]. Secondary photocurrent measured on a forward-biased junction or film with coplanar, ohmic contacts is dominated by the mobile electron density generated by the sub-bandgap optical absorption by electrons moving from defect states to the conduction band.

3.1.5 Transmittance Spectrometry

The defect density in a-Si:H depends more on how the hydrogen atoms are bonded and distributed in the material than on the total hydrogen content. Dispersive or Fourier-transformed infrared transmittance spectroscopy (FTIR) is commonly used to determine both the bonded-hydrogen content and the silicon-hydrogen bonding configurations. IR absorption measurements are much easier to make and simpler to analyze than other hydrogen detection techniques, such as hydrogen evolution (§ 3.5), nuclear resonance reactions [Ross et al. 1982], and secondary ion mass spectroscopy (§ 3.5). High-resistivity, single-crystalline Si wafers are normally used as substrates for IR absorption measurements. Float-zone-growth wafers are preferred because of their low oxygen contents, although the more widely-available Czochralski-grown Si wafers are acceptable for dual-beam measurements. Other IR-transparent crystals, such as KBr, may also be used as substrates. The spectral distribution of the absorption peaks of various silicon-hydrogen bonding modes (SiH, SiH_2, SiH_3, and $(SiH_2)_n$) has been identified [Brodsky et al. 1977, Lucovsky et al. 1979, Shanks et al. 1980]. See **Table 3-2** for the various wavenumbers corresponding to different bonding modes. To identify SiH_2 bonds, it is better to use the 875 cm^{-1} peak (bending) than the 2090 cm^{-1} peak (stretching), since the 2090 cm^{-1} peak also corresponds to the stretching of $(SiH_2)_n$ [Shanks et al. 1980] and the SiH absorption of inner surfaces of voids [Wagner and Beyer 1983]. The most common method for determining the hydrogen content of a-Si:H by converting the IR transmission spectra to absorption spectra has been the method proposed by Brodsky et al. [1977]. They used double-side-polished Si substrates for their original IR study. Most people now use single-side polished Si substrates. Langford et al. [1989a] showed that, as long as the correction for multiple reflections suggested by Brodsky et al. is used, there is no advantage in using deliberately roughened substrates or double-side-polished substrates. A new, but not yet widely accepted, set of proportionality constants between the hydrogen concentration and the integrated absorbance of the wagging and stretching modes used to determine the hydrogen content of a-Si:H from IR transmission spectra was proposed by Langford et al. [1992].

Table 3-2 **Wavenumbers for the Principal Si and Ge Bondings in a-Si:H, a-Ge:H, and Related Alloys** [Lucovsky et al. 1979, Tawada et al. 1982a, Cardona 1983, Shinar et al. 1987, Langford et al. 1990]

Group	Wavenumber (cm^{-1})	Assignment
SiH	2000	Stretching
	630	Bending
SiH_2	2080-2090	Stretching
	880	Bending-scissors
	630	Rocking
$(SiH_2)_n$	2090-2140	Stretching
	890	Bending-scissors
	845	Waging
	630	Rocking
SiH_3	2120-2140	Stretching
	905	Degenerate Deformation
	860	Symmetric Deformation
	630	Rocking
GeH	1880	Stretching
	570	Bending
GeH_2	1980	Stretching
	830	Bending-Scissors
	570	Rocking
$(GeH_2)_n$	2000	Stretching
	825	Scissors
	765	Waging
	570	Rocking
GeH_3	2050-2060	Stretching
	770, 830	Bending
SiO	1150	Surface
	950-1100	Bulk, stretching
SiH-O	2100	SiH with one O in the local cluster
$SiH\text{-}O_2$	2185	SiH with two O in the local cluster
$SiH_2\text{-}O$	2160	SiH_2 with one O atom bonded to Si
SiC	670	Stretching
$Si\text{-}CH_3$	780	Rocking or Wagging
	1250	Bending
CH_2	2850, 2910	Stretching
	1450	Bending

Table 3-2 Wavenumbers for the Principal Si and Ge Bondings in a-Si:H, a-Ge:H, and Related Alloys (Continued).

Group	Wavenumber (cm^{-1})	Assignment
CH_3	2890,2940	Stretching
	1350	Bending
SiF	830	Stretching
SiF_2	930	Stretching
SiF_4	1015	Stretching
Si(TO)	510	Activated by F
Si-Si	480	
Si-Ge	365	
Ge-Ge	265	

Langford et al. [1992] and Maley [1992] found that, for a-Si:H, the hydrogen content determined by conventional methods from the 630 cm^{-1} absorption peak is overestimated if the film thickness is less than about 1 µm. For greater accuracy, IR transmission data should be analyzed by taking the effects of optical interference into account.

Infrared absorption can also be used to identify other bonded impurities, such as O, N, Cl, and F [Rao 1963, Fang et al. 1980, Dutta et al. 1983, Lucovsky and Pollard 1983, Wu et al. 1983, Fang et al. 1985, Ray et al. 1987, Shinar et al. 1987, Langford et al. 1990]. The detection limits of IR absorption techniques is usually about 1 at.%.

Spectrophotometry in the ultraviolet and visible range is used to determine the optical bandgap of a film by means of a plot proposed by Tauc [1974], who argued that a-Si:H has a pseudo-direct bandgap; the transitions responsible for light absorption are limited only by the availability of states not by momentum considerations; and the density of extended states has a square-root dependence on energy near the band edges. In this case, the absorption follows the relation of $(\alpha h\nu)^{1/2} = C(h\nu - E_g)$, where $h\nu$ is the incident photon energy and C is a constant which depends on the material. The optical absorption coefficient, α, is defined in the equation $I(x) = I_0 \exp(-\alpha x)$, where I_0 is the incident light intensity, $I(x)$ is the light intensity remaining after traveling a distance of x in the sample. The average depth at which a particular wavelength light can travel in a material before being absorbed is $1/\alpha$ (where the light intensity drops to 1/e). The Tauc plot involves plotting the square root of the product of the absorption coefficient and the photon energy ($\alpha h\nu$) against the photon energy. The point of intersection of that function with the abscissa is the optical bandgap. The slope of this function, called the Urbach energy, can be used to characterize disorder in amorphous silicon-based alloys [Tsu et al. 1987]. Cody et al. [1982] recommend a modified Tauc model that removes the spurious thickness dependence of the derived optical gap. Jackson et al. [1985] determined the valence- and conduction-band density of states and the sub-bandgap density of states of a-Si:H and compared them to the optical bandgap determined by various methods. They found that optical determinations for the bandgap for a-Si:H tend to underestimate the mobility gap.

3.1.6 Photocarrier Grating Technique

The photocarrier grating technique, also known as the steady state photocarrier grating (SSPG) technique, is a simple and reliable method for determining the ambipolar diffusion length [Ritter et al. 1987 and Balberg et al. 1988]. It is based on the creation of a small-amplitude photocarrier grating on a uniform photocarrier background. The measurement apparatus is essentially an interferometer (**Figure 3.1.6-1**). A beam splitter diverts the laser light onto two paths with one beam 10 to 50 times stronger in intensity than the other beam. The stronger beam passes through a half-wave plate, which is used to rotate the plane of

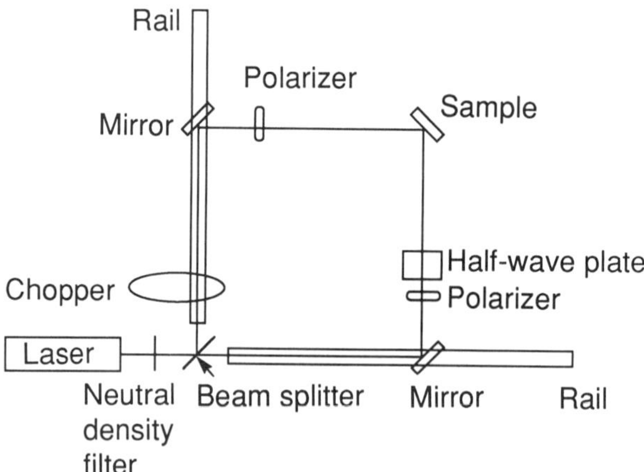

Figure 3.1.6-1 Schematic of optical components for the photocurrent grating technique

polarization. The weak beam passes through a chopper, which provides a low frequency of pulses. A steady state photocarrier grating is created by two interfering laser beams. The measurement is then repeated when the two beams are incoherent. The angle that the two beams make with respect to the sample normal determines the grating period of the interference fringes. If the grating period is much larger than the carrier diffusion length, a well-defined photocarrier grating is created in the sample; but, if the grating period is smaller than the diffusion length, an almost uniform carrier concentration will result. A voltage is applied to the sample through coplanar electrodes oriented such that the current flows perpendicular to the grating. The photocurrent caused by the chopped beam is measured using a lock-in amplifier when the beams are coherent and when they are incoherent. From a determination of the photocurrent as a function of the grating period, the diffusion length of the photocarriers can be obtained. It is convenient that the exact values for the current are not required. An accuracy of better than 10% can be obtained for diffusion lengths between 50-200 nm.

The ambipolar diffusion length in a-Si:H measured by SSPG is usually dominated by holes. Although the electrons have much higher diffusion length than holes, because of the Coulomb attraction caused by the space charge created by the slow-moving holes, the electrons and holes diffuse together. As a result, the slower moving holes dominate the ambipolar diffusion length. For characterizing a-Si:H alloys, SSPG can be used as a good complementary measurement for $\eta\mu\tau$, which is dominated by the majority carrier, electrons [Tsuo et al. 1991].

Komuro et al. [1983] used a transient photocarrier grating method to simultaneously and separately determine the diffusion coefficient and the lifetime of photogenerated carriers in a-Si:H. They used a Q-switched yttrium aluminum garnet (YAG) laser beam for the excitation light, which was split in two beams that converge on the sample to form transient photocarrier gratings, and a Kr laser beam to probe the dynamics of the transient grating produced by the excitation light. The probe light diffracted by the transient grating was analyzed by a monochromator, a photomultiplier, and a transient digitizer. They used a 1.064 μm excitation beam to study the bulk properties and a 0.355 μm excitation beam to study the surface properties. They found that the Staebler-Wronski effect in a-Si:H is a bulk effect, and optical illumination decreases the photoconductivity mainly through the decrease of the lifetime, instead of the mobility, of photoexcited carriers.

3.2 ELECTRICAL MEASUREMENTS

3.2.1 Junction Measurements

Investigations of the properties of the space-charge region that forms at a p-i, n-i, or a Schottky barrier interface by such measurements as electric field-effect, capacitance vs. voltage (C-V), isothermal capacitance transient spectroscopy (ICTS), and deep-level transient spectroscopy (DLTS) can reveal information about the density of states within the bandgap of a-Si:H alloys [see, for example, the discussions in Pankove 1984 Part C, and Tanaka 1989, Chapter 7.]. In DLTS, a voltage is applied to shift E_f in the bandgap to expose occupied defect states. Monitoring the rate electrons leave these exposed states as a function of temperature and time reveals information about the defect states. The junction recovery technique has been used to study the recombination properties of charge carriers [Snell et al. 1981, Silver et al. 1982]. ICTS has been used to determine a nearly complete experimental density of states [Jackson et al. 1985]. However, because of complexities in obtaining and analyzing the data, these junction measurement techniques are not used as standard measurements for monitoring the deposition process. Another problem is that the shape and magnitude of the density of defect states determined by DLTS do not agree with those determined by the space-charge-limited-current measurements. This problem is still unresolved. The field effect measurement, which uses a insulated gate field effect geometry of the sample, measures the defect states in the first 5 to 10 nm layer near the sample surface. This surface property may be different from the bulk.

Measuring the quantum efficiency (the ratio of collected photogenerated charge vs. the number of incident photons at a particular wavelength) vs. wavelength of a solar cell structure can give information about bandtail states [Street 1981], when below bandgap photon energies are used for the measurement. Quantum efficiency measurements under voltage bias can be used to estimate hole $\mu\tau$ products and for seeing where the losses are in a solar cell structure, and what physical causes of the losses may be [Dalal and Alvarez 1981, Catalano and Wood 1988]. Sample structures of SnO_2 / p^+ (or n^+) / i, a-Si:H / metal were used to obtain internal photoemission of carriers into both the conduction and valence bands for measuring the mobility gap of intrinsic a-Si:H and the movement of band edges with temperature [Wronski 1992]. The effective mobility gap at 300 K is found to be 1.89 ± 0.03 eV in intrinsic a-Si:H films with a Tauc gap of 1.73 ± 0.01 eV.

3.2.2 Nuclear Magnetic Resonance and Electron Spin Resonance

Nuclear magnetic resonance (NMR) is used to measure the hydrogen content and the spatial distribution of the hydrogen (whether clustered or dispersed) in amorphous silicon films [Knights 1984] and local structural arrangements, such as the shapes of microvoids. Clustered hydrogen bonded in polymeric forms is identified by a broad NMR line and dispersed hydrogen in monohydride form by a narrow NMR line. With deuteron magnetic resonance, tightly bound deuterium, weakly bound deuterium, and molecular deuterium can be identified in specially prepared films. These data can be used to learn about the hydrogen bonding and microvoid dimensions in a-Si:H. Tsai et al. [1988] found molecular non-bonding hydrogens in a-Si:H by NMR. Leopold et al. [1985] did deuteron and proton NMR and found that SiH_2 bonds do not necessarely have to be associated with voids. They could be scattered throughout the material instead of on void surfaces.

Electron spin resonance (ESR), also known as electron paramagnetic resonance (EPR), is a very useful tool for studying defect states in a-Si:H [Pankove 1984, Street 1991]. The experimental setup is similar to NMR, but at different magnetic field strengths and frequencies. Normal covalent bonds have two electrons with opposite spins and therefore zero net spin. A dangling bond in a-Si:H has a single electron with unpaired spin and can be easily detected by electron spin resonance. Three different types of localized states in a-Si:H are observed by ESR. The defect resonance at $g = 2.0055$ has been identified as silicon neutral dangling bonds. The other two resonances are attributed to band-tail electrons at $g = 2.0044$ and holes at $g = 2.012$, as shown in **Fig. 3.2.2-1**. The density of ESR Si neutral dangling bonds is between 10^{19} and 10^{20} cm^{-3} in a-Si without hydrogen and is on the order of 10^{15} cm^{-3} in high quality a-Si:H. One concern about ESR is that it measures the total dangling bond density, which is not necessarily correlated with the density of deep states that have strong effects on film properties. Another concern is that ESR measures only those defects that have an unpaired spin. Light-induced ESR, which measures the photo-induced changes in the densities of singly occupied electron states, has also been used to study the density of states in a-Si:H alloys [Boulitrop 1983, Ristein et al. 1989].

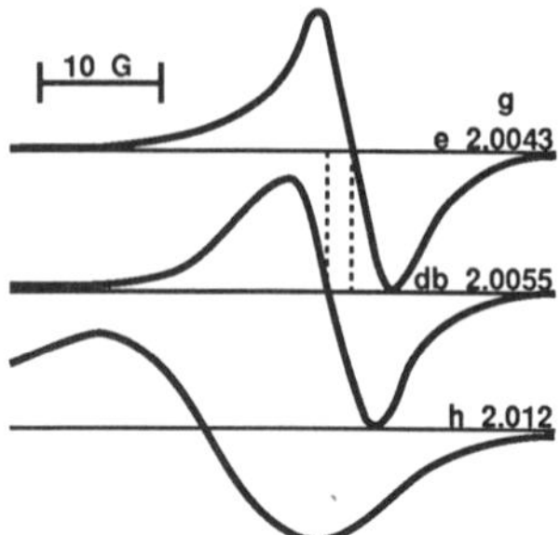

Fig. 3.2.2-1 Derivative of the ESR absorption line shapes of conduction-band-tail electrons (e), valence-band-tail holes (h), and neutral dangling bonds (db) [Dersch et al. 1983].

3.2.3 Space-Charge-Limited Current

Space-charge-limited current (SCLC) measurements are used to determine the distribution and density of defect states (DOS) of a-Si:H, typically within a few tenths of an electron-volt above E_f [Mackenzie et al. 1982, Weisfield 1983, Furukawa and Matsumoto 1983, Crandall et al. 1989]. When a voltage is applied across a sample with electrodes capable of freely injecting or absorbing one type of charge carriers (single-carrier injection), the distortion in the electric field by the build up of space charges at high current densities causes the current to become space-charge limited. An n^+-i-n^+ sample structure with an i-layer thickness of 1 μm or more and an applied voltage of up to about 5 V is usually used. Care should be taken to reduce the phosphorus contamination of the i-layer. The applied electric field should be slightly below the breakdown field of about 2 x 10^5 V/cm. The current-voltage characteristic of an n^+-i-n^+ device is linear at the low-voltage region but becomes superlinear at the high-voltage region as a result of the injection of excess electrons from the n^+ contacts into the i-layer. The space-charge-limited current is proportional to V^2. The DOS can be calculated from the dark current-voltage characteristics in the high-voltage region [Orton 1984]. The densities of defect states below E_f in intrinsic a-Si:H has been measured by SCLC of holes using p^+-i-p^+ sample structures [Dawson et al. 1991].

3.2.4 Hall Mobility Measurements

The Hall method, which measures voltage build up by carriers drifting in an electric field and a magnetic field applied perpendicular to each other and to the direction of the voltage measurement, is widely used to determine the carrier mobility and concentration of semiconductors [Putley 1960, ASTM 1992]. However, undoped a-Si:H and even some doped a-Si:H materials are very resistive, making Hall mobility measurements on a thin film difficult to perform. It is also difficult to interpret the data because the effective mass theory does not apply to amorphous semiconductors. The Hall mobility, μ_H, which is different from conductivity mobility and drift mobility [Dresner in Pankove 1984], in a-Si:H is in the range of 0.1 to 1 cm^2/Vs. This is very small compared to the 10^4 cm^2/Vs Hall mobility of single crystal Si. The sign of μ_H in a-Si:H disagrees with the sign of the thermoelectric power and the temperature dependence of μ_H is small.

3.2.5 Thermopower Measurements

If a temperature gradient exists in a conductor or semiconductor, an electric potential gradient, which is proportional in magnitude to the temperature gradient, also occurs. The determination of the sign of this thermoelectric effect allows classification of a semiconductor as either n- or p-type. In addition, the temperature dependence of the thermopower provides information about the transport mechanism (extended state conduction vs. hopping in tail states) and the energy distribution of the density of localized states [Jones et al. 1976, Overhof and Beyer 1983]. However, the resistivity of intrinsic a-Si:H is too high for thermopower measurements to be useful.

3.3 PHOTOELECTRIC MEASUREMENTS

3.3.1 Photo- and Dark Conductivity Measurements

Because they are both important and easy to measure, photo- and dark conductivities (σ_l and σ_d) are usually among the first set of properties measured for intrinsic and doped a-Si:H. The σ_d of amorphous silicon without hydrogen is usually higher than 10^{-7} S cm^{-1}. Adding hydrogen reduces σ_d by passivating the dangling-bond defects. The σ_d

for high quality, intrinsic a-Si:H with an E_g around 1.75 eV is in the 10^{-10} to 10^{-11} S cm^{-1} range. The AM1 σ_l for such an intrinsic layer is in the 10^{-5} to 10^{-4} S cm^{-1} range. The photosensitivity of an a-Si:H film is usually expressed by the AM1 photo-to-dark conductivity ratio (σ_l/σ_d). For a high quality film with an E_g around 1.75 eV, this ratio is larger than 1 x 10^6. When measuring the conductivities of a-Si:H alloys, the film thickness should be between 0.5 to 1 μm. If the layer is too thin, the measurement will be influenced strongly by the bending of energy bands at the surface; if the layer is too thick, photons will not penetrate deep enough into the film for accurate σ_l measurement.

The photoconductivity relates to the photogeneration, transport, and recombination of electrons and holes. σ_l normalized against the actual amount of light absorption is equal to the product of quantum efficiency, mobility, and recombination lifetime ($\eta\mu\tau$) for majority carriers [Fritzsche 1980]. For undoped a-Si:H, σ_l and $\eta\mu\tau$ values are dominated by electrons above the mobility edge. $\eta\mu\tau$ is a better a-Si:H property parameter to use than σ_l because $\eta\mu\tau$ corrects for the optical absorption differences due to the variations in bandgap, film thickness, and surface reflection. When reporting $\eta\mu\tau$, one should mention E_a, E_g, and the illumination intensity and spectrum. The photosensitivity (σ_l/σ_d) of an a-Si:H film is not sensitive to the position of E_f, whereas $\eta\mu\tau$ depends strongly on E_a—a shift of E_f toward the midgap reduces $\eta\mu\tau$. Thus, photosensitivity is a better quantity to compare films than is $\eta\mu\tau$, unless we also know the E_g and E_a of the films.

The temperature dependence of σ_l relates to charge carrier transport properties which reflect the influence of film quality on transport mechanisms. Spectrally-resolved σ_l at the photoconductivity edge measures the valence band tail state density [§ 3.1.4]. The conductivities are usually measured by means of a coplanar electrode geometry [Pankove 1984, Part B]. It has been noted that conducting surface space-charge regions sometimes dominate the conductance of a-Si:H films [Fritzsche 1980, Tanielian 1982, Ye et al. 1988]. Gas adsorption may cause accumulation or depletion of charge carriers in a conduction channel beneath the sample surface. For example, H_2O on the a-Si:H surface was found to act as an electron donor and Br_2 as an electron acceptor [Abelson and de Rosny 1983].

The darkconductivity activation energy (E_a) can be determined from the σ_d-versus-temperature function, using the equation $\sigma_d = \sigma_0$ x exp $-(E_a/kT)$, where E_a approximately equals E_c - E_f for electron conduction or E_f - E_v for hole conduction [Fritzsche 1980]. The above approximation is true if the mobility is only a weak function of temperature and if the presence of deep levels does not alter the occupation of band states too severely. The dominant high-temperature (above 300 K) conduction mechanism in a-Si:H is conduction by electrons in extended states above the mobility band edge, E_c. The conductivity activation energy is routinely used to estimate the Fermi level position. For high quality intrinsic a-Si:H with a 1.75 eV E_g, the material is often slightly n-type with $E_a = 0.70$ eV or higher.

The transport properties of a-Si:H are also influenced by the interaction of the charge carriers with localized defect states. This interaction is called *trapping* if the charge transfer involves thermal activation of trapped charge from a defect site to a conduction band, in which the charge diffuses to the next defect site. It is called *hopping* if the transport is by charge carrier tunneling directly between localized defect sites. The interactions between charge carriers and defects can cause a packet of charge propagating across a sample to appear as a large dispersion in the arrival time of the carriers at the far electrode [Scher et al. 1991]. The dispersive (time-dependent) photocurrent in a-Si:H can provide a measure of the density of localized states near the mobility edge, such as the photocurrent transient spectroscopy [Monroe et al. 1981, Stoddart et al. 1988] (see also § 3.3.3).

3.3.2 Surface Photovoltage Measurements

The constant surface photovoltage (SPV) technique [Goldstein et al. 1982, Moore 1982, McMahon and Konenkamp 1982, Kumar and Agarwal 1985] is widely used to determine minority-carrier diffusion lengths and collection widths in a-Si:H. In this technique, a sample with a p-n or Schottky-barrier junction at the surface is illuminated with monochromatic light. The light intensity is adjusted to achieve a constant photovoltage for different wavelengths of light. By extrapolating the straight line of the light intensity vs. the absorption length of the monochromatic light one gets the diffusion length of the semiconductor. A bias light is usually used during a SPV measurement to reduce the depletion layer width and thus reducing the effects of electric field on the measurement result. SPV is a very useful tool for monitoring the electronic quality of a-Si:H films produced under various deposition conditions. Usually, a liquid Schottky contact is applied to the surface of an i/n/metal a-Si:H structure. The n-layer is about 30 nm thick. The intrinsic a-Si:H layer needs to be thicker than 1.5 μm. Intrinsic a-Si:H deposited on n-type crystalline silicon may also be used for SPV measurement. Typical minority-carrier diffusion length values for high quality, as-deposited, intrinsic a-Si:H are 0.2 to 0.8 μm. In addition to the minority-carrier diffusion length, the space charge density in the dark can be determined by SPV.

3.3.3 Time-of-Flight Measurements

The time-of-flight (TOF) measurement is a time-resolved transport measurement that yields information on the charge-carrier drift mobilities, drift length, and band-tail states [Pankove 1984, Vol. C, Scher et al. 1991]. The drift length, $\mu_e\tau_e + \mu_h\tau_h$, is the mean distance a charge moves in the direction of a electric field before trapping and subsequent recombination. Usually, a p-i-n or Schottky device is illuminated by a short-wavelength laser pulse on one surface with a transparent or semitransparent electrode, and the photogenerated carriers are swept across the sample thickness by a reverse-bias electric pulse. The incident photons should be mostly absorbed in a depth much smaller than the sample thickness. For this measurement to be successful, the sample should have a high photosensitivity and an excess carrier lifetime, longer than the transit time across the thickness of the sample. In good quality a-Si:H p-i-n devices, the entire i-layer has electric field across it. Thus, the drift properties of charge carriers measured by TOF are more important cell quality indicators than diffusion properties [Crandall 1983, Crandall and Balberg 1991]. For good quality a-Si:H, $\mu_e\tau_e$ is larger than 10^{-6} $cm^{-1}V^{-1}$ and $\mu_h\tau_h$ is larger than 10^{-7} $cm^{-2}V^{-1}$. Marshall et al. [1984] have measured a room-temperature electron mobility of 2.5 $cm^2V^{-1}s^{-1}$ and a hole mobility of 1.5×10^{-2} $cm^2V^{-1}s^{-1}$ for a-Si:H devices.

3.4 MECHANICAL MEASUREMENTS

3.4.1 Adhesion

Although the adhesion of a-Si:H alloys is an important property, especially for films thicker than one micron, quantitative studies of the adhesion property are not available. Adhesion depends on the cleanliness and the texturing of the substrate surface. The evolution of hydrogen for a-Si:H alloys deposited at low substrate temperatures can cause loss of adhesion and cracking of the film. In general, adhesion is a difficult parameter to measure in all thin film systems and most techniques often give values of limited accuracy.

3.4.2 Stress

Most a-Si:H films have large amounts of intrinsic stress, which sometimes causes the deposited film to flake off the substrate. Internal stress has been suspected as the cause of instabilities in the electrical properties of a-Si:H [Stutzmann 1985b]. The intrinsic stress can be deduced from measurements of the radius of the film-stress-induced curvature of a film deposited on a flexible substrate [Kurtz et al. 1986, Hasegawa et al. 1992].

3.4.3 Thickness

The most reliable thin-film thickness measurement is with a stylus profiler, or profilometer, [Kane and Larrabee 1978], which can measure step heights as small as 10 nm. It is difficult to create a sharp step for stylus profiler thickness measurements by masking part of the substrate during deposition. Satisfactorily sharp steps can usually be achieved by removing part of the film by tape liftoff or scraping after deposition. The thickness of a-Si:H films may also be estimated from interference fringes of uv/visible transmittance measurements using the equation, thickness = $[2n(1/\lambda_1 - 1/\lambda_2)]^{-1}$, where n is the index of refraction, and λ_1 and λ_2 are the wavelengths of two adjacent peaks.

3.4.4 Density

Two factors determine the mass density of a-Si:H: the net hydrogen concentration and the presence of microvoids. The most direct way to measure the density of a-Si:H films is to measure the weight and volume of the film. Such measurements show that a-Si:H films made at high substrate temperatures have a mass density of about 2.3 g/cm^3, essentially identical to that of crystalline silicon (2.35 g/cm^3) [Smith 1983]. Floatation properties of a-Si:H have also been used for density measurements [Menna et al. 1987]. In this method, the a-Si:H film is first deposited on an Al foil. The Al foil is then dissolved in a 10% HCl solution. The remaining a-Si flakes are put in a solution of Zinc Bromide ($ZnBr_2$) in water. The concentration of $ZnBr_2$ in water is adjusted until the solution's density matches that of the a-Si:H film and the a-Si:H flakes suspends in the solution. Using a crystalline Si sample, having a density of 2.33 g/cm^3, Menna et al. found that the error in this floatation technique is less than 1%. The saturated density of $ZnBr_2$ in water is 2.4 g/cm^3. This sets the upper limit of the density that can be measured. For materials with higher density, such as a-SiGe:H alloys, a solution with higher density is needed.

3.5 CHEMICAL ANALYSIS MEASUREMENTS

Determination of the elemental composition of a thin film, for determining either film stoichiometry or impurity content, is one of the most important aspect of film characterization [see, Czanderna 1975, Kane and Larrabee 1978, Czanderna and Hercules 1991]. **Table 3-3** is a summary of the relative properties of selected surface analysis methods. It is from a tutorial paper by L.L. Kazmerski [1988] that focuses on the chemical and compositional analyses aspects for polycrystalline and amorphous thin films.

Secondary ion mass spectrometry (SIMS) is probably the most sensitive technique for determining trace impurities in thin films [Evans and Blattner 1978]. In SIMS, the sample is bombarded by a beam of ions (usually cesium, oxygen, or argon), which sputters and ionizes sample atoms. SIMS can also be used to obtain a depth profile of impurity distributions in a thin-film device structure. These "secondary ions" are then analyzed by their charge to mass ratio in a mass spectrometer. Detectability limit of SIMS can be as small as 10^{15} atoms/cm^3. Other techniques used for a-Si:H films include Auger electron spectroscopy (AES) [Burnham et al. 1987], x-ray photoelectron spectroscopy (XPS), and electron probe microanalysis (EPMA) [Smith and Hinson, 1986]. In AES, the sample is bombarded by an electron beam. The electron beam knocks out electrons in the inner shells of the sample atoms. The vacant positions in the inner shells are replaced by electrons in the outer shells. In doing so, the outer electrons release enough energy to eject other electrons out of the atom with energies characteristic of the atom. AES is a somewhat nondestructive method for determining the elemental composition of the sample. However, it gives no information about the local bonding, and it cannot be used to determine hydrogen content because hydrogen has only one electron and the Auger process requires two. An Ar or Xe ion beam for sputtering may be used in combination with AES for depth profiling. The detectability limit for most elements of interest in AES depth profiling is about 0.1 at.%. XPS measurements have been used to estimate the valence- and conduction-band density of states of a-Si:H [Jackson et al. 1985]. In XPS, the energy of the electrons ejected due to x-ray absorption is related directly to the binding energy, which is determined

Table 3-3 Summary of the relative properties of selected surface analysis methods

	AES	ELS	SIMS	XPS	UPS	ESD
Probe	electron	electron	ion	X-ray	UV	electron
Species detected	electron	electron	ion	electron	electron	ion
Spatial resolution	≈10 nm	>5 nm	<0.5 μm	≈10-10^3 μm	10^3 μm	≈100 nm
Depth resolution	0.5 - 6 nm	0.5 - 6 nm	>0.3 nm	0.5 - 6 nm	0.5 - 6 nm	≈0.3 nm
Detection sensitivity	0.1 at.%	0.1 at.%	<0.0001 at.%	0.1 at.%	0.1 at.%	---
Quantifiability	Medium	---	Difficult	Very good	---	---
Mapping resolution	Excellent	Very good	Good	Poor	---	---

AES, Auger electron spectroscopy; ELS, or EELS, electron energy loss spectroscopy; SIMS, secondary ion mass spectrometry; XPS, X-ray photoelectron spectroscopy, also known as ESCA (electron spectroscopy for chemical analysis); UPS, UV photoelectron spectroscopy; ESD, electron stimulated desorption.

by chemical surrounding at the atom. This makes XPS suitable for measuring chemical states, in addition to elemental identification, of the material. Electron probe microanalysis (EPMA) was used to determine the fluorine content in a-Si:H:F with absolute errors less than 1.0 at.% [Langford et al. 1989b]. Total-yield photoelectron spectroscopy has been used to detect defect states (§ 3.1.4) [Winer et al. 1988] and conduction- and valence-band-tail density of states [Aljishi et al. 1990] within the first approximately 10 nm of the surface of a-Si:H samples. It is found that all a-Si:H and a-SiGe:H films possess purely exponential conduction- and valence-band-tail densities of states [Aljishi et al. 1990]. The conduction band tail for a-Si:H is typically 35 meV or less, whereas the valence band tail is typically 45 meV or more.

Hydrogen evolution is a simple method of determining the total, instead of only bonded as in IR absorption, hydrogen content of a-Si:H alloys. In this method, a sample is placed in a small vacuum chamber which is then evacuated, sealed, and heated; the pressure in the chamber rises as hydrogen diffuses out of the sample, and the pressure can be converted to the amount of hydrogen. A mass spectrometer is sometimes attached to the chamber to verify that the pressure rise is entirely due to hydrogen and not vacuum leaks or outgassing. For a-Si:H, the rate of hydrogen evolution as a function of temperature has two distinct peaks: one at about 350°C from the less tightly bonded dihydride (SiH_2) and polymeric ($(SiH_2)_n$) groups, and the other at about 500°C from the more tightly bonded monohydrides. [Wilson et al. 1984, Sakka et al. 1989].

3.6 MICROSCOPIC MEASUREMENTS

3.6.1 Scanning Electron and Transmittance Electron Microscopy

The maximum magnification of a conventional optical microscope is about 2000. For a scanning electron microscope (SEM) and a transmission electron microscope (TEM), the magnification can be as high as 400,000 and 1,000,000, respectively. Although the SEM is a very powerful tool for studying the surface morphology, crystal structure, and electrical properties of polycrystalline silicon [See, for example, Tsuo et al. 1984], it has had limited use in studies of a-Si:H, except for the surface morphology [Fritzsche 1980] and columnar growth structure [Knights and Lujan 1979]. The reason for this is that high energy electrons induce microscopic defects in a-Si:H [Yacobi 1984] and therefore affect the electronic properties that are being measured. In addition to the problem of electron-beam-induced metastable defects in a-Si:H, electron-beam-induced current (EBIC) is not useful as a defect characterization tool for a-Si:H p-i-n devices because the electron current density is often so high that it is dominated by space-charge-limited current not by hole transport from the back electrode. There is also the problem of the uncertainty in the value of the electron-hole pair generation function in a-Si:H alloys [Najar et al. 1991]. Since amorphous silicon does not have a crystalline structure, electron channeling patterns (ECP) cannot be used to study the structure of amorphous films. However, electron channeling can be used to detect the existence of microcrystalline Si in a-Si:H depositions [Tsai et al. 1988]. TEM has been used to study the degree of crystallinity in μc-Si:H and μc-SiC:H alloy films [Chen et al. 1992].

3.6.2 Scanning Probe Microscopy

The use of mechanically scanning probes has made possible two types of commercially available, high-resolution probe microscopes—scanning tunneling microscopes (STM) and scanning atomic force microscopes (AFM). Both are nondestructive and can directly reveal the atomic structure of surfaces. The STM is based on the principle that electrons can tunnel through the potential barrier between a nanometer-scale probe tip and a conductive sample surface if the distance between the two is close enough (within a few angstroms). An AFM, which is essentially a stylus profilometer with atomic-scale resolution, can be used on non-conductive surfaces. The tip radius of a AFM is less than 40 nm. The resolution of an AFM is slightly less than STM.

STM has been used to study the morphology of a-Si:H surface with atomic scale resolution [Wiesendanger et al. 1988, Kazmerski 1989, Gallagher et al. 1992]. Kazmerski obtained spectroscopic STM images of phosphorus-doped a-Si:H films that show the existence of $(SiH_2)_n$ polymeric structures and phosphorus incorporation. He also obtained images of high-resistivity, intrinsic a-Si:H using AFM with indications of hydrogen bonding and surface defects. Gallagher et al. have shown that glow discharge-deposited, intrinsic a-Si:H grows uniformly, without islanding, on atomically clean crystalline Si and GaAs.

3.6.3 Small-Angle X-Ray Scattering

Small-angle x-ray scattering (SAXS) [Brumberger 1967] measures electron density fluctuations on a size scale from 1 to 100 nm in a-Si:H alloys. The electron density fluctuations are related to mass fluctuations. In addition to the volume fraction of the microstructure component that produces the observed x-ray scattering, the size, shape, and orientation of the density fluctuations can be obtained by studying the inversion of the scattering intensity versus angle. However, because the number of electrons for hydrogen atoms is low (only one electron per atom), SAXS cannot tell the difference between clustered hydrogen atoms and voids. A thickness of about 15 μm or more of a-Si:H films (films can be stacked) are needed for sufficient signal-to-noise ratio in SAXS measurements. Care should be taken when interpreting SAXS samples grown on Al foil substrates, because the voids in Al foil may propagate into the a-Si film during film deposition and Al can induce crystallization in a-Si:H at temperatures as low as 185°C.

SAXS measurements can be used to study the effect of various parameters such as substrate temperature, deposition method, and addition of dopants on: 1) the void volume fraction (vvf), 2) size of the voids, and 3) shape of the voids. The presumed desirable direction for material optimization is to reduce the void volume fraction and the size of the voids. The best SAXS data so far have been on a-Si:H films deposited from silane at 250°C. **Table 3-4** shows the effect of changing various parameters on the size and volume of microvoids [Williamson et al. 1989, Mahan et al. 1991a, Jones et al. 1992].

Small-angle scattering of neutrons [Postol et al. 1980, Chenevas-Paule et al. 1985] and electrons [Craven et al. 1985] have also been used to estimate the distribution of voids in a-Si and a-Si:H.

Table 3-4 Parameters Affecting Microvoids

Material	Change	Void Volume Fraction	Void Size	Void Shape
a-Si:H	Ts > 250°C	slight decrease	no change	unknown
a-Si:H	Ts < 250°C	increase	little increase	columnar
a-Si:H	rf sputter	2x increase	larger	unknown
a-SiGe:H	Ge addition	increase	larger	elongated and oriented in the growth direction
a-SiC:H	C addition	increase	little change	spherical

4
CONVENTIONAL GLOW DISCHARGE DEPOSITION PROCESSES FOR AMORPHOUS SILICON-BASED ALLOYS

There are three major categories of plasma-assisted thin-film deposition methods: (1) plasma-enhanced chemical vapor depositions (PECVD), (2) RF or DC plasma sputter depositions, and (3) plasma-activated reactive physical vapor depositions. PECVD is the most popular deposition method for a-Si:H alloys and is the subject of discussion for chapters 4 to 9. Sputter depositions can be either direct or reactive. These will be discussed in § 12.2. Plasma-activated reactive PVD for a-Si:H use a plasma to hydrogenate evaporated silicon atoms. These techniques will be discussed in § 12.1. For any plasma-assisted deposition methods, there are two types of variables: (1) plasma-related variables, such as electron and ion densities, energies, and distributions, and (2) process-related variables, such as gas flow rates, neutral and reactive gas partial pressures, and the substrate temperature. Common methods for diagnosing plasma-related variables will be discussed in § 4.2. However, these plasma-diagnostic methods are usually costly to set up and time consuming. They are not widely used for process control. Many of the process-related variables will be discussed in Chapter 6. They are easy to measure and control and often vary from system to system. The chemical reaction paths for a-Si:H depositions in conventional glow discharge methods will be discussed in Chapter 7. In addition to gas-phase and surface chemical reactions, ion and electron bombardment of the growing film surface is also an important concern and will be discussed in Chapter 7.

4.1 PLASMA-ENHANCED CHEMICAL VAPOR DEPOSITION

Plasma-enhanced (or -assisted) chemical vapor deposition is also called *glow discharge deposition* because of its visible luminosity of the plasma glow region, which is mainly the result of the de-excitation of emitting molecular and atomic species contained in the plasma. Glow discharge is sustained by inelastic electron-impact processes which are initiated by electrons that have acquired sufficient energy from the electric field as a result of successive elastic collisions with gas molecules. The field can be either direct current (DC), RF, or microwave frequency. Overviews of the process have been given by Chapman [1980], Mort and Jansen [1986], Garscadden [1990], and Boeuf et al. [1990]. Whereas most RF glow discharge is done at a frequency of 13.6 MHz, research has also been done for the frequency range 60-144 MHz [Curtins et al. 1987a,b, Oda et al. 1988, and Chatham and Bhat 1989]. The gas-phase reaction of glow discharge reduces the substrate temperature required for film deposition compared to thermal CVD (pyrolysis), which depends completely on thermally-induced gas-surface interaction. Externally applied electrical energy is used to provide the activation energy for gas decomposition in the glow discharge deposition process. This allows separate controls of the substrate temperature and electric potential. This flexibility is a very important advantage of glow discharge over many other deposition techniques, such as thermal CVD.

Major steps of a plasma-enhanced CVD process include source gas diffusion, electron impact dissociation, gas-phase chemical reaction, radical diffusion, and deposition. When a silane (SiH_4) plasma is used in a glow discharge deposition process, the inelastic electron-impact process leads to reactive neutral species, such as SiH, SiH_2, SiH_3, Si_2H_6, H, and H_2, and ionized species, such as SiH^+, SiH_2^+, SiH_3^+, and so on [Griffith, 1980]. Some of these ions and free radicals diffuse to the substrate's surface, and are accompanied by a multiplicity of secondary reactions, e.g., ion-molecule, photon-molecule. The reaction of ions and free radicals with or their adsorption onto the substrate surface is followed by the process by which these species or their reaction products are incorporated into the growing film or are re-emitted from the surface into the gas phase. The substrate temperature during deposition is usually between 200° and 300°C, and the feed-gas pressure is between 0.1 and 1 torr.

Glow discharge has become the most common technique for depositing a-Si:H alloys and devices. Plasma-assisted gas decomposition reduces the substrate temperature required for the deposition process.

This lower substrate temperature makes it possible for sufficient hydrogen to be incorporated during the deposition. Under certain conditions, plasma-induced ion bombardment of the film during deposition tends to improve the quality of the film. Because glow discharge is a highly complex process, optimization of the deposition parameters has been chiefly empirical. However, this empirical optimization of the film's electronic properties has become increasingly more difficult as device structures have become more complex. Thus, it has become imperative to better understand the growth process and its effects, through the material's structure and chemistry, on the electronic properties of these films [Knights 1984]. Over the years, much research has been done to learn how plasma conditions relate to the quality of the film, and the literature on this subject is vast. This review summarizes some of the aspects of glow discharge relevant to producing high quality a-Si:H as well as aspects that have deleterious effects on film quality, and it elucidates the effects of various deposition parameters and other factors to provide a better understanding of the growth process.

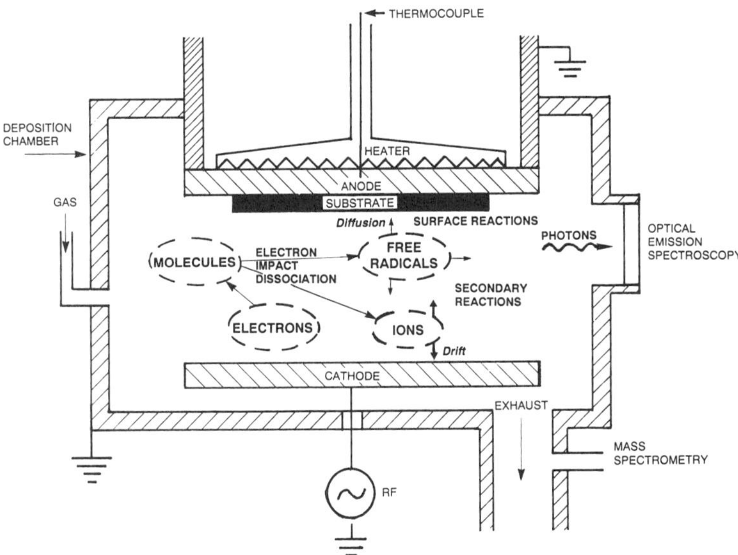

Figure 4-1 Schematic representation of the plasma-enhanced chemical vapor deposition process

There are four stages in the formation of hydrogenated amorphous silicon films from an electrical discharge in silane as shown in **Figure 4-1**. The first is the primary reaction between electrons and silane, which results in a mixture of ions and free radicals. The second stage is the transport of these species to the substrate's surface, which is accompanied by a multiplicity of secondary reactions, e.g., ion-molecule, photon-molecule. The third stage is the reaction of ions and free radicals with, or their adsorption onto, the surface of the substrate. The fourth stage is the process by which these species or their reaction products are incorporated into the growing film or are re-emitted from the surface into the gas-phase [Knights 1985].

There are at least three causes for poor quality hydrogenated amorphous silicon films: 1) ion bombardment, 2) microparticles, and 3) growth resulting from radicals having high sticking coefficients [Gallagher 1987a,b,c]. Dominant SiH_3 radicals in the plasma are essential for high quality a-Si:H films grown from both monosilane [Gallagher and Scott 1987a] and disilane [Gallagher et al. 1989].

Ion bombardment can have positive effects on the film's quality. Gallagher [1987] believes that ion bombardment energy in the range of 15 to 200 eV is beneficial to a-Si:H film deposition. Roca i Cabarrocas [1989] measured ion flux and ion energy in an RF glow discharge deposition system with an electrostatic energy analyzer. They found that ion bombardment at moderate energy (< 70 eV) does not induce any serious degradation of the electronic properties of a-Si:H films. Vepřek et al. [1989] proposed that one of the beneficial effects of ion-bombardment in a plasma deposition is reduced hydrogen surface coverage. They believe that the reactive sticking coefficients of silane plasma radicals depend on the surface coverage of hydrogen. Decreasing surface hydrogen coverage increases the deposition rate. Ion bombardment during deposition may also break up polymeric chains and increase the percentage of monohydride bonds.

Ion bombardment effects are greater for silane diluted by inert gases than for undiluted silane. Hydrogen dilution does not have such a strong effect as that caused by inert gases, however, because of hydrogen's lower mass. Ion bombardment of a substrate can be minimized or eliminated in a triode reactor geometry or by biasing the substrate appropriately. However, such an arrangement reduces the deposition rate compared with that of an unbiased case. Methods for minimizing or avoiding ion bombardment are discussed in sections 5.1 and 5.2.

Microparticulates are particles 2 to 30 nm in diameter that can be observed by light scattering. Roth et al. [1984] have observed such particulates in light-scattering experiments in which dust was not observed by unaided visual inspection of the plasma. The microparticles are caused by ion nucleation and neutral nucleation. Ion nucleation is severe in low-field regions within a reactor. Under certain circumstances, ions can remain in low-field regions of the plasma long enough to grow into microparticulates [Gallagher 1986a]. Neutral nucleation is more dominant than ion nucleation in high-field regions of the plasma. Microparticulates cause voids to form in a-Si:H films, and this has been observed to correlate with poor film quality. It appears that plasma confinement, magnetically or by a grounded shield, can reduce high-field regions. This could explain the better a-Si:H film quality at high deposition rates that is obtained with confined plasmas. Plasma confinement is discussed in section 5.3.

There is considerable evidence that structural inhomogeneity (specifically, voids) exists in much of the material that has poor electronic properties [Knights 1984, Carlson 1986]. Although there are strong correlations between the presence of structural inhomogeneity and high levels of electrically active defects, no direct connections between the electrical and structural defects have been made [Knights 1984].

Radicals with high sticking coefficients exacerbate both the microscopic and macroscopic shadowing that results in columnar growth of a-Si:H film. Columnar growth in turn results in inferior electronic properties. The SiH_3 radical has a low sticking coefficient and high surface migration; as a result, SiH_3 yields good quality films. Silane depletion in a reactor reduces the availability of SiH_3 and increases the relative abundance of other (sticky) radicals, thus causing poorer film quality [Gallagher 1987].

4.2 PLASMA DIAGNOSTICS

Diagnostic measurements of the plasma are needed for better process control and for understanding the plasma deposition processes [see, for example, the book by Auciello and Flamm 1988]. Basic properties of a plasma include the distributions of luminous intensity, electrical potential, electric field, space charge density, and current density, and the properties of plasma species including their natures, concentrations, kinetic lifetimes, spatial distributions, velocity distributions, and internal temperatures. Plasma diagnostic methods include electric probe and various spectroscopic techniques.

Spectroscopic plasma diagnostics can be carried out in three different modes: emission, absorption, and scattering. They may be used separately, in sequence, or simultaneously. The emission mode, optical emission spectroscopy (OES), uses emission spectra to measure the intensities of photons that permeate a plasma-filled space. The absorption mode, optical absorption spectroscopy (OAS), uses photons from an external radiation source to measure the absorbance of plasma species. We can also monitor laser-induced fluorescence (LIF) and optogalvanic [Murnick et al. 1989] effects. The scattering mode uses a laser beam to perform spontaneous or coherent Raman spectroscopies, such as, coherent anti-Stokes Raman spectroscopy (CARS). The optical diagnostic techniques provide means for *in-situ*, generally nonintrusive examination of the plasma. **Table 4.2-1** lists the plasma species in a SiH_4 deposition plasma that can be detected by several optical spectroscopies [Venugopalan 1989].

Table 4.2-1 Plasma species in a SiH_4 deposition plasma that can be detected by the listed optical spectroscopies.

Method	Detectable Species
Optical Emission Spectroscopy, OES	H^*, H_2^*, Si^*, SiH^*
Optical Absorption Spectroscopy, OAS	SiH, SiH_2, SiH_4
Laser-Induced Fluorescence, LIF	Si, SiH
Coherent Anti-Stokes Raman Spectroscopy, CARS	SiH_2, SiH_4

* = excited state

4.2.1 Optical Emission Spectroscopy

The plasma glow is caused by inelastic collision of high-speed electrons with atoms followed by spontaneous emission of photons by the excited atoms. Optical emission spectroscopy (OES) in the visible and ultraviolet regions of the spectrum is commonly used to determine the presence of neutral and ionic species in plasmas [Kampas 1983, Jansen et al. 1985]. Typical excited species in a SiH_4 plasma that can be identified are Si and SiH [Knights 1985], H [Bauer and Bilger 1981], H_2 [Matsuda and Tanaka 1982], and impurity species such as N_2 and Cl [Knights 1985]. The main advantage of OES is that it does not perturb the plasma, and it can also be used as a monitor to achieve reproducible plasma conditions. However, OES can provide only semiquantitative data on concentration trends and excitation processes. Selwyn [1986] discusses in detail actinometry, spatial resolution, temporal resolution, and optical emission end-point detection and provides an extensive reference list.

Jasinski [1990] developed laser-based techniques to obtain absolute rate constants for the reactions of mono-silicon hydride radical species, SiH, SiH_2, and SiH_3, with stable molecules such as silane, hydrogen, and disilane. This technique uses a pulsed ultraviolet eximer laser photolysis to generate silicon hydride radicals and high resolution laser spectroscopy to detect specific radicals directly in a time-resolved manner.

4.2.2 Optical Absorption Spectroscopy

Optical absorption spectroscopy (OAS) measures the absorbed energy from a beam of light that passes through the plasma. By operating in the wide wavelength regions of vacuum UV, UV, visible, and IR, it can be used to identify a large number of stable molecules in a gas. Wormhoudt et al. [1983] describe infrared (IR) tunable diode laser absorption spectroscopy for sensitive and selective detection of a variety of species. Representative sensitivity is 10^{-3} for a 10^{-4} cm^{-1} laser line width. They provide a listing of minimum detectable densities for stable molecules by IR absorption spectroscopy. Jasinski et al. [1984] describe the detection of SiH_2 in direct-current (DC) glow discharge plasmas from silane and disilane using frequency-modulated IR absorption spectroscopy.

4.2.3 Laser-Induced Fluorescence

A particle excited by photon absorption may exhibit fluorescence by undergoing radiative transitions by which photons of lower energies than the absorbed photons are emitted. Since fluorescence times are considerably shorter than 1 μs, it is necessary to use laser pulse widths of the order of nanoseconds. Laser-induced fluorescence (LIF) has been used to measure the neutral species distribution in a plasma. It offers greater spatial resolution, higher sensitivity, and better selectivity than optical emission spectroscopy and can be used to probe ground-state species directly in state-specific quantum levels [Roth et al. 1984, Selwyn 1986]. The sensitivity is in the 10^6 to 10^8 cm^{-3} range. Using LIF in tandem with OES provides an especially powerful diagnostic tool, because the two techniques can provide both an overview of plasma conditions (through OES) and detailed quantitative information on a selected species (by LIF) [Selwyn 1986]. Selwyn provides a detailed description of LIF and a good list of references [Selwyn 1986]. Moore et al. [1984] measured plasma electric fields with high spatial resolution ($\leq 10^{-4}$ cm^3) using LIF. Gottscho and Mandich [1985] described the use of LIF to measure time-varying plasma phenomena - in particular, plasma formation and decay rates, concentration gradients, and electric fields. Laser-induced fluorescence can detect species complementary to those detected by coherent anti-Stokes Raman spectroscopy [Hata and Tanaka, 1985], such as Si and SiH.

4.2.4 Coherent Anti-Stokes Raman Spectroscopy

Coherent anti-Stokes Raman spectroscopy (CARS) is excellent for measuring neutral-molecule spacial distribution even in the presence of a strong background plasma emission [Hata and Tanaka 1985]. A Raman scattering is a light scattering process where a photon get scattered with a very small loss or gain of energy (12400 cm^{-1} = 1 eV) [Tobin 1971]. In spectroscopy, the difference between the photon energies of the inelastically scattered light and the incident light is called a Stokes shift when the difference is due to phonon emission and an anti-Stokes shift when the difference is due to phonon absorption. The origin of the phonon component is the electron-phonon interaction.

CARS can be used to measure, *with good spatial resolution*, SiH_4 and SiH_2 concentrations and the gas-phase temperature profile during a silane glow discharge process [Shing et al. 1987]. The second harmonics of an Nd:YAG laser and a dye laser can be used as light sources for CARS. Details are provided by Hata et al. [1983a and 1983b]. The gas-phase temperature profile of a SiH_4 discharge was found by Hata et al. to be determined by the temperatures of chamber walls and electrodes with little effect from the discharge. It is unfortunate that SiH_3, the radical normally attributed to good quality a-Si:H film depositions, cannot be detected by CARS.

4.2.5 Mass Spectroscopy

Mass spectroscopy is used to measure ions and neutral species in the plasma. A commercially available quadrupole mass spectrometer or residual gas analyzer (RGA) [Goldfarb 1986], consisting of an ionizer, a quadrupole mass sorter, and a detector, is usually used for this purpose because of its high sensitivity, rapid scanning, good resolution, linear mass scale, and small size. For high-vacuum studies, the electron multiplier version of RGA detector is used; for 10^{-3} Torr vacuum to high pressure operations, the Faraday cup detector version of RGA detector is used. Using a magnetically confined electron-beam ionizer and a quadrupole mass spectrometer, Robertson and Gallagher [1986] achieved high sensitivity (discrimination of one radical in 10^6 SiH_4 molecules or 10^9 radicals per cm^3), which is essential in identifying low-abundance species. It should be noted that the generation rate for neutral radicals exceeds the ionization rate by more than an order of magnitude in typical operating conditions [Jansen et al. 1985].

4.2.6 Electric Probe Techniques

An electrostatic Langmuir probe can be used to monitor plasma density and electron energy distribution [Chapman 1980]. However, proper interpretation of probe data is complex, and the probe may become a power drain or even a contamination source for the plasma.

4.3 FEED-GAS ANALYSIS

Contaminant impurities can have a significant effect on the electronic properties of films [Knights 1984], especially nitrogen, oxygen, carbon, and chlorine. Minute quantities of other impurities in the feed gas are also suspected to affect the electronic properties. Gas chromatography/mass spectroscopy (GC/MS) is used by most a-Si:H solar cell manufacturers for selecting a feed gas and controlling purity. The detectability limits of a GC/MS system can reach 100 ppb (10^{-7}) or better for most impurities in silane and disilane. A strong correlation has been found between undesirable gas impurities and the surface photovoltage diffusion length of deposited a-Si:H films [Carlson et al. 1984b].

In a CVD process, the problem of gas purity usually does not arise from the quality of the "delivered" gas from the supplier. It is the quality of the "reaction" gas—the gas that has passed through the storage and delivery subsystems of the deposition system into the reactor—that is to blame. An on-site gas purity analysis technique is helpful to ensure the quality of the reaction gas. Besides GC/MS, a quadrupole residual gas analyzer (RGA) [Goldfarb 1986], equipped with a vacuum source and an adjustable orifice sample inlet, can be used for this purpose.

5
DESIGN OF GLOW DISCHARGE DEPOSITION REACTORS

A plasma deposition system, like any chemical vapor deposition system, usually consists of several functional subsystems, such as the reactor, the process gas storage, metering, and delivery subsystems, and the high-vacuum pumping system, the process gas pumping system, the effluent gas disposal system, and the RF or DC circuits. For p-i-n device fabrications, single-chamber reactor systems [Böhm et al. 1985], load-locked single-chamber reactor systems [Dickson et al. 1986], and multichamber reactor systems [Ohnishi et al. 1983] have been used. The use of multiple chambers with gas locks between them in a reactor design is believed to be important in the deposition of high-efficiency p-i-n devices because they minimize cross-contamination from one layer to the next and increase the throughput of device deposition. In designing a reactor chamber, one needs to consider such factors as the electrode geometry, the gas flow pattern, and the heater design. Other significant design features are hot walls, small electrode distances, electrode size, elimination of high-field regions, frequency selection (e.g., microwave frequency versus radio frequency), and bias control. All such design or process features can affect the film quality. Because there are numerous variations and engineering considerations in the design of a deposition system, we will limit our discussions in this chapter to those factors that we think relate directly to the physics and chemistry of film growth.

The reactor electrode design is important because it affects the reactions between electrons and gas molecules, the transport of species generated in the plasma to the substrate, and secondary reactions at the substrate, and thus can affect the resulting quality of the film. Most modern a-Si:H deposition systems use a flat-bed reactor design with internal electrodes [Semiconductors and Semimetals, A, 1984]. Tube reactors with external electrodes or with an inductively coupled discharge were used in the early stages of a-Si:H research [Chittick et al. 1969]. However, tube reactor systems are difficult to scale up for industrial production. More recent work with inductively coupled glow discharge systems show that the deposition rate, conductivity, and dihydride content are strongly affected by the distance between the substrate and the inductive coil [Yokota et al. 1992]. A triode geometry of electrodes with the proper bias can minimize ion damage to growing films. Plasma confinement, by grounded screens or magnetic fields, minimizes film contamination from walls, reduces film deposition on walls, and can provide a more uniform electric field. Reactors with two electrodes, a powered or RF-driven electrode and a grounded electrode, are called diode reactors, and those with three electrodes are called triode reactors.

5.1 GEOMETRY

5.1.1 Diode Geometry

First, we will discuss some of the nomenclature pertaining to these reactors. In diode reactors powered by a radio frequency or other AC frequency electric field, the electrode containing the substrate on which film is to be deposited is typically the grounded electrode. The combined area of the grounded electrode and the grounded chamber walls or plasma confinement screens is usually much larger than the area of the powered electrode. As a result of this asymmetric geometry and the fact that the electron mobility is much larger than the ion mobility [Chapman 1980, Kasper et al. 1992], the two electrode regions will have different amounts of potential drop, which accelerates ions toward the electrodes and electrons away from the electrodes. The larger potential drop appears at the powered electrode, which is more negative with respect to ground than the grounded electrode, as shown in **Figure 5.1-1**. This is the reason why the powered electrode is often called the cathode and the

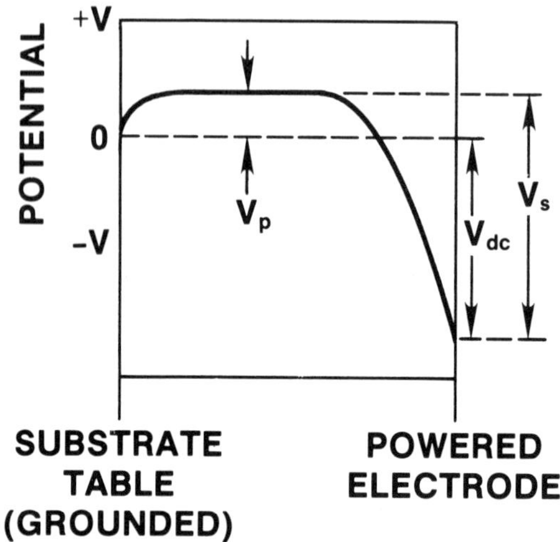

Figure 5.1-1. Spatial distribution of average potential in RF-powered diode reactor [Chapman 1980]

grounded electrode is called the anode. In Figure 5.1-1, V_p = plasma potential, V_{dc} = average self-bias, and V_s = sheath potential. The ground potential is always negative with respect to the plasma potential, and hence the substrate's surface on the grounded electrode suffers bombardment by the positive ions. The powered electrode is even more negative with respect to the plasma and thus subjected even more to positive ion bombardment than the anode. The plasma glow region is the place where the potential is near the plasma potential. This is where most of the inelastic electron-impact gas dissociation takes place. The dark region between the boundaries of the glow region and the electrodes is usually called the sheath or dark space. Increasing power increases the sheath voltages but decreases the sheath widths. For a DC glow discharge process, the potential distribution is similar to that shown in Figure 5.1-1. The glow region also has a positive plasma potential. Almost all of the applied potential is across the cathode sheath and, typically, only about a 10-20 V potential drop is across the anode sheath.

Cathodic deposition rates (for DC and RF) when asymmetric electrodes are used are always higher than anodic rates because of the negative self-bias. This is consistent with the fact that the generation of neutral radicals takes place relatively close to the cathode both for DC and RF discharges [Jansen et al. 1985]. At comparable power levels, the cathodic deposition rate is higher for DC discharge than for RF discharge. In DC glow discharge, the substrate is usually placed on the cathode. In RF glow discharge, the substrate is normally located on the anode. An example of an industrial diode deposition chamber for about 0.7 m^2 substrates is shown in Figure **5.1-2.**

In diode reactor designs, ions produced in DC and RF glow discharges are accelerated in high-field regions of the plasma sheath and bombard the growing film surface. To avoid damage from positive ion bombardment, a cathode screen is sometimes added near the substrate for DC glow discharge (proximity DC glow discharge) [Chapman 1980]. Films grown on the cathode experience more ion bombardment than films grown on the grounded anode. Cathode films generally contain less hydrogen than anode films; in addition, a smaller fraction of the hydrogen is in the form of dihydride or polyhydride groups [Mort et al. 1986, p. 46]. This fact is attributed to the effect of ion bombardment on the dihydride in the growing film [Kampas 1984]. For a low silane concentration in inert gas (but not for undiluted silane), Collins and Cavese [1987] found a higher-density microstructure at the anode than that at the cathode. Lucovsky et al. [1979] found more polymeric $(SiH_2)_n$ groups for anode films and more isolated SiH_2 groups for cathode films. A comparison of DC and RF excitation methods for anodic and cathodic depositions is shown in **Table 5.1-1**.

5.1.2 Triode Geometry

Using a screen or triode configuration largely eliminates ion bombardment of the growing film [Gallagher 1986]. In a triode geometry, two different arrangements of the electrodes may be involved, as shown by **Figures 5.1-3** and **5.1-4**. In Figure 5.1-3, the third electrode contains the substrate on which the film is grown. The bias voltage is applied between the anode and the third electrode [Ando et al. 1984]. In this arrangement, enhanced ion transport to the substrate results from a negative bias of the substrate with respect to the space plasma potential. In Figure 5.1-3, the DC discharge is between the anode and the cathode.

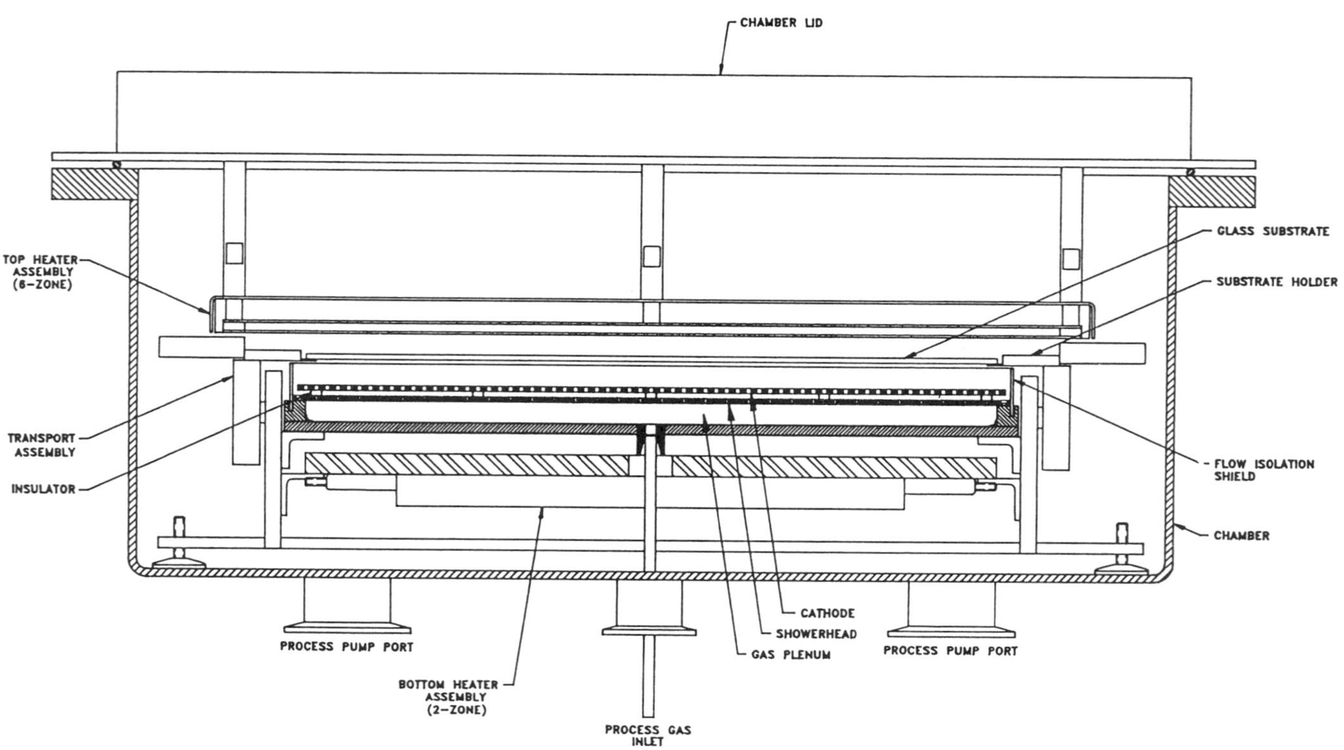

Figure 5.1-2 Industrial Diode Glow Discharge Deposition System [Oswald and O'Dowd 1992]

TABLE 5.1-1. A Comparison of DC and RF and Anodic and Cathodic Deposition

	DC	RF
Anodic Deposition		Normal substrate location
	Less ion bombardment than on cathode	Less ion bombardment than on cathode
	Higher hydrogen content than on cathode	Higher hydrogen content than on cathode
Cathodic Deposition	Normal substrate location	Lower deposition rate than DC
	Higher deposition rate than on anode	Higher deposition rate than on anode
	Fewer polyhydrides than on anode	Fewer polyhydrides than on anode

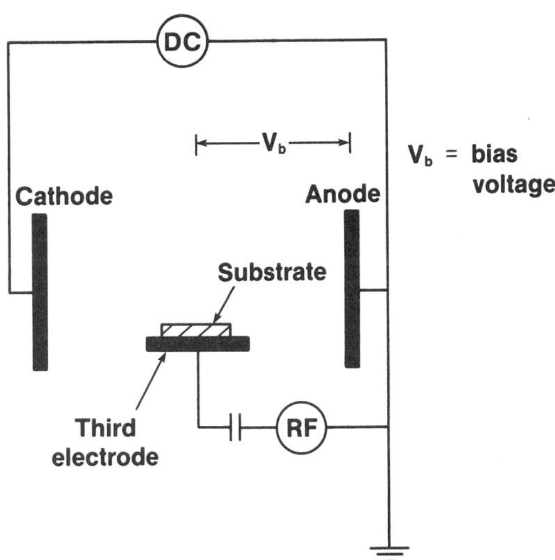

Figure 5.1-3. Triode reactor geometry with substrate on third electrode [Ando et al. 1984]

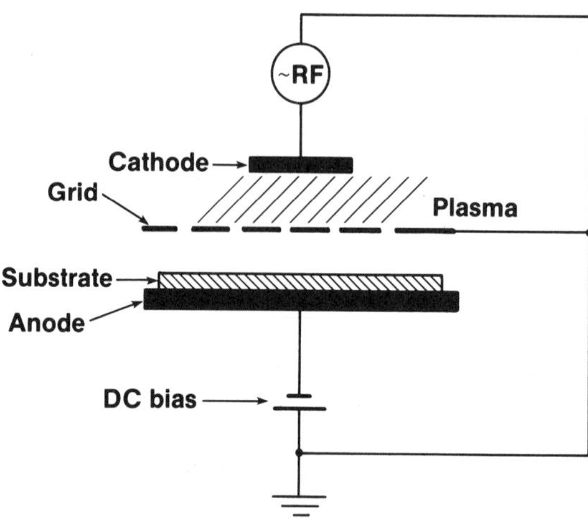

Figure 5.1-4. Triode reactor geometry with grounded third electrode grid

Figure 5.1-4 shows an RF triode arrangement in which the third electrode is a grid that is grounded [Gallagher 1987] and the substrate is located on the anode that can be DC-biased relative to the grid. With any such properly biased triode system, there is less ion bombardment (10^{-2} to 10^{-3}) than with a diode system. For triode glow discharge with a negatively biased substrate relative to the grounded grid, as shown in Figure 5.1-4, the plasma is formed between the grid and the RF electrode. Depending on the bias voltage, ions (especially the heavier ones) can be extracted from the plasma by the negative potential applied to the substrate. Thus, SiH_3^+ ions can reach the substrate, whereas the lighter H^+ ions stop at the grid.

In the absence of a triode screen, the fraction of initially produced radicals reaching the substrate before reacting are largely independent of the reactor chamber pressure [Gallagher 1987]. Very transparent screens, used at a small distance from the substrate, with a fine mesh (to avoid coating inhomogeneities) can be used to obtain relatively high deposition rates in triode discharges [Gallagher 1987]. The screen transparency controls the deposition rate [Gallagher 1986]. A modification to Figure 5.1-4 is the DC-proximity glow discharge system mentioned in which the anode is the powered electrode and the grid is the cathode, and the substrate is located 1 to 2 cm beyond the screen and is thus protected from ion bombardment [Pankove 1984, Vol.B, p.43].

The triode reactor geometry is much less common than diode geometry because it has the drawbacks of low deposition rate due to film deposition on the grid and film flakes from the grid falling onto the sample substrate.

5.1.3 Electrode Spacing

The electrode spacing in a plasma deposition reactor affects the deposited film's quality because it limits the distance that radicals travel before reaching the growing film's surface and it determines the glow-discharge-sustaining voltage through Paschen's law [Brodsky 1977, Ishihara, S., et al. 1987a], which states that the glow-discharge-sustaining voltage is a function of the product of the process gas pressure (P) and the electrode spacing (d), as shown in **Figure 5.1-5.**

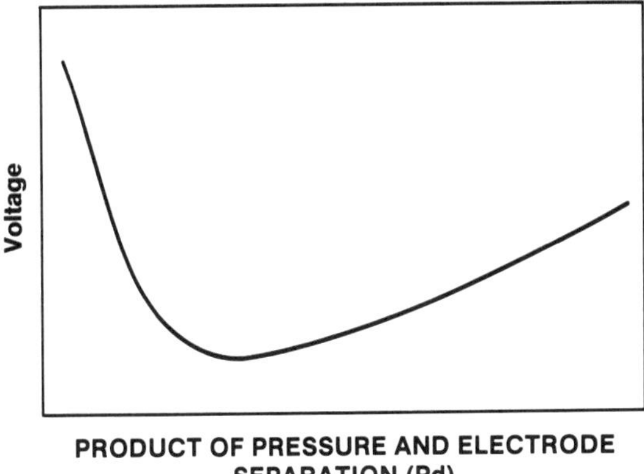

Figure 5.1-5. A schematic illustration of Paschen's law for the voltage V needed to sustain (or initiate) a glow discharge in a plasma [Brodsky 1977]

A voltage slightly above the glow discharge-sustaining voltage is typically used to deposit high quality a-Si:H. To achieve that voltage, the product of pressure and electrode distance must be selected appropriately. Since increasing the gas pressure increases the possibilities of gas-phase ion nucleation, it would seem that a larger electrode spacing would be preferred to achieve a given Pd-product. On the other hand, a small electrode spacing is desirable to minimize the distance that radicals must diffuse before reaching the substrate (or the residence time) to avoid excessive gas-phase nucleation which causes powder formation (plasma polymerization). To reduce the dust formation, a-Si:H deposition is usually done with Pd-product on the left of the Paschen minimum. Electrode spacings of 1 to 5 cm are typically used in a-Si:H depositions. Spacings of less than 1 cm are undesirable because they can cause nonuniform film deposition over the substrate area.

5.2 BIAS CONTROL

In both the diode and the triode geometry, a bias voltage applied to the substrate on which film deposition takes place relative to the plasma potential can have significant effects on film properties. Since glow-discharge deposition is caused mainly by neutral radicals (see section 7), the application of a bias voltage does not affect the type and sticking properties of the deposition precursors. A bias voltage mainly affects the film properties by controlling the extent and energy of ion and electron bombardment during film growth. A moderate negative substrate bias, either externally applied or self-induced by the nonsymmetrical electrode geometry (as discussed in §5.1.1) improves the structural and electrical properties of the deposited a-Si:H [Knights and Lujan 1979, Potts et al. 1981, Turner et al. 1983, Roca i Cabarrocas et al. 1991]. This occurs because moderate ion bombardment, caused by the negative bias, can remove defect sites associated with SiH_2 bonding and removes weakly-bonded species from the surface. Drevillan et al. [1983] showed that, in pure silane multipole DC discharge depositions, an increase in ion (mainly SiH_3^+) bombardment energy favors the formation of high-density homogeneous and isotropic films correlated with the predominance of monohydride bondings. However, excessive ion bombardment [Gallagher 1987] and electron bombardment [Bhat at al. 1983] can have damaging effects on the film's quality. In the following paragraphs, the various effects of an applied bias voltage on (a) the deposition rate, (b) the spin density, (c) the dihydride-to-monohydride ratio, and (d) the doping efficiency are discussed for various electrode geometries.

5.2.1 Effects of Bias Control on Deposition Rate

Shimada et al. [1984] describe a symmetric, vertically stacked, diode-electrode geometry RF glow discharge system using hot walls. The reactor-chamber walls are made of fused quartz. There are no self-biasing electrodes. In this system, when 100% silane is used, the deposition rate decreases from 2.3 nm/s at -110 V DC bias to 1.9 nm/s at no bias, and to 1.5 nm/s at +80 V bias.

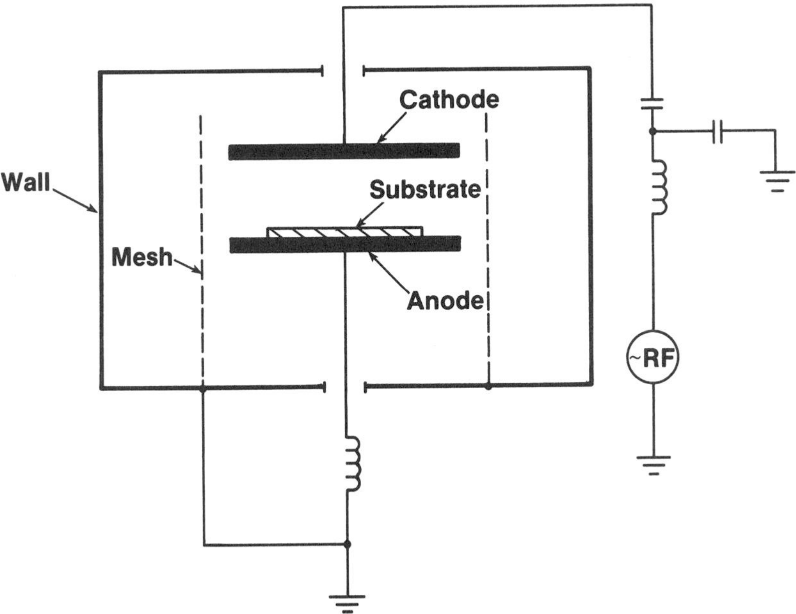

Figure 5.2-1 Diode reactor geometry with biased substrate [Hattori et al. 1984]

Hattori et al. [1984] discuss another diode glow discharge system (see **Figure 5.2-1**). They investigated two voltage biasing methods: in one, the substrate is biased relative to the reactor wall that is grounded and a closed circuit is formed to the reactor walls; and in the other, the circuit is closed via the power electrode that is grounded via an inductor. The deposition rate for silane increased from 0.2 nm/s at -100 V to 0.45 nm/s at no bias with both biasing methods. Neither the spin density nor the photosensitivity was affected much by either a positive or negative bias voltage on the substrate.

Roca i Cabarrocas et al. [1991] show a decrease in deposition rate with increasing negative bias at 100 mtorr but no change at 30 mtorr.

Aozasa et al. [1986] describe a triode glow discharge system, as shown in Figure 5.1-2. The deposition rate decreases from 1 nm/s at -500 V to 0.3 nm/s at 0 V. When the negative bias of the substrate relative to the anode was moderate (-50 to -150 V), the film quality improved, as indicated by the dihydride-to-monohydride ratio, because of the promotion of SiH incorporation or a reduction in SiH_2 incorporation, or both. On the other hand, a -100 V negative bias resulted in a large increase in the dark conductivity (one to five orders of magnitude, depending on the discharge current) and a decrease in the photosensitivity at high discharge current.

5.2.2 Effects of Bias Control on Spin Density

Knights [1979b] showed that, in depositions from silane in a diode glow-discharge system, the effect of a bias voltage on spin density depends on the RF power density. At a low power density, there is no difference in spin density between biased and unbiased conditions. At a high RF power density in the unbiased condition, the spin density increases by a factor of 100 in comparison to the low power condition, whereas the spin density remains constant in the negatively biased condition.

Similarly, Hattori et al. [1984] found for diode glow discharge (see Figure 5.2-1) at a low power density (40 mW/cm^2) that varying the bias voltage over the range of -100 to +100 V had no appreciable effect on the spin density, which remained at about 10^{16} cm^{-3}.

Roca i Cabarrocas et al. [1991] observed a 2/3 reduction in bulk density of states going from no bias to -50 V bias at 100 mtorr but an insignificant reduction at 30 mtorr. At the same time the valance-band tail became sharper. Going beyond -50 V gradually increased both parameters again.

5.2.3 Effect of Bias Control on Dihydride/Monohydride Ratio

Hotta et al. [1981] studied bias voltage effects on hydrogen content (and found very little effect) and hydrogen incorporation as a dihydride or monohydride in films from monosilane in a diode RF glow discharge with a DC cross field (the dihydride to monohydride ratio increased from 0.5 to 0.85 as the bias voltage increased from 0 to +100 V). They concluded that a negative bias favors monohydride bonding, whereas a positive bias favors the formation of dihydride bonds.

Aozasa et al. [1986] reported for a triode geometry (Figure 5.1-2) that a film's quality (for deposition from silane) is enhanced by a negative bias up to an optimum bias level. A further increase in the negative bias (beyond -200 to -300 V) leads to a deterioration in quality. With an adequate negative bias, SiH incorporation is promoted but SiH_2 incorporation is depressed, which results in an improvement in the film quality. This effect is more pronounced at a substrate temperature of 180°C than at 270°C, because films deposited at lower temperatures have a higher SiH_2 concentration in the absence of a bias voltage.

5.2.4 Effects of Bias Control on Doping Efficiency

Hotta et al. [1981, 1982] studied the effects of the substrate bias voltage in a cross-field plasma deposition system. In addition to effects on hydrogen bonding, they also noticed that the boron impurity and the doping efficiency were strongly affected by the plasma voltage. Doping efficiency refers to the percentage of dopants that is electrically activated. In Si, fourfold-coordinated, rather than five or threefold-coordinated, phosphorus or boron are active dopants. A 100-V positive bias voltage enhanced the dark conductivities of boron a-Si:H samples by more than three orders of magnitude, compared with samples deposited with no applied bias.

Alvarez et al. [1985] also reported on the effects of substrate bias on doping efficiency (a positive bias improves the phosphorus incorporation and a negative bias improves the boron incorporation) and conductivity (a positive bias enhances the conductivity of phosphorus-doped films and a negative bias enhances the conductivity of boron-doped films). The last effect is the opposite of that reported by Hotta et al. [1981].

5.3 PLASMA CONFINEMENT

Electron and ion motions in a plasma can be influenced easily by a magnetic field. Taniguchi et al. [1980] found that a magnetic field applied to a plasma column generally lowers the electron temperature, reducing the diffusion loss of charged particles in the direction perpendicular to the magnetic field, and it affects the dihydride and monohydride content in films from silane. A magnetic field across the plasma acts as an ion filter. It can exclude light H^+ ions from the substrate that tend to etch the deposited a-Si. However, heavier ions, such as Ar, can proceed to the substrate and remove hydrogen bonded to the surface, especially hydrogen bonded as SiH_2.

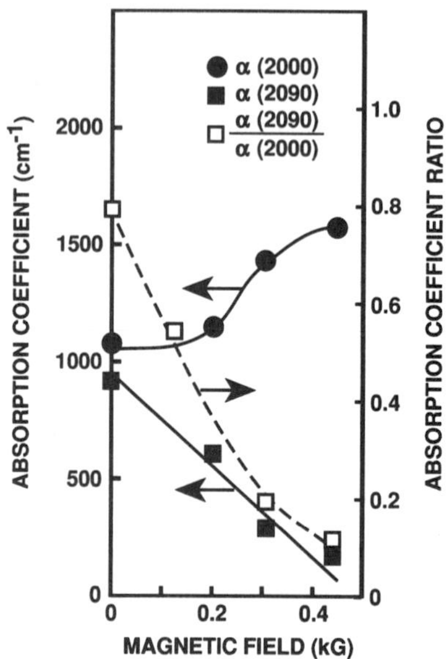

Figure 5.3-1. Infrared absorption coefficients in a-Si:H prepared from silane by RF discharge vs. magnetic field strength [Pankove 1984]

A magnetic field parallel to the electric field (e.g., by ring-magnet around a tubular reactor) confines the positive plasma column and lowers the electron temperature [Pankove 1984]. Increasing the magnetic field reduces the dihydride-to monohydride ratio (see **Figure 5.3-1).**

Ohnishi et al. [1987] describe a method of plasma confinement in which electromagnets are located under the cathode, creating a magnetic field between the cathode and anode. They call it the controlled plasma magnetron method (see section 8.4). The diffusion of electrons toward the walls is suppressed by the magnetic field and results in an increase in the electron density in the plasma and an increase in the deposition rate, because of the high radical concentration due to electron confinement.

Mejia et al. [1985] reported on the effect of an external magnetic field in microwave (2.45 GHz) excited plasmas undergoing electron cyclotron resonance (ECR), using a 10% SiH_4 and 90% H_2 gas composition. The photosensitivity of the resulting film increased by five orders of magnitude for a tripling of the magnetic field. The electron temperature was about 10 eV with ECR, compared with 1 to 2 eV with glow discharge. The ions were well confined by the magnetic field and did not impinge on the growing film.

Plasma confinement can also be achieved by using grounded screens instead of a magnetic field. These screens confine the RF power. Hamasaki et al. [1984] reported that grounded mesh confined plasmas from silane yield films with a better thermal stability than that of nonconfined plasmas and also increase the film deposition rate significantly.

5.4 HOT WALLS

In a conventional CVD system, only the substrate is intentionally heated. In a hot-wall CVD system, hot reactor walls are used to increase the density of the material deposited on the walls, so that fewer impurities are absorbed by the material on the walls when the reactor is opened to change the substrates. This minimizes the contamination of subsequent films by outgasing from the walls. This also reduces particulate generation from the plasma walls. This features are especially useful for single-chamber deposition systems that have no load-lock chamber.

Hot-wall, plasma-enhanced CVD from undiluted silane was reported by Boulitrop et al. [1985]. The resulting films were similar to those deposited by conventional glow discharge. The defect density was studied as a function of film thickness, deposition temperature, and pressure.

Hot-wall chemical vapor deposition of a-Si:H from disilane and higher silanes for thin-film transistors has been investigated by Ahn et al. [1991]. they found the electronic properties to be as good as those from silane by cold-wall CVD.

Muramatsu et al. [1987] reported achieving a conversion efficiency of 10.3% at a deposition rate of 1.5 nm/s and 11.2% at 0.4 nm/s using a hot-wall symmetric plasma CVD reactor for the i-layer from monosilane. Photosensitivity for the i-layers was about 10^6 and constant for a deposition-rate range from 0.1 to 3 nm/s.

A modified cylindrical-geometry hot-wall CVD system with a center electrode and substrates around the periphery that allows high deposition rates (2.5 nm/s) has been reported by Zhang et al. [1990]. However, the increase in deposition rate was accompanied by a decrease in the photosensitivity.

5.5 CONTAMINATION FROM IMPURITIES

External vacuum leaks, virtual leaks, and pump oil back-diffusion into the reactor chamber can introduce unwanted impurities into the depositing film. Virtual leaks may include water vapor, fluorocarbon cleaning fluids from freshly cleaned vacuum components, and trapped air or doping gas. A residual gas analyzer (RGA) may be used to distinguish the type of leaks - real, virtual, or pump oil. For an a-Si:H glow discharge deposition vacuum system, the leak rate should be maintained at a level below 1×10^{-5} torr·liter/s (1 torr·liter/s = 79 sccm). Good quality a-Si:H films typically contain less than 5×10^{18} cm^{-3} oxygen, 5×10^{18} cm^{-3} carbon, and 5×10^{17} cm^{-3} nitrogen [Carlson et al. 1985].

Under high base-vacuum conditions (10^{-9} torr background pressure), levels of 2×10^{18}, 2×10^{18}, and 1×10^{17} cm^{-3}, for oxygen, carbon, and nitrogen, respectively, have been achieved, giving films with an ESR spin density of 2×10^{15} cm^{-3} and a space charge density of 5×10^{14} cm^{-3} [Tsuda et al. 1987b]. But, for unknown reasons, the dark and photoconductivities and the photosensitivity of the resulting high base-vacuum films by Tsuda et al. were not as good as those achieved for the best a-Si:H glow discharge films deposited under normal base-vacuum (10^{-6} torr) conditions.

Tsai et al. [1984] studied the defect density and photostability of ultra-high-vacuum-deposited, undoped a-Si:H films in which the oxygen and nitrogen contents were varied by four to five orders of magnitude. They found that these impurities cause rapid increases in the defect density when they exceed a concentration of about 10^{20} cm^{-3} for oxygen and about 10^{19} cm^{-3} for nitrogen.

Carlson et al. [1985, 1987] studied the influence of oxygen, carbon, nitrogen, chlorine, and fluorine impurities on the space-charge width and surface-photovoltage diffusion length of a-Si:H deposited by DC glow discharge. These impurities were introduced by mixing gases such as CO and F_4 in SiH_4. Deteriorations in film quality were observed in all cases they studied. For example, adding 2% nitrogen to the silane gas, they observed about a factor of five reduction of the electron diffusion length and a factor of six reduction of the space-charge width in the resulting films.

Nakano et al. [1987] showed the important effects of an ultra-high vacuum (10^{-9} torr or better) and correspondingly low oxygen, nitrogen, and carbon impurities on ESR spin density (2×10^{15} cm^{-3}), Urbach edge (<50 meV), hole diffusion length (1 μm), and space-charge density (5×10^{14} cm^{-3}). They also described the effect of the dihydride content on spin density ($3x10^{16}$ cm^{-3} at 10^{21} Si-H_2 bonds per cm^3) and fill factor (0.58 at 10^{21} Si-H_2 bonds per cm^3) after exposure of the device to light and the effect of the oxygen content on degradations in photoconductivity due to light exposure.

Morimoto et al. [1990] investigated the effects of C ($2x10^{17}$ - $2.4x10^{19}$ cm^{-3}), O ($2x10^{18}$ - $3x10^{20}$ cm^{-3}), and N ($2x10^{16}$ - $1.2x10^{20}$ cm^{-3}) impurities prepared by several glow discharge systems. They found that the dark conductivity, the activation energy, and the density of charged dangling bonds are closely correlated with the N and/or O impurity content.

6
GLOW DISCHARGE DEPOSITION PARAMETERS HYDROGENATED AMORPHOUS SILICON

Because there are numerous variations in designs of glow discharge deposition systems, the optimum deposition parameters do vary from system to system and are often empirically optimized for individual systems. In this chapter, we present only some general trends reported in the literature for high quality a-Si:H film growth. Power density, substrate temperature, chamber pressure, feed-gas concentration, and flow rate are the primary deposition parameters affecting film quality. Typical a-Si:H deposition rates are 0.1-0.5 nm/s. The effects of these parameters at higher deposition rates have been reported by Luft [1988b].

A summary of the effects of the primary deposition parameters for glow discharge is given in **Tables 6-1 and 6-2.** Table 6-1 is for a-Si:H films deposited from (mono)silane and Table 6-2 is for those deposited from disilane. The tables show the effects of increasing the primary deposition parameters on the deposition rate (DR), the defect density (N_s), the total hydrogen content (C_H), the band gap (E_g), the dihydride content (SiH_2), the dihydride-to-monohydride ratio (SiH_2/SiH), the photoconductivity (σ_l), the photo-to-dark conductivity ratio (σ_l /σ_d),

Table 6-1 Summary of the Effects of Various Primary Deposition Parameters in Glow Discharge Deposition of a-Si:H Films from Monosilane.

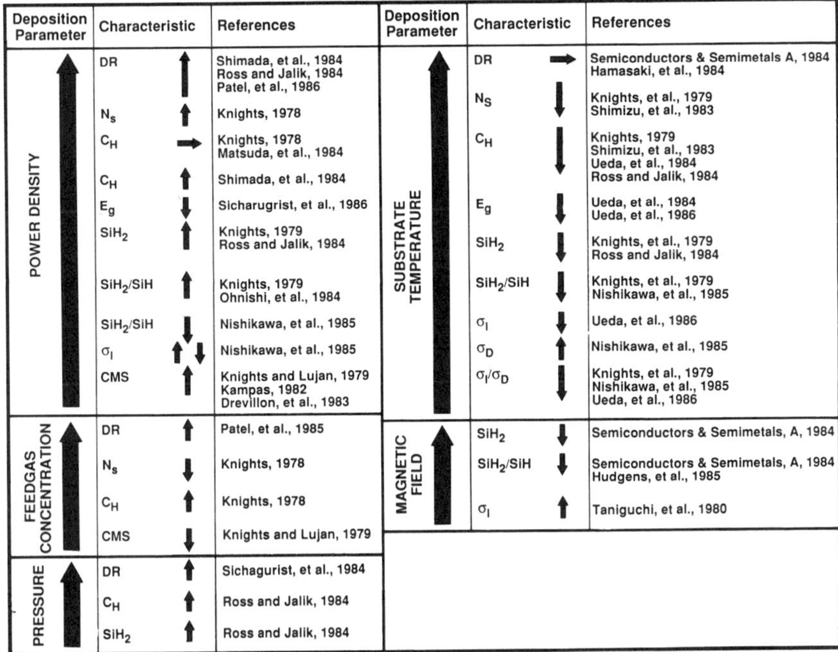

Deposition Parameter	Characteristic		References
Power density ↑	DR	↑	Shimada, et al., 1984; Ross and Jalik, 1984; Patel, et al., 1986
	N_s	↑	Knights, 1978
	C_H	→	Knights, 1978; Matsuda, et al., 1984
	C_H	↑	Shimada, et al., 1984
	E_g	↓	Sicharugrist, et al., 1986
	SiH_2	↑	Knights, 1979; Ross and Jalik, 1984
	SiH_2/SiH	↑	Knights, 1979; Ohnishi, et al., 1984
	SiH_2/SiH	↓	Nishikawa, et al., 1985
	σ_I	↑↓	Nishikawa, et al., 1985
	CMS	↑	Knights and Lujan, 1979; Kampas, 1982; Drevillon, et al., 1983
Feedgas concentration ↑	DR	↑	Patel, et al., 1985
	N_s	↓	Knights, 1978
	C_H	↑	Knights, 1978
	CMS	↓	Knights and Lujan, 1979
Pressure ↑	DR	↑	Sichagurist, et al., 1984
	C_H	↑	Ross and Jalik, 1984
	SiH_2	↑	Ross and Jalik, 1984
Substrate temperature ↑	DR	→	Semiconductors & Semimetals A, 1984; Hamasaki, et al., 1984
	N_S	↓	Knights, et al., 1979; Shimizu, et al., 1983
	C_H	↓	Knights, 1979; Shimizu, et al., 1983; Ueda, et al., 1984; Ross and Jalik, 1984
	E_g	↓	Ueda, et al., 1984; Ueda, et al., 1986
	SiH_2	↓	Knights, et al., 1979; Ross and Jalik, 1984
	SiH_2/SiH	↓	Knights, et al., 1979; Nishikawa, et al., 1985
	σ_I	↓	Ueda, et al., 1986
	σ_D	↑	Nishikawa, et al., 1985
	σ_I/σ_D	↓	Knights, et al., 1979; Nishikawa, et al., 1985; Ueda, et al., 1986
Magnetic field ↑	SiH_2	↓	Semiconductors & Semimetals, A, 1984
	SiH_2/SiH	↓	Semiconductors & Semimetals, A, 1984; Hudgens, et al., 1985
	σ_I	↑	Taniguchi, et al., 1980

and the columnar microstructure (CMS) in terms of increasing, decreasing, first increasing and then decreasing these attributes, or leaving them constant. The effects shown in the tables are for the parameter ranges normally used in high quality film depositions. The effects of deposition parameters are often interrelated and cannot be expressed as a simple straight-line relationship. Readers should see the references listed in the table and the discussions in later sections of this chapter for further details.

6.1. POWER DENSITY

In the glow discharge of monosilane, the deposition rate increases monotonically with radio frequency (RF) power density until it is limited by the gas flow rate (i.e., the deposition becomes gas-phase-limited instead of reaction-rate-limited) [Shimada et al. 1984, Ross and Jaklik 1984, and Patel et al. 1986]. The disadvantages of increasing the deposition rate by increasing the RF power include poor film quality and powder formation in the deposition chamber. Hydrogen dilution (at a constant gas flow rate) reduces the deposition rate; i.e., the rate decreases with increasing dilution [Knights 1979, Hamasaki et al. 1983 & 1984, Patel et al. 1985]. In disilane depositions, the dependence of the deposition rate on RF power is similar [Matsushita et al. 1984, Kumeda et al. 1985, Wiesmann et al. 1987].

Table 6-2 Summary of the Effects of Various Primary Deposition Parameters in Glow Discharge Deposition of a-Si:H Films from Disilane.

Deposition Parameter	Characteristic	References
POWER DENSITY ↑	DR ↑	Matsushita, et al., 1984 Kumeda, et al., 1985 Wiesmann, et al., 1987 Bhat, et al., 1987
	N_S ↓	Kumeda, et al., 1985
	C_H ↓	Wiesmann, et al., 1987
	C_H ↑	Matsuda, et al., 1983
	E_g ↑→	Wiesmann, et al., 1987 Bhat, et al., 1988
	SiH_2 ↓	Wiesmann, et al., 1987
	SiH_2/SiH ↓	Ohnishi, et al., 1984 Wiesmann, et al., 1987
	σ_I ↑	Ohnishi, et al., 1984 Fukuda, et al., 1984
	σ_D ↓→	Matsuda, et al., 1983 Wiesmann, et al., 1987
PRESSURE ↑	DR ↑↓	Vanier, 1986 Bhat, et al., 1988
	E_g ↑	Bhat, et al., 1988
	σ_I ↑	Wiesmann, et al., 1987
	σ_I ↓	Bhat, et al., 1988
	σ_D ↓	Bhat, et al., 1988
	σ_I/σ_D ↓	Bhat, et al., 1988

Deposition Parameter	Characteristic	References
SUBSTRATE TEMPERATURE ↑	DR ↑→	Fukuda, et al., 1984 Bhat, et al., 1988 Wiesmann, et al., 1987
	N_S ↓	Kumeda, et al., 1985
	C_H ↓	Ross and Jalik, 1984 Kumeda, et al., 1985 Wiesmann, et al., 1987
	E_g ↓	Wiesmann, et al., 1987 Bhat, et al., 1988
	SiH_2 ↓	Ross and Jalik, 1984 Wiesmann, et al., 1987
	SiH_2/SiH ↓	Wiesmann, et al., 1987
	SiH_2/SiH ↑→	Bhat, et al., 1988
	σ_I ↑	Ohnishi, et al., 1984 Fukuda, et al., 1984 Bhat, et al., 1988 Wiesmann, et al., 1987
	σ_D ↑	Ohnishi, et al., 1984 Bhat, et al., 1988
	σ_I/σ_D ↑	Wiesmann, et al., 1987
	σ_I/σ_D ↓	Bhat, et al., 1988
FEEDGAS CONCENTRATION ↑	DR ↑	Wiesmann, et al., 1987
	C_H ↑	Wiesmann, et al., 1987
	SiH_2/SiH ↑	Wiesmann, et al., 1987
	SiH_2/SiH →	Bhat, et al., 1988

For films from monosilane, the spin density as a function of RF power depends on the deposition conditions, such as electrode bias [Knights 1979]; in the unbiased condition, the spin density increases more than two orders of magnitude as the power is increased by a factor of 30 [Knights 1979]. For films from disilane, the spin density decreases (at 200°-300°C substrate temperatures) with increasing RF power [Kumeda et al. 1985].

According to Knights [1979], the hydrogen content in films from silane/argon mixtures first increases and then decreases (but remains essentially constant in the range of 14-17%) with increasing RF power (4-30 W). Ross and Jaklik [1984] showed the total hydrogen content to increase monotonically with increasing power. In films from disilane, the total hydrogen content decreases with increasing power density according to Kumeda et al. [1985] and Wiesmann et al. [1987], but it increases according to Matsuda et al. [1983], while the dispersed hydrogen content increases [Kumeda et al. 1985].

The effect of RF power on the optical bandgap in films from monosilane glow discharge is first an increase in the bandgap and then a decrease as the power level increases [Sichagurist et al. 1986]. The increase in the optical bandgap with RF power at low power levels is due to the increase in the hydrogen content, since the band gap increases linearly with increasing hydrogen content [Cody et al. 1981, Tsuo et al. 1987]. At very high power levels, the formation of microcrystalline silicon causes a sharp decrease in both the hydrogen content and the optical bandgap [Tanaka et al. 1981, Kamiya et al. 1981, Hata et al. 1981]. In films from disilane, the bandgap remains essentially constant [Wiesmann et al. 1987] or increases [Bhat et al. 1988] with increasing power density.

In films from monosilane, an increase in the power density increases the dihydride content [Tanaka et al. 1980, Ross and Jalik 1984] and the dihydride-to-monohydride ratio [Knights et al. 1979, Ohnishi et al. 1984]; in films from disilane, the opposite is true [Ohnishi et al. 1984, Wiesmann et al. 1987].

According to Nishikawa et al. [1985] the dark and photo-conductivities of films from monosilane first decrease and then increase significantly with increasing RF power. Films from disilane show an increase and then saturation in conductivities with increases in the RF power [Ohnishi et al. 1984, Fukuda et al. 1984]. Wiesmann et al. [1987] report different behavior for helium-diluted films from disilane; depending on the helium dilution, the photo- and dark conductivities either decrease or remain constant with increasing power density.

Increasing the power density also increases the microcolumnar structure in deposition from a silane/argon mixture, according to Knights and Lujan [1979]. A columnar microstructure is undesirable because it results in poor electronic film properties. Kampas [1982] found that a high power density leads to a columnar microstructure and a high spin density in glow discharge-deposited films from a silane/argon mixture. The growth of such microstructures is associated with the presence of voids and $(SiH_2)_n$ polysilane groups [Drevillon et al. 1983].

For high quality a-Si:H depositions, the supplied power is usually maintained at a level slightly above the minimum value needed to sustain the plasma. For a pure silane or disilane RF glow discharge deposition, this discharge-sustaining value is usually less than 0.1 W/cm^2. It is about 20 mW/cm^2 for a pure silane plasma with a 1.9 cm electrode spacing and a 0.7-torr pressure.

6.2 SUBSTRATE TEMPERATURE

During a-Si:H deposition, substrate temperature affects the hydrogen elimination and the reconstruction of atoms after the deposition precursors, such as SiH_3, have arrived on the growing film surface. It is generally agreed that the optimum deposition temperature for glow discharge deposition of a-Si:H from monosilane or from disilane is between 200° and 300°C. The valence band tail (Urbach) parameter and neutral dangling-bond density of a-Si:H both show a minimum at the deposition temperature of about 250°C [Smith and Wagner 1987]. The deposition rate for films from monosilane, with a few exceptions [Qiao et al. 1985, Ishihara et al. 1987b] is reported to be relatively independent of the substrate temperature [Longeway 1984]. This means that the sticking coefficients of the deposition precursors are relatively independent of the substrate temperature, and the deposition rate is limited by the supply of the deposition precursors from the plasma. For films from disilane, Fukuda et al. [1984] show a constant deposition rate with increasing substrate temperatures at a high RF power density, and an increase in the deposition rate with increasing substrate temperatures at low RF power. Wiesmann et al. [1987] also show an increase in the deposition rate with temperature. Bhat et al. [1988] describes an essentially constant deposition rate with increasing temperatures over a power-density range of 0.03 to 0.09 W/cm^2.

The total hydrogen content decreases with increasing substrate temperatures in films both from monosilane [Knights 1979, Shimizu et al. 1983, Ueda et al. 1984 & 1986, Qiao et al. 1985] and disilane [Ross and Jalik 1984, Kumeda et al. 1985, Wiesmann et al. 1987], but the dispersed hydrogen content increases. Similarly, the optical gap of films from silane [Ueda et al. 1984 & 1986] and from disilane [Wiesmann et al. 1987] decreases with increasing substrate temperatures due to the decreased hydrogen content. Values of the optical bandgap of a-Si:H depend on hydrogen content, following the empirical relation $E_g = 1.56 + 1.27C_H$, where C_H is the atomic fraction of bonded hydrogen [Ross and Jaklik 1984, Tsuo et al. 1987]. This equation is for glow discharge-deposited a-Si:H deposited at high temperature or dehydrogenated by annealing. For sputtered material, E_g = 1.45 for C_H = 0.

The spin density decreases with increases in the substrate temperature both in films from monosilane [Shimizu et al. 1983] and disilane [Kumeda et al. 1985] when the substrate temperature is less than 300°C. This is probably related to the higher number of microvoids and polyhydride bonds in films deposited at temperatures lower than 200°C [Soule et al. 1981]. The number of microvoids decreases with increasing temperatures. Weitzel et al. [1981] found that a-Si:H films glow discharge deposited at substrate temperatures above 250°C had densities close to that of crystalline silicon and were impervious to water.

In films from both monosilane and disilane, the polyhydride and dihydride contents [Knights et al. 1979, Ross and Jalik 1984, Wiesmann et al. 1987] and the dihydride-to-monohydride ratios [Knights et al. 1979, Wiesmann et al. 1987, Nishikawa et al. 1985] decrease with increasing substrate temperatures. However, Bhat et al. [1988] found an increasing or constant dihydride-to-monohydride ratio, when the dihydride was measured by the 850 cm^{-1} absorption. In films from silane an increase in the substrate temperature decreases the SiH content [Drevillon et al. 1983]. Films from silane by RF glow discharge show no dihydride bonds above a 220°C substrate temperature. However, films grown at a substrate temperature higher than 350°C (the onset of hydrogen evolution from the film) contain insufficient hydrogen to passivate dangling-bond defects. At substrate temperatures above 550°C (the onset of microcrystalline silicon growth), the deposited films become polycrystalline with no detectable hydrogen content. In films from disilane, a high RF power or a substrate temperature above about 250°C also eliminates dihydride bonds.

In glow discharge of monosilane, Ueda et al. [1986], Nishikawa et al. [1985], and Knights et al. [1979] describe a decrease in photosensitivity with an increase in temperature. Ueda et al. [1986] show that the photosensitivity decreases with increasing substrate temperature from 10^5 to 2 x 10^3 as the temperature increases from 350° to 450°C; from 300° to 350°C the sensitivity is approximately constant. Both the photoconductivity and dark conductivity of films from disilane increase with increasing substrate temperatures [Wiesmann et al. 1987, Bhat et al. 1988, Ohnishi et al. 1984, Fukuda et al. 1984]. Whereas Wiesmann et al. [1987], Ohnishi et al. [1984], and Fukuda et al. [1984] found an increase in photosensitivity, Bhat et al. [1988] found a decrease in photosensitivity with increasing temperature.

6.3 FEED-GAS CONCENTRATION

The composition and structure of a-Si:H are determined by both feed-gas concentration and deposition parameters. As we discussed earlier, SiH_3 is generally believed to be the deposition precursor responsible for high quality a-Si:H films. Films produced in the silane depletion regime, where there is a lack of SiH_3 radicals, have more defects than those produced in the silane excess regime [Hirose, in Pankove 1984]. Thus, a high silane-versus-diluent-gas ratio should be beneficial to high quality film deposition. A high gas pressure, a high gas flow rate, and low power also help to ensure that the deposition is taking place in the silane excess regime.

Patel et al. [1985] found an increase in the glow discharge deposition rate when the silane concentration in the feed-gas was increased from 10% to 35%. Reporting on glow discharge from helium-diluted disilane, Wiesmann et al. [1987] also showed an increase in the deposition rate with increasing disilane concentrations in helium.

The spin density of films from monosilane decreases sharply with increases in the silane concentration in argon [Knights 1979]. Increasing the atomic weight of the diluent gas also increases the defect density of the film. Helium and neon produce less microstructure and fewer active defects than do argon and krypton [Knights et al. 1981].

The hydrogen content of a-Si:H films from monosilane increases with increases in the silane concentration for RF glow discharge films [Knights 1979]. In films from disilane, the total hydrogen content increases with increases in the disilane concentration (20-50%) in helium [Wiesmann et al. 1987].

Knights and Lujan [1979] found that a low concentration of silane in argon leads to a high columnar microstructure and a higher defect density, except when the substrate is negatively biased. Kampas [1982] reported that a low silane concentration in helium leads to a columnar microstructure.

Doping gas can affect the deposition rate of a-Si:H. Adding PH_3 to SiH_4 lowers the deposition rate by roughly 10%. Adding B_2H_6 to SiH_4 increases the deposition rate by roughly 50%. Adding B_2H_6 to SiH_4 also lowers the activation energy for thermal CVD of a-Si:H [Collins and Cavese 1988].

6.4 PRESSURE

The effect of gas pressure on the deposition rate in the glow discharge deposition of both monosilane [Sichanugrist et al. 1984] and disilane [Vanier 1986, Bhat et al. 1988] is the same. In the supply-limited region, the deposition rate is proportional to the pressure; in the power-limited region, it remains constant. Bhat et al. [1988] shows first an increase and then a slight decrease in the deposition rate as the undiluted disilane pressure is increased. In the low-pressure regime, the supply of SiH_3 radicals is depleted more easily, there is a higher chance for SiH and SiH_2 radicals to reach the film's growing surface, and the ion bombardment of the growing film's surface is more severe. For high quality film deposition, the high pressure, power-limited regime is favored. However, the pressure should be below the level that causes gas-phase polymerization, which results in the accumulation of yellow powders inside the glow discharge reactor [Brodsky 1977]. For pure silane RF glow discharge deposition, this usually means keeping the gas pressure below 1 torr. The optimum pressure for deposition also depends on the reactor electrode spacing, as shown by Paschen's law [§ 5.1.3].

Ross and Jaklik [1984], on the other hand, found that in RF and DC glow discharge at pressures in the range of 0.1 to 0.5 torr (that is, below the pressures for plasma polymerization) of films from monosilane, the total hydrogen content and the dihydride content increase with increasing pressure. Hudgens et al. [1985] reported an increase in the dihydride/ monohydride ratio for microwave glow discharge, as the pressure is increased from 50 to 200 mtorr, and a decrease in the photoconductivity in films from monosilane. Bhat et al. [1988] describes an increase in the optical gap with increasing pressure (300 to 500 mtorr) for undiluted disilane.

Wiesmann et al. [1987] report that the photoconductivity increases with increasing pressure in films from helium-diluted disilane, whereas Bhat et al. [1988] reports a decrease with increasing pressure not only in the photoconductivity but also in the dark conductivity and the photosensitivity with undiluted disilane.

6.5 GAS FLOW RATE

The gas flow rate is another important primary deposition parameter, but it is not one that is directly comparable from one system to another. The flow rate is inversely related to the residence time, the average time that a gas molecule spends in the plasma. This time determines the probability that the molecule will be dissociated and incorporated into the growing film. The residence time in turn affects the depletion of the gas. At a given pressure and power density level, the longer the residence time, the greater the depletion of the feedstock gas. This depletion appears to be the critical parameter affecting film quality. Hirose et al. [1981] studied the SiH and SiH_2 stretching modes as a function of SiH_4 and H_2 flow rates and found the incorporation of monohydride bonds to be predominant as well as the elimination of dihydride bonds in high-flow-rate regimes.

Whereas gas flow rates are usually reported, such additional information as the volume between the electrodes and the fraction (f_d) of the feed gas being deposited on the substrate are not reported. This generally makes it impossible to determine the gas depletion from published data.

In the absence of any deposition on the substrate from the feed-gas, the residence time (t_R) of a gas molecule is

$$t_R = P_1 V_1/(P_2 Q_g) ,$$

here P_1 is the chamber pressure, P_2 is the feed-gas pressure, Q_g is the feed-gas flow rate, and V_1 is the volume between the electrodes.

With deposition on a substrate, the gas flow rate at the partial pressure in the reactor chamber is

$$Q_{out} = Q_{in} (1 - f_d)$$

$$k\, n_e = Q_{in}\, f_d/[V_1 (N_g)_o (1 - f_d)] ,$$

where k is the total rate constant of the electron impact dissociation, n_e is the electron number density, and $(N_g)_o$ is the initial silane number density [Shing et al. 1987].

As the flow rate increases, the depletion decreases [Shing et al. 1987] and the deposition rate increases [Sichanugrist 1984 & 1986]. Shing et al. [1987], however, found the opposite; namely, that as the flow rate increases, the deposition rate decreases, and as the depletion increases, the deposition rate increases.

6.6 MAGNETIC FIELD

A magnetic field can be used to confine the plasma, influence the electron density and temperature, and modify the glow discharge deposition process without changing other deposition parameters. Increasing the applied magnetic field perpendicular to the substrate's surface during RF glow discharge deposition from silane has been shown to increase SiH, decrease SiH_2 [Pankove 1984, Hudgens et al. 1985] (see Figure 5.3-1) and increase photoconductivity [Tanaguchi et al. 1980]. Sugai et al. [1985] used a toroidal magnetic field to confine a DC hot tungsten-filament-generated silane plasma (DC toroidal discharge) and correlated the film properties with the ion and radical fluxes to the substrate. Ohnishi et al. [1987] placed electromagnets behind the cathode of RF-generated silane plasma (the controlled plasma magnetron method) to produce high quality a-Si:H films at high deposition rates (up to 1.5 nm/s).

6.7 FREQUENCY

Both AC (mainly RF) and DC glow discharges have been used in high quality a-Si:H depositions. Although the RF glow discharge is used more often, there has been no conclusive evidence as to which method produces better quality films. RF glow discharge deposition is usually done at a frequency of 13.56 MHz. This is the allotted frequency of the U.S. Federal Communication Commission and international communication authorities. RF-excited plasma is popular because it causes no charging of insulating or electrically isolated surfaces inside the reactor, and it is believed to be more efficient than DC excitation in promoting ionization and sustaining the discharge. At low AC frequencies, the glow discharge is discontinuous, and the electrodes successively take opposite polarities. At frequencies above about 100 kHz, the discharge can be maintained almost continuously, and the ions in the plasma, because of their heavy mass, cannot instantaneously follow the excitation field. Consequently, only electrons can instantaneously respond to the RF field and gain sufficient energy to ionize the gas [Chapman 1980]. Models of RF discharges at frequencies greater and smaller than the ionic plasma frequency have been studied by Pointu [1987].

Curtins et al. [1987a,b] studied the glow discharge deposition rate as a function of excitation frequency from 25 to 150 MHz and found that it increases with frequency to reach a maximum at about 70 MHz, the rate being about four times higher than at the standard 13.56 MHz frequency. The film quality parameters at 70 MHz were comparable to those at lower frequencies. Above 70 MHz the deposition rate fell off again. The authors claim that there are lower internal stresses in films produced at 70 MHz than at 13.56 MHz. At 70 MHz the plasma glow is self-confined, eliminating the need for other confinement.

Microwaves with a frequency of 2.45 GHz have been used for gas decomposition in plasma deposition of a-Si:H [Kato and Aoki 1985, Mejia et al. 1983]. In the microwave frequency range, even the electrons cannot respond in phase with the high-frequency electric field. This results in low efficiencies of energy transfer from the electric field to the plasma, and consequently, a higher applied power is needed to maintain the plasma than those used in the RF discharges. The increased power raises both the gas and substrate temperatures and is detrimental to the quality of the deposited film. Electron cyclotron resonance has been used to enhance the energy transfer in microwave-generated plasma, and this resulted in a high a-Si:H film deposition rate [Kato and Aoki 1985]. ECR-generated argon or hydrogen plasmas have been used to generate a remote silane plasma (ECR CVD) for a-Si:H film deposition (see Chapter 9).

6.8 SUBSTRATE EFFECTS

The electrical and optical properties of a-Si:H films are not strongly dependent on the substrate material as long as the deposition temperature is not high enough to cause significant diffusion from substrate into the film. However, there are reports that the microstructure of a-Si:H depends on the surface roughness and conductivity of the substrate [Canillas et al. 1990, Hishikawa et al. 1990]. The most popular substrate for a-Si:H-based devices is transparent conducting oxide (TCO)-coated glass. Most a-Si:H film characterization measurements are done with film deposited directly on Corning Code 7059 glass. Stainless steel substrates are also used for some solar cell applications.

An important consideration for choosing substrates is the thermal expansion coefficient. For example, an a-Si:H film can be peeled off very easily when it is deposited on fused silica, which has a very different thermal expansion coefficient than a-Si:H. Corning code 1724, 1729, 7059, and 7740 glasses all have similar thermal expansion coefficients to a-Si:H (**Fig. 6.8-1**). The 7059 glass is the most widely

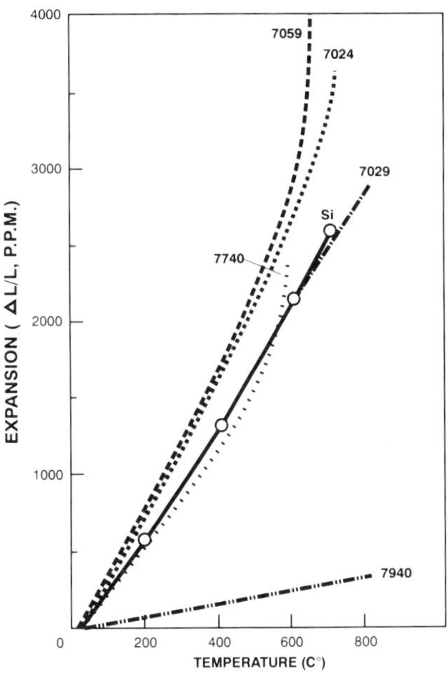

Figure 6.8-1 Linear expansion coefficient vs. temperature for Si and several glass substrates by Corning Glass Works [Troxell et al. 1987].

used and can tolerate process temperatures up to 600°C. The Corning 7059 is an alkali-free barium aluminoborosilicate glass with the approximate composition of 49 wt.% SiO_2, 10 wt.% Al_2O_3, 15 wt.% B_2O_3, 25 wt.% BaO, and 1 wt.% As_2O_3. The 1729 glass (alkaline earth aluminosilicate) has the best thermal expansion fit with Si from 0 to 700°C and it can tolerate process temperatures up to 800°C (viscosity strain point = 799°C, anneal point = 855°C) [Troxell et al. 1987]. The thermal expansion properties of a-Si:H depend on the hydrogen content. Jansen et al. [1987] reported that, for an a-Si:H film deposited at 250°C with a density of 2.0 g/cm^3 and 20 at.% of hydrogen, the linear thermal expansion coefficient is 4.4×10^{-6} °C^{-1} and the biaxial elastic modulus is 150 GPa. (These are averaged values measured between 30 and 100°C. For crystalline Si, the average linear thermal expansion coefficient is 2.5×10^{-6} °C^{-1}.)

Popular TCO coatings for a-Si:H substrates include indium tin oxide (ITO), fluorine-doped tin oxide (SnO_2:F), and zinc oxide (ZnO). Sakai et al. [1990] reported that highly textured SnO_2-coated glass may cause stripe-like defect regions in the p-i-n solar cell deposited on it and degrades the V_{oc}. The atomic hydrogen in a silane plasma may cause chemical reduction of tin oxide coating on glass, especially under high hydrogen dilution conditions. ZnO-coated glass is less affected by this reduction problem. Indium tin oxide-coated glass can cause indium contamination of a-Si:H at normal film deposition temperature and is seldom used as substrates [Eicke and Bilger 1988]. SiO_x formation at the TCO and a-SiC:H interface is more serious a problem, by an order of magnitude, for SnO_2 than for ITO and ZnO [Eicke and Bilger 1988]. SiO_x at the interface causes series resistance problems. Alkali-containing glasses may dope the a-Si:H with alkali impurities that act as donors [Carlson and Wronski 1985b]. Soda-lime glass because of its high optical transmission and low cost is often preferred for industrial production of a-Si:H solar cells. However, electrically active impurities in the soda glass substrate may diffuse into the film or get sputtered into the plasma, forming compounds like Na-Si and K-Si, and then get trapped in the film [Koo et al. 1988]. A thin layer of deposited stable oxide is necessary to passivate soda glass substrates for subsequent a-Si:H depositions.

Table 6.8-1 Properties of Corning Code 7059 Glass [Bocko et al. 1989]

Working Point	1160°C	Density	2.76 gcm^{-3}
Soft Point	844°C	Liquidus Viscosity	1 x 10^6 poise
Anneal Point*	639°C	Acid Solubility	13 mg/cm^2
Strain Point**	593°C	Refractive Index	1.5333 (589.3 nm)
Expansion (0 to 300°C)	4.6 x 10^{-6}/°C	Dielectric Constant	5.84 (20°C, 1 MHz)
Expansion (25°C to set point)	5.05 x 10^{-6}/°C	Log Resistivity	13.1 (250°C) 11.0 (350°C)

* anneal point = temperature at which viscosity is 10^{13} poise
** strain point = temperature at which viscosity is $10^{14.5}$ poise

For substrate temperatures lower than 400°C, metals such as Cr, Ti, V, Nb, Ta, and Mo have all been used to make a-Si:H solar cells without causing significant impurity problems [Carlson and Wronski 1985b]. For Fe substrates the a-Si:H deposition temperature should be

330°C or less. Because the eutectic temperature (the crystalline-liquid transformation temperature that melting occurs at the interface followed by crystallization) for Al in contact with evaporated a-Si is about 275°C, the deposition temperature for a-Si:H on Al substrates should be less than 275°C. Amorphous Si or Ge in contact with certain metals appears to crystallize at temperatures lower than the usual crystallization temperature. Herd et al. [1972] found that in simple eutectic systems vacuum-evaporated a-Si crystallizes at 0.72 (and Ge at approximately 0.65) of the eutectic temperature expressed in degrees Kelvin. The crystallization or silicide formation temperatures (whichever is lower) for a variety of a-Si-metal and a-Ge-metal systems found by Herd et al. are: 540°C for Si-Ag (crystallization), 335°C for Si-Al (cryst.), 186°C for Si-Au (cryst.), 175°C for Si-Cu (β-Cu-Si), 500°C for Si-Cr (Cr_5Si_3), >800°C for Si-Mo, 320°C for Si-Ni (θNi_2Si), 220°C for Si-Pd (Pd_2Si), 280°C for Si-Pt (Pt_2Si), 210°C for Ge-Al (cryst.), 125°C for Ge-Au (cryst.), and 25°C for Ge-Cu (Cu_5Ge). Maa and Lin [1979] found that, depending on the device structure, annealing temperatures as low as 120°C could cause Al-induced crystallization in electron-beam-evaporated a-Si films. Some metals, e.g. Au and Cu, appear to be unsuitable as substrates due to interdiffusion and silicide formation at low temperatures. Interfacial reactions between a-Si and Pd films have also been observed [Hentzell et al. 1985]. The growth of Pd_2Si from 200 to 235°C is diffusion controlled with an activation energy of 1.25 eV. Enhanced crystallization of remaining a-Si in contact to Pd_2Si occurs between 400 and 500°C. Poor adhesion, which is evident in the form of small blisters in the a-Si:H films, has been observed on Ag (perhaps due to an oxide layer) and on highly polished single-crystalline semiconductors, such as Si, Ge, and GaAs [Carlson and Wronski 1985b]. Stainless steel or Mo substrates for Si need a buffer layer to prevent the diffusion of Fe into silicon or the reaction of Mo with Si. The buffer layer might be RuO_2 [Jia & Anderson 1990] or TiN [Pramanik & Jain 1990].

To make good contacts between metal substrates and the a-Si:H film, the work function of the metal needs to be considered. For example, high work function metals such as Pt and Pd make good contacts to p-type layers, while low work function metals such as Mg, Al, Ti, Cr, and Mo make good contacts to n-type layers [Sze 1981]. Ti followed by Ag is often used as contact to the n-layer of an a-Si:H solar cell deposited on glass because it combines the low work function of Ti and the high optical reflectivity of Ag. TCO and metal combinations may also be used to enhance the optical reflection of solar cell back contacts [Banerjee and Guha 1991]. The optical reflectance values of common a-Si:H/metal contacts are: 0.70 for Al, 0.94 for Ag, 0.22 for Ti, 0.75 for 4 mm Ti/Ag, and 0.95 for ITO/Ag [Catalano in Kanicki 1991]. Aluminum, having both low work function and good optical reflectivity, is normally used for contact to n-type a-Si:H. In addition to barrier height, it is important to study the ability of a metal to make a good low-resistance electrical contact to an a-Si:H film by examining the value of the specific contact resistance. All contacts having contact resistance below 0.5 ohm-cm^2 are acceptable, in general, for a-Si:H TFT structures to be used in flat panel LCD's [Kanicki 1988].

Flexible and light-weight polymer films, such as polyimide, polyethylene terephthalate, and transparent polyether sulphone, have been used as substrates for a-Si:H solar cells [Ishikawa et al. 1985, Jacobson et al. 1987, Nakatani, et al. 1989, Nath et al. 1990, and Kishi et al. 1991]. These films are usually coated with a TCO or metal layer for electrical contact before a-Si:H deposition. Because of outgassing from the polymer substrates, lower deposition temperatures than glass or stainless steel substrates are required. This can usually be achieved by using hydrogen dilution to reduce the temperature required for good quality a-Si:H alloy film depositions.

The intrinsic stress in a-Si:H films depends strongly on the substrate that the film is deposited on [Stutzmann 1985, Kurtz et al. 1986]. Stutzmann found that for an a-Si:H film deposited on a Corning 7059 glass substrate, the intrinsic stress is about 400 MPa (compressive) in the first 0.5 μm of the film. This stress decreases for a-Si:H further away from the substrate, and eventually reaches zero for a-Si:H deposited more than 3 μm away from the glass substrate. For a-Si:H films deposited on thin, easily deformed Al substrates, Stutzmann found the intrinsic stress to be less than 100 Mpa. Infrared absorption data indicate that some ordering of the a-Si:H structure may be occurring on crystalline Si substrates at deposition temperatures as low as 300°C [Carlson and Wronski 1985].

The electrical and optical properties of a-Si:H may vary as a function of film thickness. When the surface mobility is high and/or the sticking coefficient is low during deposition, the surface roughness of a-Si:H decreases with increasing film thickness [Collins 1988]. Raman backscattering spectra of glow discharge-deposited a-Si:H films were found to depend on the material of the substrate (glass, Si, and stainless steel) and on the thickness of the films [Schubert and Bauer 1990, Hishikawa et al. 1990]. Stress originated from the a-Si:H/substrate interface is induced into the film and relieved as film growth propagates. Schubert and Bauer found that the interface stress is highest for a-Si:H (deposited at 247°C) on ITO, less on 7059 glass and on ZnO, and not detectable on Al. In this study, ITO, ZnO, and Al films are deposited on 7059 glass. Hishikawa et al. found that a-Si:H films deposited on crystalline Si substrates by RF glow discharge at 160 to 200°C seem to have smaller bond-angle distortion and a larger effective force constant of the Si-Si bonds than the films deposited on quartz or glass substrates. *In-situ* IR phase modulated ellipsometry study of the interaction between growing a-Si:H (or a-SiC:H) and Corning 7059 glass substrate showed evidence of boron incorporation into an interface layer [Blayo and Drévillon 1990].

6.9 DEPOSITION RATE

Amorphous silicon solar cells and modules are typically produced at a deposition rate of 0.1-0.3 nm/s. Thus, the deposition of a typical 400 nm-thick intrinsic amorphous silicon layer requires 22-66 minutes. Increasing the deposition rate is believed to be one way to reduce manufacturing time and, hence, manufacturing cost. For instance, increasing the deposition rate to 1-2 nm/s reduces the deposition time of a typical intrinsic amorphous silicon layer to about 3-7 minutes, a duration that is commensurate with the time needed for other process steps.

There are various potential approaches to increasing the deposition rate for good quality material: (1) by selecting a different feed gas than silane, (2) by using a modified or different deposition method that will improve the film quality at high deposition rates, (3) changing the reactor geometry, or (4) changing the parameter space for deposition. A thorough review of the effect of various deposition parameters on the deposition rate and film quality has been made by Luft [1988].

6.9.1 Powder Formation

Since 1980, researchers have attempted to achieve higher deposition rates in preparing amorphous silicon films. The simplest way to achieve higher deposition rates is to increase the power density. Increasing the power density, using undiluted silane as a feed stock, results in gas-phase polymerization and attendant microparticulate or powder formation, which results in poor film quality [Knights et al. 1981] and greater light-induced degradation. In attempts to avoid this problem, disilane or other higher silanes have been used as a feedstock [Scott et al.

1980, Ogawa et al. 1981]. However, powder formation at high power densities also occurs with disilane [Vanier 1986]. Increasing the frequency for glow discharge increases the power density before onset of powder formation; the threshold power is nearly tripled by increasing the frequency from 20 MHz to 80 MHz [Tscharner et al. 1992].

Gallagher [1988] has speculated that ion nucleation in low-field regions of a plasma is severe, since ion-molecule reactions cause a rapid growth in the size of dust particles, and only the ion drift to the surface in the electric field can compete effectively with the rapid growth. Thus, any ions formed in the low-field region can grow to a large size before reaching the surface. As they grow, the drift becomes slower and the reaction rate faster. Small dimensions between electrodes minimize the time that elapses between species generation and their deposit on the substrate and thus minimize the time the species are available for polymerization in the gas phase. Similarly, elimination of high-field regions, e.g., by plasma confinement, minimizes neutral nucleation. In general, gas-phase reactions can be reduced by increasing the feed-gas flow rate, reducing the pressure, or both. As a result of such preventive measures, the significant problem of powder formation at high-rate deposition can be controlled. Researchers have been able to reach deposition rates up to 2 nm/s with reasonable efficiencies without powder formation.

6.9.2 Parameter Space

The deposition rate can be increased by higher power density, higher feed gas flow or concentration, higher pressure, higher frequency, and higher magnetic field strength. Increasing the partial pressure of the feed gas is not a desirable alternative for increasing the deposition rate, because it increases the gas-phase reactions and thereby encourages dust formation. Increasing the feed-gas concentration to increase the deposition rate is desirable. High-rate deposition of a-Si:H films has been done with undiluted gases with no dust formation. Using undiluted gases has the additional advantage of reducing the ESR spin density at a given power level. Dilution of feed gases is still needed for high quality a-SiGe:H deposition, but for different reasons.

The effects of deposition parameters such as power density, substrate temperature, feed gas concentration, pressure, magnetic field strength, and power frequency have been evaluated for both silane and disilane as feed gases [Luft 1988b]. At high deposition rates the silicon dihydride content, bonded hydrogen content, defect density, amount of voids, and number of strained bonds increase and the diffusion length and life time decrease due to trapping and recombination. The photoconductivity and sub-bandgap absorption (at 1.2 eV) as well as the Urbach energy decrease with increasing deposition rate in the range of at least 0.01-0.1 nm/s. The increase in bonded hydrogen content gives an increase in the optical bandgap and a decrease in the absorption coefficient. The increase in ESR spin density correlates with an increase in clustered hydrogen and a decrease in dispersed hydrogen. All this leads to lower short-circuit current density, fill factor and efficiency.

Good quality intrinsic films, meaning films with a high photosensitivity and a low dihydride/monohydride ratio and a low density of states, can be produced at deposition rates of 2 nm/s. To obtain high quality films at high deposition rates, all the precautions for high quality film deposition at low rates must be observed: a high predeposition vacuum; low O, N, and C impurities; clean feed gases; confined plasmas; a low deposition pressure; minimization of the low-field region; optimized deposition parameters; and no (or only low energy) ion bombardment (achieved by a triode geometry or proper bias voltage). Because of the interactions that occur among the deposition parameters, they must be optimized for each set of deposition conditions, generally by trial and error.

6.9.3 Feed-Gas Selection

There is not much difference in device performance whether (mono)silane or disilane are used as i-layer precursor. The defect density $g(E_f)$ at 2 nm/s is slightly lower with silane (5×10^{15} cm^{-3}) than with disilane (2×10^{16} cm^{-3}). In films from monosilane, most of the incorporated hydrogen is bonded to silicon. Films made from disilane have a large amount of hydrogen that is not bonded with silicon (7%-12%, depending on deposition power), and this may be reflected in the dark conductivity and photoconductivity of the films. As a consequence, the two kinds of films exhibit different structural and electronic properties [Matsuda et al. 1983]. Properties of a-Si:H deposited from disilane are in general not as good as those of films from monosilane. The best density of states, $g(E_f)$ (measured by space-charge-limited current), is approximately 3.5×10^{16} cm^{-3} V^{-1} for films from disilane deposited at a rate of 2 nm/s [Bhat et al. 1988] versus 5×10^{15} cm^{-3} V^{-1} for films from monosilane. Optimum material properties are obtained at higher deposition temperatures for disilane than for monosilane films [Bhat et al. 1988]. Higher deposition temperatures increase the rate of surface reactions and thereby the rate of surface equilibration and reduce the hydrogen content slightly, thus reducing the optical band gap and thereby increasing the short-circuit current while at the same time reducing the open-circuit voltage.

6.9.4 Alternative Deposition Methods and Modified Reactor Geometry

Other avenues to increase the deposition rate without deteriorating the film quality have been sought in modifications of the reactor equipment. For instance, Nishikawa [1985] found that one way to increase the deposition rate from monosilane was to modify the deposition equipment to ensure plasma confinement. Better reactor design avoids dust formation because of improved reactor geometries (avoidance of high-field regions), smaller electrode spacing (reducing low-field regions), and by confining plasmas (lower power densities for a given deposition rate).

Some new deposition methods, especially those that generate an abundance of hydrogen radicals, such as microwave plasma CVD or ECR-CVD, seem promising in improving film quality at high deposition rates. The reason appears to be a better microstructure as a result of hydrogen etching during film growth. Radicals impinging on the growing surface must be given time to move around to find the lowest energy level that will give the most homogeneous structure (minimal voids and dangling bonds) and not get buried under further depositions before this minimum energy level has been achieved. An abundance of hydrogen radicals might accomplish this as enhanced atomic hydrogen etching during deposition may have beneficial effects on the film properties.

6.9.5 Device Performance

Efforts to increase the cell efficiency of devices produced at high deposition rates have been twofold: (1) by increasing the charge generation, that is, the short-circuit current density, and (2) by increasing the charge collection under forward bias, that is, the fill factor. However, these efforts are not different from those being pursued for low-deposition-rate devices. The charge generation can be increased by reducing the band gap (for example, by increasing the substrate temperature during deposition). In device fabrication, there is a limit to how high the substrate temperature can be increased because of interlayer effects such as the diffusion of dopants or other impurities from one layer to the next. The charge collection can also be improved by increasing the substrate temperature during deposition to reduce the density of states (at least for cells produced from disilane).

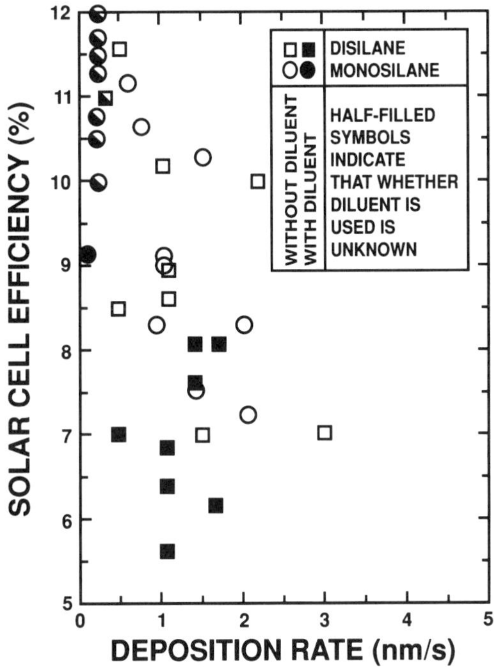

Figure 6.9-1 Small-Area Solar-Cell Initial Efficiencies as a Function of Deposition Rate when Monosilane and Disilane are the Feed-Stock Gases [Luft 1988b]

The best combinations of a high deposition rate and high initial cell efficiency in laboratory-size cells prepared by RF glow discharge are 10.3% efficiency for a cell deposited at 1.5 nm/s from monosilane [Muramatsu et al. 1987] and 10.2% efficiency at 1 nm/s in a 1 cm^2 cell from disilane [Tanaka, M., et al. 1988]. As the deposition rate increases, the efficiency decreases. **Figure 6.9-1** illustrates small-area solar-cell initial efficiencies as a function of deposition rate. Except at the lowest deposition rates, we can see that the best efficiencies have been achieved from undiluted feed gases. At the lowest rates, it is not known whether the feed gases were diluted or not. As the deposition rate increases, the

Staebler-Wronski degradation increases (see **Figure 6.9-2**).

For a-Si:H devices deposited by glow discharge at rates of 0.5-2 nm/s, it appears to make no difference in device efficiency whether silane or disilane is used as a feed-gas. However, there is some evidence that fluorine-containing gases (such as SiF_4) in the presence of atomic hydrogen will result in HF that might provide beneficial etching of strained or weak bonds [Shibata et al. 1987a].

Part of the higher efficiencies that have been achieved in a-Si:H films at all deposition rates are device-related; that is, they are the result of a better p-i interface, light-trapping, etc. For instance, modulating the gas flow rate and reducing the RF power at the beginning of the i-layer deposition can avoid plasma damage to the p-i interface [Azuma et al. 1987].

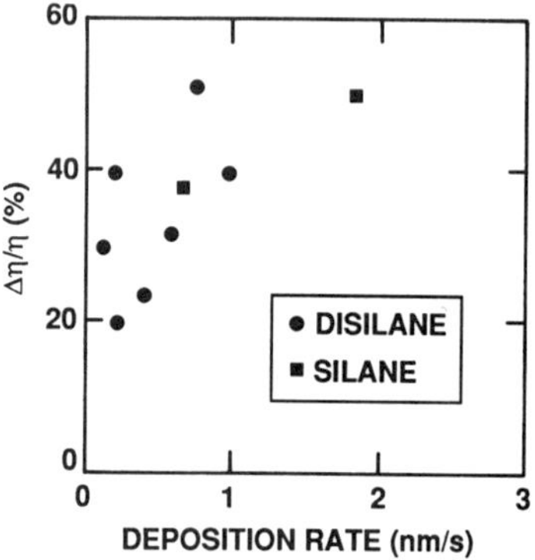

Figure 6.9-2 Light-Induced Degradation as a Function of Deposition Rate for single-junction p-i-n devices after 200 hours of Continuous Illumination.

7
GLOW DISCHARGE DEPOSITION REACTION CHEMISTRY FOR HYDROGENATED AMORPHOUS SILICON

The growth of a-Si:H and its alloys depends partially on the ions and neutral radical species in the gas phase, partially on the reactions at the film surface, and partially on the reactor system, e.g., whether DC or RF glow discharge systems are used.

There are four stages in the formation of a-Si:H films from a silane plasma: (1) electron impact dissociation of the feedgas, (2) transport of species to the substrate surface accompanied by particle reactions, (3) adsorption of species at the growing film surface, and (4) reactions and reconstructions that result in the final film.

Examples of electron impact processes in silane plasmas are:

$$
\begin{aligned}
e + SiH_4 \text{-->} \quad & SiH_4 + e \\
& SiH_2 + H_2 + e \\
& SiH_3 + H + e \\
& SiH + H_2 + H + e \\
& SiH_2^+ + H_2 + e + e \\
& SiH_3^+ + H + e + e \\
& SiH_3^- + H \\
& SiH_2^- + H_2
\end{aligned}
$$

Examples of first-order heavy particle reactions in silane plasmas are:

$SiH_4 + H \rightarrow SiH_3 + H_2$
$SiH_4 + SiH_2 \rightarrow Si_2H_6$
$SiH_4 + SiH_3 \rightarrow Si_2H_5 + H_2$
$SiH_4 + Si_2H_6 \rightarrow Si_nH_m$

Neutral radicals dominate the deposition under conditions that produce high quality a-Si:H films. The dominant radical is SiH_3. For a variety of discharge conditions, the total SiH_3 flux to the surface equals the silane depletion and is consistent with the film growth rates. Longway (1) et al. [1984] concluded the same based on scavenging of radicals by NO. Other experiments confirming the SiH_3 dominance are measurements of radical species passing through the surface of a substrate hole [Robertson et al. 1983 & 1986], and laser-induced-fluorescence measurements of discharge radicals [Itabashi et al 1988, and Matsuda & Goto 1989]. The deposition by SiH_2 is negligible because of the high rate coefficient (k_2) for the $SiH_2 + SiH_4 \rightarrow Si_2H_6$ reaction [Inoue & Suzuki 1985, Jasinski & Chu 1988].

7.1 GENERAL EFFECTS OF THE GAS-PHASE AND PLASMA SPECIES

For DC discharges in silane and silane/noble gas mixtures it has been found that the cathode is bombarded by ions with sufficient energy (>50 eV) to do considerable structural damage to the growing film. In addition, resonant charge exchange between noble gas ions and atoms produces an energetic neutral flux to the cathode that can exceed the energetic ion flux and do considerable film damage.

The deleterious effects of these energetic ion collisions can be reduced by placing the substrate behind a screen cathode (DC proximity glow discharge). In typical DC proximity discharge the energetic ions reaching the film surface are $<10^{-5}$ of the film growth Si and the reactive radicals Si, SiH, and SiH_2 that are deleterious to film quality are almost completely reacted before reaching the substrate, thereby further improving the film quality.

In RF discharges between parallel-plate electrodes, the cycle-average sheath voltage accelerating ions toward the substrate is much smaller than at DC cathodes. Thus, although the ion and neutral bombardments that occur at the DC cathodes also occur at RF electrodes, the bombardment energies are correspondingly smaller and much less film damage occurs. The energetic ion bombardment depends on the relative size of the powered and substrate electrodes and on the pressure-electrode distance product. At high p x d levels (0.5-1 torr cm), particulates form.

Ions are a minor contributor to a-Si:H film growth. This is because in DC discharges where about 25% of all electron collisional dissociations of SiH_4 result in an Si ion, each ion produces on an average 20 neutral radicals on its way to the substrate as a result of ion-molecule collisions in the high-electric-field region. In RF discharge typically only 1 ion is produced for 20-30 dissociations [Gallagher et al. 1990].

The species in the gas or plasma are believed to have a definite impact on the quality of the a-Si:H film, which is defined by the photosensitivity, the density of states, the dihydride content, and the microstructure. Inhomogeneities in the films, such as columnar morphology and microvoids, are caused by certain species in the gas phase, especially radicals that have high sticking coefficients. The feed-gas type, feed-gas impurities, and diluents used all affect film quality by means of the species created in the gas phase. The effects on film properties of various gas or plasma species and the effects of feed-gas diluents are discussed, as are reactions at the film's surface and controls on the surface growth. In general, diluents tend to increase structural inhomogeneities and thus reduce the film's quality. There are indications that SiH_3 in the gas phase as a result of the dissociation of both silane and disilane is the species responsible for good quality a-Si:H films.

The species in the plasma are determined partially by the source gas, by the diluent, by species created by gas dissociation, and by species sputtered from the growing film surface. The initial step in a glow discharge is the dissociation of silane or disilane by electron impact. Electron-impact dissociation creates both neutral and ionized species. Neutral dissociation has a lower threshold energy than dissociative ionization. Typically, in RF deposition plasmas, the ratio of neutral to ion products ranges from 50 to 100 [Kushner 1987]. The excitation and dissociation processes in a plasma are balanced by reaction and recombination processes of the species. The term "radicals" is used to indicate the entire group of chemically active species in a plasma. Ions and radicals may go through some secondary reactions among themselves in the plasma before reaching the substrate's surface and partaking in film growth. Reactions among the neutral and charged components of the plasma are called homogeneous. Reactions between the plasma species and the substrate are called heterogeneous [Selwyn 1986].

As mentioned in Chapter 5, section 5.1, a large portion of the plasma is at the plasma potential and is essentially field-free (the glow region). Most of the voltage drops are at the sheath (the dark space) regions near the electrodes. High energy electrons are accelerated by the externally applied field and directed by the sheath voltage toward the glow region, so the sheath acts both as the electron source and as the energy source for the glow. Most of the electron-impact dissociation happens in the glow region, and this region serves as an ion source for the sheath. In SiH_4 plasma, the products of electron-impact dissociation include SiH_n (n = 0 - 3), H, H_2, and SiH_n^+ (n = 0 - 3) [Semiconductors and Semimetals, A, 1984].

The radicals Si, SiH, SiH_2, and SiH_3 have all been suggested as film precursors for silane discharges. However, Weakliem et al. [1983], Longeway [1984], and Gallagher [1986] determined that, for low-power-density DC glow discharge from silane, SiH_3 is by far the dominant radical leading to film deposition. Gallagher [1988] found that to also be true for RF glow discharge of silane. He also concluded that $<98\%$ of neutral radical deposition of a-Si:H is by SiH_3 for typical deposition pressures (>0.1 torr at 240°C). This is because, at normal a-Si:H depositions at low power and high pressure (0.1-1 torr), Si, SiH, and SiH_2 radicals react with each other and with SiH_3 before reaching the substrate surface. Disilane is also formed by surface recombination. The rate of formation of Si_2H_6 in silane RF glow discharge has been observed by mass spectroscopy to undergo a transition from one level to another higher level at 40-60 mW/cm^2 power density. Higher levels of disilane in the plasma from silane lead to poorer film quality. Ions are not considered as main precursors of film growth because the ion flux density is too small to account for the deposition rate. In a diluted silane discharge, depending on the diluents, different radicals are probably produced, but there are no measurements of this. When silane is significantly depleted, the relative abundance of SiH_3 decreases and light radicals increase [Gallagher 1988]. This depletion of SiH_3 may be the reason for the poor quality of films from depleted feedstock gases.

There have been fewer investigations of disilane than of silane glow discharge. Longeway et al. [1984] concluded that the film precursors from disilane DC discharge are SiH_3 and Si_2H_5 and that the difference in film properties between silane- and disilane-prepared films can be explained in part by noting that the film precursor in silane glow discharge (at low power) is primarily SiH_3, with no significant contribution from Si_2H_5.

Gallagher [1987] described test results that might explain why the quality of a-Si:H films produced from silane DC and RF glow discharges deteriorates with increases in power, while that of films from disilane improves. He showed that, in low power silane glow discharge (which yields an excellent quality film), almost all the deposition is by the radical products of silane dissociation. But at higher power densities, a major fraction of the silane turns into disilane, which is then dissociated into other depositing species. Conversely, in low-power disilane discharges, the latter species dominate the deposition, which are generally found to yield films of lower quality. At higher power densities, an increasing fraction of disilane is initially dissociated into silane. A part of this silane is subsequently dissociated to produce the film. Thus, at high power a large fraction of the deposition from a disilane discharge is actually from silane. This may account for the increase, with increasing power density, in the quality of films produced from disilane.

Gallagher [1987] also speculated, on the basis of silane-germane glow-discharge plasma measurements, that the primary Si deposition process for producing a-Si:H films is largely eliminated in silane-germane mixtures. The low sticking coefficient and high surface migration associated with SiH_3 deposition, which are believed essential to the homogeneity and quality of a-Si:H films, are virtually missing. This has important implications for a-SiGe:H films. Although this is only true at low power-to-flow ratios, at high power-to-flow ratios and subsequently severe depletion of the GeH_4 feed gas (where SiH_3 deposition again occurs), the conditions correspond to those that normally yield poor a-Si:H films from silane discharges.

Kampas [1984] and Longeway [1984] describe the chemical reactions and enthalpies of formation for species from silane. Chatham and Gallagher [1985] present silane ion-molecule reactions and measured rate coefficients for DC silane glow discharges.

The deposition rate of films from silane is proportional to the [SiH] emission intensity of the plasma, but for disilane the deposition rate varies as the square root of the emission intensity [Bhat et al. 1988]. The [SiH] or [H] emission intensity divided by the gas flow rate is proportional to (RF energy)^m [Fukuda et al. 1984]. As shown in **Figure 7.1-1**, a large [H]/[SiH] emission-intensity ratio for the plasma yields poor films from silane, that is, films with a large (2090/2000 cm^{-1}) absorption-coefficient ratio (silicon dihydride-to-monohydride ratio, [$(SiH_2)/(SiH)$]) [Ohnishi et al. 1984]. Nakayama et al. [1985] provide similar data for SiH_2F_2 and for a mixture of SiF_4 and SiH_4 on the relationship between the [H]/[SiH] emission-intensity ratio and the

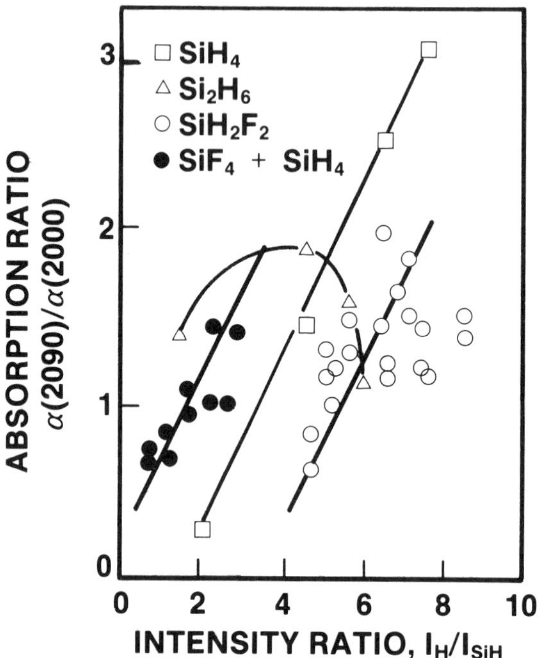

Figure 7.1-1. Ratio of absorption coefficients at 2090 cm^{-1}/2000 cm^{-1} vs. emission intensity ratio I_H/I_{SiH} [Ohnishi et al. 1984, Nakayama et al., 1985]

2090/2000 cm^{-1} absorption-coefficient ratio. For films from disilane, Ohnishi et al. [1984] show first an increase and then a decrease in the absorption ratio (2090/2000 cm^{-1}) with an increasing intensity ratio of [H]/[SiH].

7.2 EFFECTS OF DILUENTS

Knights et al. [1981] observed a general trend of increases in the defect density and microstructural inhomogeneity with increases in the atomic weight of the diluent gas for glow discharge from silane/inert gas mixtures. Collins and Cavese [1987] investigated the effect of silane concentration in both hydrogen and inert gas diluents and found evidence that as the dilution ratio is increased there is an increase in the volume fraction of voids in the a-Si:H network; that is, the density of the a-Si:H film decreases as dilution increases. Similarly, Tsai et al. [1986] found that at low silane concentrations in argon or helium (RF diode glow discharge), the deposition has a physical-vapor-deposition-like component that produces defective material. In this subsection we review the effects of hydrogen, helium, and argon as diluents for silane and disilane. It will be apparent that the results are not totally consistent. Whereas there is some indication that hydrogen dilution results in better a-Si:H film quality, some of the data show the opposite to be the case. There is generally an improvement of a-SiGe:H with H or He dilution (see Sections 2.3.3 and 2.4.1).

7.2.1 Effect of Hydrogen Diluent

Hydrogen is a reactive diluent in glow discharge that can scavenge deleterious radicals. Hydrogen can act as an inhibitor of surface deposition and can result in silicon etching. Shirafuji et al. [1985] describe the effect of hydrogen dilution of silane on the optical gap, the activation energy, the hydrogen content, and the photoconductivity. A greater dilution results in a lower optical gap, activation energy, total hydrogen content, and photoconductivity measured at 100 K (but not when measured at room temperature). Chaudhuri et al. [1984] found that the quality of the a-Si:H film, as evident in the monohydride to dihydride ratio, for example, improved with increasing hydrogen dilutions until a H_2/SiH_4 ratio of 0.25 was achieved, and then the quality started to deteriorate.

Maessen et al. [1987] investigated the role that hydrogen dilution plays in hydrogen incorporation in glow discharge films from silane. They found that 70% of the hydrogen in an a-Si:H film originates from SiH_x compounds and not from the hydrogen diluent.

Nakayama et al. [1985] found that glow discharge of SiH_2F_2 and of SiF_4 + SiH_4 with a mixture of hydrogen and helium as the diluent decreases the dark conductivity of the film as the hydrogen content of the diluent mixture is increased **(Figure 7.2-1).** This occurred at a deposition-rate range of 0.28 to 1.1 nm/s, a substrate temperature (T_s) of 300°C, a pressure of 0.3 torr, and power of 150-400 W. These results would indicate that hydrogen dilution is better than helium dilution (however, see Section 7.2.2 for Patel's results).

Highly diluted silane in hydrogen in a glow discharge deposition process can cause extensive etching of the growing film's surface by atomic hydrogen. Because atomic hydrogen selectively etches strained and weak bonds [Kampas 1984, Schmitt 1983], this can result in the

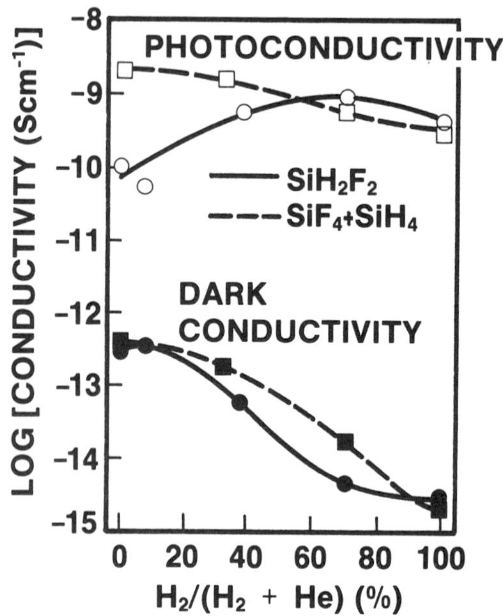

Figure 7.2-1. Dark and photoconductivities vs. $H_2/[H_2+He)$ at 200 W RF power [Nakayama et al. 1985]

deposition of microcrystalline rather than amorphous silicon. Collins and Cavese [1987] found by ellipsometry that, in RF glow discharge of silane diluted in hydrogen at concentrations less than 0.1, there is a hydrogen etching effect that leads to a thickness-dependent microstructure for cathode a-Si:H films that is absent in anode films. Hanaki et al. [1987] found that high hydrogen dilutions and a high RF power caused microcrystalline silicon to form in their p-type a-SiC:H depositions.

Possible paths for hydrogen dilution or hydrogen plasma exposure to improve a-Si:H film properties include:

1) by preferential etching of strained and weak bonds [Tsai 1988]. H-plasma etches a-Si:H at a rate of ~0.03 nm/s. However, H-plasma does not seem to etch a-Ge:H.
2) by altering the reactions of radicals in the plasma before they reach the growing film surface.
3) by hydrogen atoms sticking to the growing film surface and thus decreasing the reactivity of the surface and increasing the diffusion length of the adsorbed precursors [Asano et al. 1990].
4) by hydrogen atoms diffusing into the growth zone [Shibata et al.

1987] and causing the reconstruction of the Si-Si network structure in this zone.

5) by increasing the diffusion length of carriers [Catalano 1989 & 1990].

a-Si:H p-i-n solar cells deposited from SiH_4 without and with (1:1 to 10:1) hydrogen dilution have the same initial performance and the same degraded performance after light-soaking [Bennett and Tu 1990]. Subsequent data from solar cells and modules deposited from hydrogen diluted silane indicate that their light-induced degradation is reduced. Kishi et al. [1991] showed SiH_2/SiH vs. hydrogen dilution (a ratio of 0.12 at 0 hydrogen to a ratio of 0.03 at 2.5:1 H_2/SiH_4), photosensitivity increased from 10^3 at 0 hydrogen to 10^6 at 5:1 H_2/SiH_4, for films deposited at 180°C.

The major effects of diluting SiH_4 with H_2 in glow discharge a-Si:H are a pronounced decrease of the amount of $(Si\text{-}H_2)_n$-chains (R^* at 250°C decreasing from 0.34 without H to 0.04 with H-dilution), an increase of the macroscopic density (up to 90% of the density of crystalline silicon), an increase in the refractive index, a decrease of the optical bandgap, and a decrease of the total amount of incorporated hydrogen in the film (at T_s=250°C a reduction from 18 at.% to 13 at.%) [Meiling et al. 1990a and 1990b].

For a-SiGe:H alloys, hydrogen dilution significantly increases the photosensitivity for several deposition methods (see section 2.3.3, Table 2.3.3-2). More importantly, Ge-Ge clustering and the dihydride bonding fraction are lower and the $\mu\tau$-product and diffusion length are higher for films made with hydrogen dilution.

a-SiC:H films deposited from silane-methane diluted in either H_2 or Ar are of superior quality to those deposited from silane/methane with no dilution. The a-SiC:H material deposited with hydrogen dilution is denser and has fewer voids, it has lesser dihydride bonding, lower Urbach energy, and lower absorption coefficient at 1.2 eV and higher photoconductivity, photosensitivity, and diffusion length (see Section 2.4.1).

7.2.2 Effect of Helium Diluent

Nakayama et al. [1985] reported on the glow discharge of a-Si:H from SiH_2F_2 and from SiF_4 plus SiH_4 diluted with hydrogen and helium (at 300°C, 0.3 torr, 150-400 W, and a 0.28-1.1 nm/s deposition rate), and they observed that a higher helium ratio results in a higher dark conductivity. Consequently, helium dilution yields poorer films than does hydrogen dilution (see **Figure 7.2-1**).

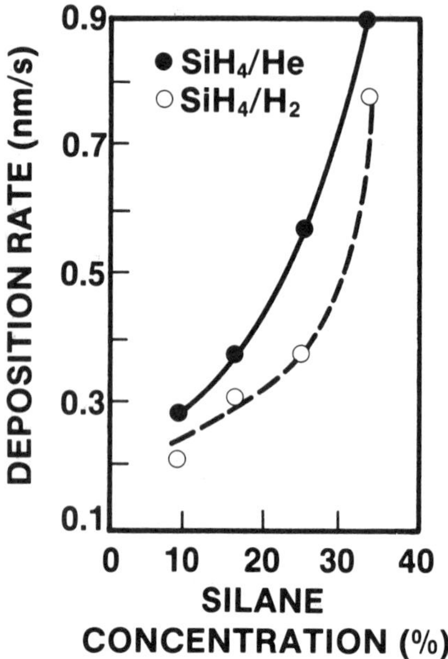

Figure 7.2-2. Deposition rate vs. silane fraction for hydrogen and helium dilution [Patel et al. 1986]

Patel et al. [1986], on the other hand, showed that, in RF glow discharge of a-Si:H from silane, helium dilution produces films with better properties than does hydrogen dilution and that a deleterious powder formation starts at a higher power density with a helium dilution than it does with a hydrogen dilution. Also, helium dilutions result in slightly higher deposition rates than do hydrogen dilutions at equal power densities (see **Figures 7.2-2 and 7.2-3**). Helium and neon dilutions resulted in better films than argon or krypton dilutions did.

In studying the glow discharge of silane, Collins and Cavese [1987] reported that, for a given silane concentration, the density deficit (low-density microstructure) is largest for films produced with helium and smallest for argon dilutions. Thus, dilution in argon resulted in better quality films.

Roca i Cabarrocas [1989] describes effects of helium dilution and bias voltage on a-Si:H film quality. Ion bombardment at 25-75 eV is efficient in reducing the disorder and the defect density of a-Si:H films.

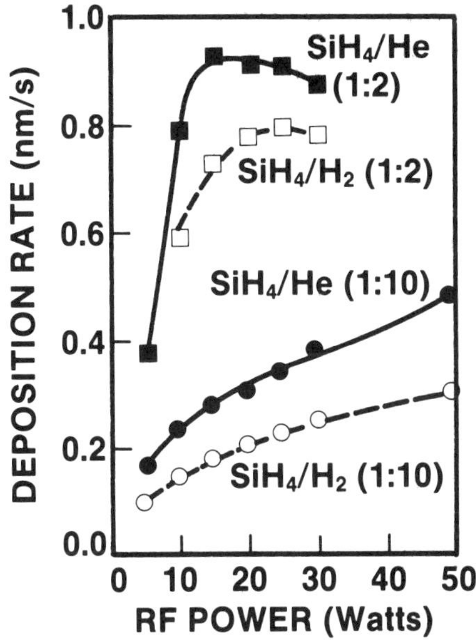

Figure 7.2-3. Deposition rate vs. RF power for hydrogen and helium dilution [Patel et al. 1986]

The electron and hole collection has an optimum at 50 eV ion energy. The saturated light-induced defect density of these films does not show any correlation with the substrate bias indicating that ion bombardment does not increase the susceptibility of the films to the creation of defects by light-soaking [Roca i Cabarrocas et al. 1990].

Tsuo et al. [1991b] found helium dilution of silane/germane gas mixtures more effective than hydrogen dilution in improving a-SiGe:H (see also Section 2.3.3).

7.2.3 Effect of Argon Diluent

Tanaka et al. [1980] reported that argon in silane hinders H_2 effusion and the thermal rearrangement of the local hydrogen environment. Films grown on the cathode had an 8 at.% Ar content versus 1 at.% for films grown on the anode.

Robertson and Gallagher [1986] showed that, in silane-argon mixtures, the relative abundance of monosilicon radicals other than SiH_3

decreases with increases in the fraction of silane in the mixture. Thus, a higher silane concentration leads to higher quality films.

Kushner [1986] reported that diluting silane with argon increases the deposition rate at a constant RF power because of an increase in the electron temperature, an increase in the rate of dissociation, and a decrease in the power lost to vibrational excitation.

Collins [1986] found by ellipsometry that diluting silane in argon results in a more extensive initial growth period during which the density of Si-Si bonds in the growing film is significantly less than that in the bulk of thick films. The conditions under which extensive initial nucleation effects are observed correspond to those in which columnar morphology is evident from scanning electron micrographs.

7.3 REACTIONS AT THE GROWING FILM SURFACE

Glow-discharge deposition onto substrates placed on electrodes is caused by a mixture of chemical and physical processes involving both ions and radicals generated in the plasma. The type and reaction probability of radicals arriving at the growing film's surface have very important effects on the final quality of the film. Since the main precursors in high quality a-Si:H deposition are the SiH_3 radicals, and the resulting film contains only about 20% hydrogen or less, the hydrogen elimination process must occur at the surface of the growing film along with silicon-network building and hydrogen incorporation. Hydrogen atoms are eliminated from the surface in the form of H_2 or SiH_4 [Scott et al. 1983]. Ion bombardment of the growing film's surface also affects the film quality (see Section 2.1.1).

Hamasaki et al. [1983] concluded (from pulsed-plasma-deposition studies) that a-Si:H growth takes place through heterogeneous reactions of adsorbed chemical species on the substrate's surface and that the gas-phase polymerization of radicals plays no important role in the deposition. Species are adsorbed onto the substrate. Depending on the surface mobility and desorption rates, these species migrate until they encounter a sufficiently attractive potential to prevent further migration [Knights 1984]. Gallagher [1988] found that radicals with a high sticking coefficient to the growing film's surface (i.e., those radicals that insert easily into SiH surface bonds) can cause poor film quality. Collins and Cavese [1987], describing glow discharge from silane/hydrogen mixtures, stated that the added energy imparted to the film's surface by ion bombardment (when the substrate is located on the cathode) may enhance the convergence of clusters (2-3 nm laterally) in the initial growth stage by providing added energy for hydrogen-elimination reactions, cross-linking, or Si incorporation by surface diffusion of the dominant SiH_3 radical. Surface diffusion of the SiH_3 radicals promotes the CVD-like deposition process, which Tsai et al. [1986] have shown to be associated with high quality a-Si:H film growth.

The reactions between radicals and the growing a-Si:H surface have been studied by Gallagher [1986] and Robertson and Gallagher [1986]. The reaction enthalpies and a simple set of rate equations have been derived. A silicon surface is nonreactive when it is saturated with hydrogen. During a silane glow discharge deposition process, hydrogen molecules compete with silicon and slow down the silicon incorporation reaction. Reactive surface sites are created during deposition by the release of hydrogen. The transition from hydrogen-rich (H/Si greater than 2) to hydrogen-poor (H/Si less than 0.2) a-Si:H film occurs in a single monolayer [Gallagher and Scott 1987]. Collins and Pawlowski [1986] found from in situ ellipsometry that a-Si:H from silane coalesces to bulk density at a thickness of about 5.5 nm.

Some radicals are formed by sputtering from the growing film's surface. The primary radicals at the surface in glow discharge of undiluted silane are SiH_3 (85%), Si_2H_4 (7%), Si_2H_4 (4%), and Si_2H_5 (4%). Nearly all of the Si, SiH, and Si_2H_n are produced by ion sputtering of the surface [Gallagher and Scott, 1987]. In glow discharge of silane or silane-argon mixtures, disilane (Si_2H_6) is formed primarily at the surface and sputtering is a significant source of radicals near the cathode [Robertson and Gallagher 1986]. Longeway et al. [1985] suggest that a significant fraction of the disilane found in the plasma from silane is produced by the surface combination of SiH_3 radicals. Weakliem et al. [1983] reached the same conclusion.

SiH_3 radicals exhibit a good deal of surface migration and have low sticking coefficients [Gallagher 1986]. At high power densities, there may be less surface migration, more radicals with a higher sticking coefficient, or both. A significant fraction of the lighter SiH_n radicals may be favored at higher powers, and these lighter radicals may be incorporated immediately upon a single surface collision [Robertson and Gallagher 1986]. Collins and Cavese [1987] speculated that when a silane/argon mixture is used at a high power, the dominant radicals may be SiH and SiH_2, which incorporate immediately upon reaching the surface; that is, they have a high sticking coefficient. On the other hand, SiH_3, which is dominant in pure silane plasma at a low power, requires a dangling-bond site for incorporation, and the surface dehydrogenation step limits this incorporation. Thus, it has a lower sticking coefficient. The columnar morphology observed for silane/argon films is attributed to a low surface mobility coupled with physical shadowing, which would cause any microscopic roughness to become more pronounced over time. Device-quality material, prepared under conditions leading to the lowest defect densities, shows some evidence of columnar microstructure, but there are no observable voids nor is there a low-density intercolumnar phase. The surface mobility of adsorbed precursors as well as the local mobility of larger structures may contribute to the knitting of the initial growth microstructure.

Knights [1985] reported that nuclear magnetic resonance (NMR), specific heat measurements, and infrared spectroscopy have shown both covalently bound and trapped molecular hydrogen in a-Si:H. The molecular hydrogen amounts to a fraction of an atomic percent and appears trapped in small voids. Transmittance electron microscopy (TEM), scanning electron microscopy (SEM), and small-angle neutron scattering have all provided direct evidence of structural inhomogeneities. Knights believes that these inhomogeneities result from nucleation at discrete sites, followed by the growth but incomplete coalescence of islands, which produces interstitial voids. The propagation of this process into the columnar form observed has been ascribed to a physical shadowing mechanism, similar to that seen in physical vapor deposition. For example, Tsai et al. [1986] postulated that plasma deposition under device-quality conditions is a CVD-like process that is rate-limited by surface reactions and that results in conformal step coverage; that is, trench side walls are covered by the deposition process. The transition from a chemical vapor deposition to a more physical-vapor-deposition-like process occurs at low silane concentrations and when the RF power is increased.

Using NMR, Boyce et al. [1985] identified molecular hydrogen (0.25 at. %) in voids with a mean radius of 2.5 nm in a-Si films deposited from silane. Molecular hydrogen cannot break weak Si-Si bonds or compensate Si dangling bonds and therefore cannot participate in the bonding reconstruction that occurs on the internal surfaces of hydrogenated microvoids upon illumination or annealing [Carlson 1986]. Gleason et al. reported [1986] that they used multiple-quantum NMR to study the clustering of hydrogen in a-Si:H. For device-quality films (8 to 20 at. % hydrogen and approximately 10^{15} dangling-bond defects per cubic centimeter), they found the average cluster size to be six protons and the intercluster distance to be 1.0-1.4 nm. The concentration of these 5-7 atom defects increases with increasing hydrogen content until, at a very high hydrogen content, the clusters are replaced by a continuous network of silicon-hydrogen bonds.

Kushner [1986] explained the higher hydrogen content observed in films produced from silane and disilane with increases in the power density and deposition rate by the burial of bonded hydrogen on the surface before silicon-to-silicon interconnection and hydrogen elimination can occur. [Note, however, that the hydrogen content is reduced with increases in RF power [Kumeda et al. 1985] for disilane depositions but remains constant for monosilane [Knights 1979]. Collins and Cavese [1987] attribute the difference between films grown from undiluted and hydrogen-diluted silane to the presence of a buried interface layer in the latter. In films from undiluted silane, such buried layers were not observed.

7.4 INTENTIONAL CONTROL OF THE GROWING FILM SURFACE

There are many variables in a glow discharge, such as deposition precursors, density of atomic hydrogen or fluorine for etching, ion and electron bombardments, and photon excitation.. However, traditional RF and DC glow discharge deposition processes allow very little control of the growing surface. The deposition parameters that can be used to deposit high quality films are interrelated, and they lie in a very narrow range. Although a substrate voltage bias can be applied (Section 5.2) and independently controlled, its effects are not yet well understood and it is seldom used in a well-controlled fashion in film deposition. Most RF glow-discharge systems utilize the electrode self-bias effect caused by the unsymmetrical electrode areas and the difference in mobility between electrons and ions to enhance the film's properties. However, this is usually done rather randomly, without first making a systematic study of the film properties as a function of the electrode geometry. The exact ratio of the grounded and powered electrode areas is almost never reported in the literature when a deposition system is described.

The process of a-Si:H film deposition by glow discharge is actually a balance of etching and deposition. In addition to the effects of physical sputtering by energetic ions, chemical etching occurring at the film's surface during growth can also affect the film's quality. Kampas [1984] noted that atomic hydrogen can both create and passivate dangling bonds. Under conditions that produce a large flux of atomic hydrogen on the film's surface, in comparison to the flux of silicon-containing species, etching of the film will become important. The silicon atoms most likely to be etched away by hydrogen are those in positions that resulted in strained or weak bonds. It is well known that weak Si-Si bonds can serve as charge-carrier recombination centers, and the breaking of such weak bonds deteriorates the film quality by increasing the dangling-bond density; this may also be the cause of the light-induced instability commonly observed in a-Si:H [Stutzmann 1985]. One of the explanations [Oda et al. 1986] of the benefit of incorporating fluorine in a-Si:H:F and a-SiGe:H:F alloys is that fluorine serves as an etchant during deposition, etching off weak bonds. It is thus very likely that well-controlled chemical etching during amorphous-silicon-based alloy film deposition could result in films with lower defect densities and fewer voids. However, this aspect of the glow discharge deposition process has not been systematically investigated to date.

In recent years, many efforts have been made to try to intentionally influence the growing film's surface in a plasma deposition. Such efforts might improve our understanding of the deposition process and eventually lead to better deposition techniques. For example, Tsuo et al. [1987] reported on a method that uses an etch gas to influence the growing film's surface. In this technique, a silane pulsing method is used to allow periodic etching of the growing film's surface by a xenon difluoride etch gas. Good quality, intrinsic a-Si:H films have been produced by this periodic etching and deposition method. More studies are needed to better understand the mechanisms and benefits of this process.

Another example is the hydrogen-radical-enhanced CVD (HRCVD) method proposed by Oda et al. [1986]. In the HRCVD method, RF plasma is used for the SiF_4 gas decomposition and a microwave source is used to generate atomic hydrogen. The concentration of hydrogen atoms in the chemical process and their contribution to the film-growth mechanism can be regulated. Remote-plasma-deposition methods, such as the electron cyclotron resonance CVD (ECR CVD) method [Mejia et al. 1986, Watanabe et al. 1986, Kobayashi et al. 1987, Fujita et al. 1987], also have the potential to allow more control of the deposition process and thus to produce films with properties different from those of conventional glow discharge-deposited films. For example, ECR CVD-deposited, wide-bandgap (>2 eV), p-type a-SiC:H:B materials with high conductivity values have been reported [Hattori et al., 1987].

Ohnishi et al. [1987] reported on a controlled-plasma-magnetron (CPM) deposition method, in which a magnetic field is used to enhance the radical concentration at the growing film surface of RF-plasma-deposited a-Si:H, which resulted in a high photo-to-dark conductivity ratio of 1 x 10^6 for a film deposited at a 1.5 nm/s.

7.5 MODELS OF SILANE GLOW DISCHARGE DEPOSITION

There are four stages in the formation of hydrogenated amorphous silicon films from an electrical discharge in silane [Knights 1985]. The first, the electron impact process, is the primary electron collisional dissociation of silane, resulting in a mixture of reactive species consisting of ions and free radicals. The second stage is the diffusion or drifting of these species to the substrate's surface, accompanied by a multiplicity of secondary reactions, e.g., ion-molecule, photon-molecule. The third stage is the reaction of gas or plasma species with or their adsorption onto the substrate's surface. The fourth stage is the process by which these species or their reaction products incorporate into the growing film or are re-emitted from the surface into the gas phase.

Kampas [1984] stated that the basic deposition process is the insertion of SiH_2 into a Si-H bond to produce an excited SiH_3 group bonded to a silicon. Hydrogen can then be eliminated by a variety of means. For example, SiH_3 attaches at a silicon dangling bond to form a high-internal-energy Si-SiH_3 group, from which hydrogen is again eliminated.

Toyoda et al. [1986] stated that, for a pure silane DC discharge with a combined electrostatic-magnetic deflection technique, hydrogen breaks surface Si-Si bonds, decreases the sticking probability of Si atoms, and replaces SiH bonds with SiH_2 bonds to increase the hydrogen content of the films.

Gallagher [1987] provided a model for the glow discharge deposition of a-Si:H from silane. In **Figure 7.5-1,** the electron collision processes are shown as thick arrows, the gas collisions as thinner arrows, and the transport between gas and surface as the thinnest arrows. The chemical reactions are initiated by electron collisions with silane molecules, producing ions and neutral radicals. Successive collisions of these ions with silane molecules process these ions into heavier ions, in competition with the rate of ion drift to the electrodes in the plasma sheath fields. The result of these reactions under typical conditions is a significant, but not large, fraction of heavier (multiple Si) ions. If the heavier ions were produced in larger quantities, they would continue to grow through ion-molecule interaction until they nucleated into particulates that destroy the film's quality upon deposition [Gallagher et al. 1987].

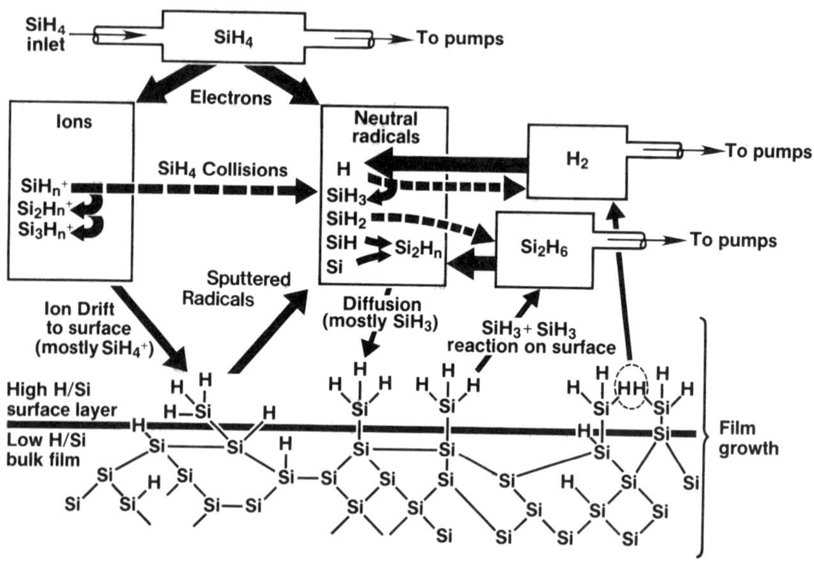

Figure 7.5-1. Model of glow discharge deposition of a-Si:H from silane [Gallagher 1987]

To avoid such nucleation, a small electrode-to-substrate distance is desirable.

The neutral radicals reacting with silane molecules undergo several processes. The most important is that the dominant neutral radical produced by the initial electron excitation—namely, atomic hydrogen—is converted into SiH_3 by reaction with silane. At the same time, at typical glow discharge deposition pressures (0.1-1 torr), SiH_2, SiH, and Si react with silane or hydrogen molecules to form disilane and disilicon radicals before reaching the substrate's surface. The result of these neutral radical reactions and the production of SiH_3 by ion reaction is the diffusion of primarily SiH_3 to the surface. Thus, the silyl radical, SiH_3, is the primary precursor of film deposition in low power, high pressure DC and RF glow discharge [Gallagher 1987]. This agrees with the finding (using laser resonance absorption flash kinetic spectroscopy) that the rate constants for reaction of silylene (SiH_2) with H_2, SiH_4, and Si_2H_6 are, respectively, 10^4, 10^2, and 10 times faster than values previously calculated from values in the literature for the Arrhenius parameters [Jasinski and Chu, 1988]. Similar results were also obtained by Inoue and Suzuki [1985]. However, for high power, low pressure (<50 mtorr) silane discharges, the dominant precursor is not necessarily SiH_3 [Kampas

1985].

Radicals sputtered from the surface by ion bombardment add to those produced by the gas chemistry and are the dominant source of radicals under some conditions. Disilane is produced in silane glow discharge by the surface reactions of two SiH_3 radicals. The primary source of hydrogen (H_2) is the conversion of two surface SiH bonds into Si-Si bonds, which releases H_2 [Gallagher 1987].

In the hydrogenated amorphous silicon surface, almost all the surface bonds of silicon are attached to hydrogen atoms. Under typical discharge conditions, only one in 10^4 or 10^5 of these silicon bonds is dangling (without a hydrogen atom). Thus, the production of the film occurs predominantly by the reaction of the SiH_3 radicals with the hydrogen-passivated silicon surface [Gallagher 1987].

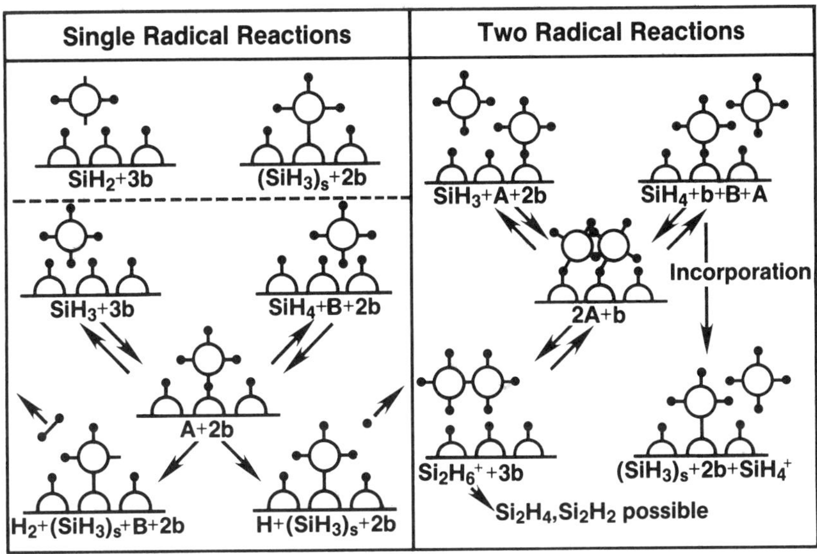

Figure 7.5-2. Model of radical-surface reactions [Gallagher 1986].

A model of the initial surface deposition is shown in **Figure 7.5-2** [Gallagher 1986a]. In this figure, silicon atoms are open circles and hydrogen atoms are filled dots. Surface SiH bonds are labeled b and dangling bonds are labeled B. The SiH_2 insertion reaction is at the top left. The SiH_3 + b --> A --> SiH_4 + B and A --> S + H or S + H_2 + B reactions are at the lower left. The two SiH_3 --> S + SiH_4 and two SiH_3 --> Si_2H_6 reactions are on the right. Here, A is an absorbed SiH_3 and S is an SiH_3 connected to the film by one Si-Si bond. Any SiH_2 or other radicals with more than one dangling bond will incorporate immediately into the hydrogen-covered silicon film. On the other hand, SiH_3 cannot incorporate directly into a SiH bond but can remove a hydrogen atom to yield SiH_4 and a dangling bond on the surface. Another SiH_3 molecule, migrating along the surface or coming directly from the gas, can then incorporate into this dangling bond. Following the initial step of incorporating SiH_3 into the surface with a single silicon-silicon bond, two SiH_3 units can make additional Si-Si bonds, releasing H_2. This process requires the two silicon atoms to have approximately the spacing and morphology of a crystal in order for the hydrogen release to become exothermic. Thus, it tends to build nearly crystalline structures [Gallagher 1987]. At high growth rates, hydrogen elimination may be impeded by the "englobement" of hydrogen-covered surfaces, which results in the formation of voids [Knights 1985].

Kushner and coworkers [Kushner 1986a, 1986b, 1987, 1988, Kampas and Kushner 1986, Hebner and Kushner 1987] have done extensive numerical modeling of RF plasma-enhanced chemical vapor deposition. Their model starts with a Monte Carlo simulation for electrons, with time and spatially varying electric fields and sheaths, to give the energy, position, and time distribution function between the electrodes. The time distribution is averaged over many cycles to provide the energy and position distributions. The excitation rate constants over position and time are calculated by convolving cross sections with the electron distribution function. From the developed electron-impact rate constants and the electric field distribution, rate equations in a one-dimensional chemistry model for ions and neutral particles are calculated as a function of position and time, considering electron impact, molecular and atomic collisions, ion drift, and diffusion. This yields radical and ion fluxes to the substrate's surface that are handled by a surface-deposition model. The surface-deposition model is an accounting (in rate equation form) of processes leading to lattice building, hydrogen incorporation, and hydrogen elimination. The rates for these processes are obtained directly or are determined parametrically by comparison with experimental data. Kushner [1988] gives all the species; electron impact excitation, dissociation, and ionization and their rate constants; neutral heavy particle reactions and their rate constants; ion-molecule reactions and their rate constants; and charged particle-charged particle reactions. This step yields deposition rates and hydrogen fractions. Calculations have been made chiefly for silane discharges, but the model can also be used for disilane discharges. The model shows how the deposition rate depends on the reactor geometry (electrode spacing), power density, gas mix, and flow rate. The model also describes conditions for dust (Si_nH_m, n > 5) formation, and predicts dust to be more likely to be neutral than charged. Gleason et al. [1987c] studied the Monte Carlo simulation of precursor diffusion on the growing film's surface before bonding to the amorphous silicon network.

Kushner [1987] found from his modeling of the glow discharge of silane that, in order for silyl (SiH_3) to be the dominant radical, its branching ratio from the electron-impact dissociation of SiH_4 must exceed 0.75, and the yield of H atoms in the branch for SiH_2 must also exceed 0.75. The deposition rate of a-Si:H is then controlled by the yield of H atoms and the subsequent generation of radicals by hydrogen abstraction from SiH_4. These conclusions require that certain assumed rate constants are met.

McCaughey and Kushner [1988] found further that film properties are critically dependent on the ratio of SiH_3/SiH_2 in the radical flux. High values for this ratio results in film properties resembling chemical vapor deposition whereas low values result in film properties resembling physical vapor deposition. Rough films (>1 nm) accordingly result from radical fluxes having high SiH_2 fractions. Surface roughness and hydrogen fraction increase with increasing growth rate and with increasing film thickness. However thin films have large hydrogen content as the result of a hydrogen-rich surface layer with a thickness approximately equal to the surface roughness. they found a correlation between porosity and hydrogen content, implying that the inner surfaces of voids are lined with Si-H configurations.

8

MODIFICATIONS OF CONVENTIONAL GLOW DISCHARGE PROCESSES AND EQUIPMENT

8.1 THE NEED FOR ALTERNATIVE DEPOSITION PROCESSES

Plasma-assisted chemical vapor deposition (CVD), also known as glow discharge deposition, has become the most common technique used in the deposition of hydrogenated amorphous silicon (a-Si:H) for semiconductor device applications. The best quality a-Si:H material to date is deposited by glow discharge, especially under high vacuum conditions that ensure low impurity levels of oxygen, carbon, and nitrogen. Such material has high photosensitivity (1×10^7), low Urbach energy (42 meV), low midgap density of states (1×10^{15} eV^{-1} cm^{-3}), and low electron spin resonance (ESR) spin density (2×10^{15} cm^{-3}). However, the conventional glow discharge deposition method has its limitations, including low deposition rates, poor quality low-bandgap (< 1.5 eV) and high-bandgap (> 2.0 eV) alloy materials, poor doping efficiencies for p- and n-type materials, and a limited selection of dopants. In addition, there has been very little real improvement in recent years in the defect and photostability properties of glow discharge-deposited a-Si:H films. Although the fundamental reasons for these difficulties are not well understood, we believe the solution lies in a better understanding and control of the a-Si:H deposition process.

In the next few chapters, we review several alternative deposition processes to glow discharge for a-Si:H-based alloys. We group these alternative processes into major categories, including (1) modifications of conventional glow discharge, (2) remote-plasma-assisted CVD, (3) photochemical vapor deposition (photo-CVD), (4) thermally-induced CVD, (5) spontaneous CVD, and (6) physical vapor deposition (PVD) methods. We also include in the review posthydrogenation methods for amorphous silicon. The material properties for the resulting a-Si:H(:F) and a-SiGe:H(:F) films are compared for these alternative deposition methods. a-Si:H films nearly as good as those produced by glow discharge have been achieved by photo-CVD. For a-SiGe:H(:F) with an optical bandgap of 1.5 eV, approximately the same photosensitivity of 1 x 10^4 has been achieved by several of the new methods discussed herein; however, glow discharge and photo-CVD films have shown about twice the photosensitivity (2 x 10^4) of that for other deposition methods. Some alternative deposition methods have achieved higher deposition rates than conventional glow discharge. None of the deposition methods discussed have improved the photostability of high quality a-Si:H.

In addition to the need to improve the charge carrier transport properties of a-Si:H and related alloys, there are other materials problems that need to be addressed by improved deposition techniques. One of the most important is that a-Si cells at present lose efficiency as a result of exposure to light, due to the so-called Staebler-Wronski effect [Staebler and Wronski 1977 & 1980], so that the stable, long-term efficiency is considerably lower than the initial efficiency. Metastable defects similar to the Staebler-Wronski defects can also be induced thermally or by electrical bias [Street et al. 1987]. Another problem is how to achieve good material properties at deposition rates high enough to allow economic film production. Presently, film quality and device efficiency decline as the deposition rate is increased to more cost-effective rates [Luft 1988b]. Yet another problem is the difficulty encountered in doping a-Si:H and related alloys. Presently, only boron and phosphorus are routinely used as dopants for a-Si:H and related alloys. Even for boron and phosphorus doped materials, the doping efficiency is very low. Important electrical properties of the doped materials, such as dangling-bond defect density and charge carrier diffusion length and lifetime, are very poor [Street 1982b, Vanecek et al. 1983, Ray et al. 1984, and Stutzmann et al. 1987].

How are these problems—relatively low solar-cell efficiency, poor alloy material quality, light-induced degradation, low deposition rates, and poor doping properties—going to be overcome? We believe that it must be by deposition processes that allow better control of the chemical composition and structure of the deposited film. The deposition method and the deposition parameters largely determine the defect properties of a-Si films, such as bonding distortions, point defects, and microstructural defects. These defects in turn determine the electronic properties of the films—such as band tails and deep defect states—that limit the ultimate photovoltaic energy conversion efficiency. These defects may also determine the magnitude of the illumination-induced degradation. Systematic studies are needed to relate the deposition to the defect properties. The need to understand the deposition conditions as a prerequisite of film property improvements is the reason for the present interest in plasma diagnostics. There is also now much effort devoted to the development of alternative deposition methods with the objective of eliminating some of the present problems. The potential of such new methods for improved control of the deposition processes and improved material characteristics will be discussed.

High quality a-Si:H films can often be deposited by slightly modifying the conventional glow discharge deposition methods, which use a DC or RF excitation source, a diode or triode geometry, and a constant feed-gas flow. Examples of such modifications include changing the electrode geometry, changing the excitation frequency, and adding separate etching cycles. Some of these methods are discussed hereafter.

8.2 FEED-GAS MODIFICATIONS

Since Sterling and Swann [1965] first reported RF glow discharge decomposition of silane for a-Si thin-film deposition, silane has always been the dominant feed-gas for depositing a-Si:H. Many other feed-gas variations have been studied recently. Some of them have been shown to increase the film deposition rate significantly over pure silane discharge. However, none has been shown to be able to deposit intrinsic a-Si:H films with photoelectric properties significantly better than films deposited by the pure silane discharge.

Dilution of silane by hydrogen, helium, or argon can significantly alter the deposition processes [see the discussions in section 7.2]. Hydrogen dilution of silane is commonly used by the a-Si:H solar-cell industry for safety and economic reasons. In addition, dilution of silane by hydrogen enhances etching during glow discharge deposition, which may have beneficial effects on film quality by preferential etching of weak bonds. Hydrogen dilution may also affect the reactions of radicals in the plasma. Hydrogen atoms sticking to the growing film surface reduces the reactivity of the surface and increase the diffusion length of the adsorbed precursors [Asano 1990]. Hydrogen atoms diffuse into the growth zone and cause the reconstruction of the Si-Si network structure in this zone [Shibata et al. 1987]. High levels of hydrogen dilution in a silane plasma makes it possible to deposit microcrystalline silicon films at low substrate temperatures (100-200°C) [Madan et al. 1979, Johnson et al. 1988, and Tsai et al. 1988&1989]. Hydrogen dilution results in photosensitivities about an order of magnitude higher in a-SiGe:H than do undiluted gas mixtures [Luft 1989]. High levels of hydrogen may cause feed-gas depletion. For example, in a SiH_4 and GeH_4 mixture, the GeH_4 decomposes faster and is more affected by feed-gas depletion causing reduction of the Ge content in the deposited a-SiGe:H alloy [Tsuo et al. 1991]. Feed-gas depletion also reduces the deposition rate and reduces gas-phase polymerization.

Higher-order silanes, mainly disilane (Si_2H_6) and trisilane (Si_3H_8), have been studied extensively for at least three reasons: (1) higher deposition rates [Ogawa et al. 1981, Kenne et al. 1984, Ashida et al. 1984, and Scott et al. 1980], (2) easier decomposition for photo-CVD processes [Inoue et al. 1983 and Kumata et al. 1986], and (3) potentially better a-SiGe:H alloy deposition using a mixture of disilane and germane [Luft 1988a].

Fluorine-containing feed-gases, such as mixtures of SiF_4 and H_2 [Madan et al. 1979, Okada et al. 1990]; SiF_4, SiH_4, and H_2 [Nakayama et al. 1983]; intermediate species SiF_2 and H_2 [Matsumura et al. 1983]; XeF_2 and SiH_4 [Tsuo et al. 1987b, Langford et al. 1987, and Weil et al. 1991]; and SiH_3F and SiH_2F_2 [Matsuda et al. 1984a and Pernisz et al. 1987] have been used to deposit a-Si:H:F films. The benefits of fluorine-containing feed-gases, if there are any, come from the fluorine passivation of dangling bonds and/or enhanced etching during deposition. Langford et al. [1989&1990] reported that larger than 1% fluorine incorporation in a-Si:H is not beneficial for photovoltaic applications. They noted that, during a-Si:H deposition, H selectively etches weak bonds whereas F indiscriminately etches, H is more mobile on the growing film surface than F, and H suppresses polymers in the gas phase whereas F induces polymers in the gas phase.

For intrinsic or doped a-Si-based alloys, such as a-SiC:H, the choice of feed-gas mixture may affect the chemical bonding structures of the dopants and alloy elements in the film. For intrinsic a-SiC:H alloys, instead of the conventional SiH_4/CH_4 or SiH_4/C_2H_4 mixtures [Tawada et al. 1982], several alternative gas mixtures have been studied. These include $SiF_2/CF_4/H_2$ mixtures [Matsumura et al. 1985]; hydrogen-diluted mixtures of SiH_4 + CH_4, C_2H_4, or SiH_3CH_3 [Matsuda et al. 1986]; and disilylmethane mixtures [Beyer et al. 1989]. Improved photoconductivities were reported with many of these alternative methods. Better film properties have been reported for p-type a-SiC deposited with $B(CH_3)_3$ as the doping gas (rather than the conventional B_2H_6) in photo-CVD and glow discharge [Tarui et al. 1987, Kuwano and Tsuda 1988, Wu et al. 1989a&b, Fieselmann and Goldstein 1989, Lechner et al. 1990]. At high boron concentrations in the solid film, $B(CH_3)_3$-doped films have less bandgap reduction than do B_2H_6-doped films. Photoconductivity, photoluminescence intensity, doping efficiency, and blue response of a-SiC films by photo-CVD are better when using $B(CH_3)_3$ instead of B_2H_6. Large-area (100 cm^2) a-Si:H cells with the p-layer deposited using $B(CH_3)_3$ gas in a glow discharge had a 10.0% active-area initial efficiency (total area efficiency = 9.0%) [Kuwano and Tsuda 1988]. An important advantage of using $B(CH_3)_3$ is the reduction of thermal decomposition that typically happens on heated substrates and reactor walls with a B_2H_6 and SiH_4 gas mixture [Collins 1988, Tsuo et al. 1989, Roca i Cabarrocas et al. 1989]. Kuwano and Tsuda [1988] found that using BF_3 as the doping gas for p-type a-SiC:H did not result in shrinkage of the E_g that is normally associated with boron doping.

8.3 SEPARATED PLASMA TRIODE GLOW DISCHARGE

The separated plasma triode (SPT) glow discharge method is a modification of the conventional triode glow discharge, in that the triode grid is grounded via a variable impedance, and the anode is not biased [Tanaka 1989]. This allows a plasma to exist not only at the cathode side of the mesh third electrode but also on the anode side, providing a high density of radical species near the substrate, which enables a high-rate deposition. In conventional triode geometry, the plasma is confined between the cathode and the third electrode. Whereas this arrangement limits the ion bombardment, it also reduces the deposition rate [Luft and Tsuo 1988]. The plasma potential (potential between plasma and grounded substrate) in the SPT method is lower than that in a diode geometry with the same electron density [Tanaka 1988]. For a-Si:H films deposited at a rate of 1 nm/s from disilane, values of $\sigma_l = 1 \times 10^{-4}$ S/cm, $\sigma_d = 2 \times 10^{-10}$ S/cm, $\sigma_l/\sigma_d = 5 \times 10^5$, and ESR spin density $= 7 \times 10^{15}$ cm^{-3} were achieved. a-Si:H solar cells with the i-layer deposited at 1 nm/s by this method had a conversion efficiency of 10.2% for a 1 cm^2 area and 7.5% for 100 cm^2 area [Tanaka 1989]. This ranks them among the highest efficiencies achieved at such a high deposition rate.

8.4 CONTROLLED PLASMA MAGNETRON (CPM) GLOW DISCHARGE

The controlled plasma magnetron (CPM) glow discharge method is a diode or triode RF plasma-assisted CVD method with electromagnets placed under the cathode to create a magnetic field between the cathode and the anode. The substrate is located on the anode. In the triode case, the third electrode is used for plasma control. The diffusion of electrons in the plasma toward the walls is strongly suppressed by the magnetic field, thus increasing the electron density in the space between the electrodes by about a factor of two at 500 gauss, compared to no magnetic field, while reducing the electron density between the electrodes and the walls. The SiH emission intensity is about seven times higher than for conventional glow discharge [Ohnishi et al. 1987]. The triode glow discharge method results in high deposition rates. The photosensitivity for a-Si:H films ranged from 1×10^6 at 0.1 nm/s to 5×10^5 at 1.5 nm/s deposition rate [Kuwano and Tsuda 1988]. A space charge density of 7×10^{14} cm^{-3} was obtained at a deposition rate of 0.7 nm/s [Ohnishi et al. 1988]. An efficiency of 10.1% was achieved at 1 nm/s deposition rate for a small-area cell, and 7.55% total-area efficiency was achieved at 1.5 nm/s deposition rate for a 100 cm^2 submodule of a-Si:H [Ohnishi et al. 1988]. These efficiencies are near the highest ever achieved at these rates. Good p-type microcrystalline SiC films with $\sigma_d = 0.7$ S/cm at $E_g = 1.8$ eV were also achieved with this method [Kuwano and Tsuda 1988]. The absorption coefficient of a-SiC fabricated by the CPM method was one order of magnitude lower than that for conventional deposited a-SiC. The CPM film had a lower hydrogen density and a higher Si-C bond density and a 0.3-0.5 eV higher band gap compared to conventional a-SiC [Nishikuni et al. 1991].

8.5 LOW- AND HIGH-FREQUENCY GLOW DISCHARGES

Glow discharges powered by DC and RF at 13.56 MHz (the allotted frequency of the U.S. Federal Communications Commission and international communication authorities) are commonly used for a-Si:H deposition. Although the RF glow discharge is used more often, there seems to be no significant difference in the quality of materials produced by these two methods. Microwaves with a frequency of 2.45 GHz have also been used for gas decomposition in plasma deposition of a-Si:H with the substrate in direct contact with the plasma [Kato et al. 1982, Mejia et al. 1983, Hudgens and Johncock 1985, Kato and Aoki 1985, Fujita et al. 1987]. Although high deposition rates have been achieved with microwave glow discharge, the photoelectric properties of the films are not as good as those of the best a-Si:H films produced by DC and 13.56 MHz RF glow discharge, due to a large increase in defect density.

Other power-source frequencies have been studied for glow discharge deposition of a-Si:H. On the low frequency side, Boulitrop et al. [1985] studied films deposited at 400 kHz and 13.56 MHz in a hot-wall glow discharge system. They found that pressure and RF have no effect on defect density and disorder. However, the Fermi level shifts toward the conduction band for deposition at 400 kHz and toward midgap for 13.56 MHz with increasing gas pressure. Matsuda et al. [1984b] studied the influence of power-source frequency on a-Si:H film properties in the range of 10 kHz-50 MHz; they found that the mechanical properties, rather than the photoelectric properties, of the film are more influenced by the frequency. High levels of compressive internal stress were observed in films deposited at frequencies lower than 500 kHz. They interpreted this effect as being caused by higher ion bombardment during deposition, because the total number of ions impinging onto the growing film surface increases drastically when the frequency decreases below 1 MHz. An even lower frequency (50 Hz) has been used by Shimozuma et al. [1989] for the deposition of a-SiN:H and a-C:H films. They believe the higher electron energy of the 50 Hz plasma, as compared to the 13.56 MHz plasma, makes it possible to deposit high quality films at low substrate temperatures.

On the high frequency side, Curtins et al. [1987a&b, 1988] studied the effects of power-source frequency in a silane glow discharge from 25 MHz to 150 MHz. They observed a strong dependence of the film deposition rate on the frequency, with the deposition rate peaking above 2 nm/s at about 70 MHz. The photoelectric properties of deposited films were found to be relatively independent of frequency. Properties of films deposited by this very high frequency (VHF) method at substrate temperatures from 50° to 320°C were reported by Ziegler et al. [1989]. Fisher et al. [1989] reported the fabrication of a 5.5%-efficient p-i-n cell using this method. Both a-Si:H and microcrystalline Si were deposited by Oda et al. [1988] using a VHF (144 MHz) SiH_4/H_2 plasma. Chatham and Bhat [1989] studied glow discharge deposition of a-Si:H from 13.5 MHz to 110 MHz. They observed that the deposition rate increases monotonically with frequency until 110 MHz, the highest frequency studied by them. This is contrary to the maximum deposition rate, at about 70 MHz, observed by Curtins et al. [1987b]. Chatham et al.

achieved 2 nm/s deposition rate at 110 MHz for disilane and 0.5 nm/s deposition rate for silane. They also found that the operating pressure decreases with increasing frequency for good quality film deposition. They believe there are two main benefits with a high frequency deposition: (1) The ion bombardment energy decreases with increasing frequency as the self-bias potential decreases from 30 to 50 V at 13.5 MHz to 10 V or lower at 110 MHz. (2) The flux of low energy ions increases with increased frequency, due to lower operating pressure. This improves the surface mobility at the deposition surface. A p-i-n a-Si:H cell with a 9.7% conversion efficiency was deposited at 1.8 nm/s using disilane at 110 MHz. Models of RF glow discharges [Wertheimer and Moisan 1985, Pointu 1986&1987] for different power-source frequencies have been presented that show reasonable agreement with experimental results. Wertheimer and Moisan [1985] proposed that some observed frequency effects should be interpreted as resulting from different electron energy distribution functions of the plasmas. Oda et al. [1990] concluded that frequencies larger than the collision frequency and close to the plasma frequency are useful in plasma processing because of the high efficiency of radical generation and the high feasibility of control of the energy and spatial distribution of electrons by means of an applied magnetic field or a DC bias. Therefore, the excitation frequency should be considered a variable parameter in optimizing plasma processing.

8.6 PULSED AND SQUARE-WAVE DISCHARGES

Pulsed discharge, instead of continuous discharge, has been studied by Yoshida et al. [1989] to obtain a high, transient electron energy in the plasma. They found that the amount and bonding configuration of a-Si:H can be controlled by hydrogen dilution and by the pulse duty. Yoshida, T., et al. [1990] found the doping efficiency for p-type a-Si(C):H films using BF_3 to be higher with pulsed discharge CVD than with continuous RF plasma. Using such p-layers, 11.7% efficient solar cell were made.

Square wave modulated discharges were studied by Verdeyen et al. [1990] and Lloret et al. [1991]. Using helium-diluted silane as the process gas, Verdeyen et al. found that the time-averaged electron density for a modulated discharge is greater than that of a conventional continuous discharge. This enhancement is a function of the modulation frequency and peak power driving the discharge. The bandgap of a-Si:H deposited from silane was found to be lower with a modulated discharge than that deposited from a continuous discharge. The amount of dust generated in a modulated discharge is also much lower. Verdeyen et al. attributed these improvements to the low level of negative ion density in a modulated discharge. Lloret et al. found that, at modulating frequencies less than 40 Hz, square-wave-discharge-deposited a-Si:H films have shown a significant increase of porosity and surface roughness than continuous-wave RF discharge films. At higher frequencies (40 to 4000 Hz), the material density and roughness of the two types of films are similar.

8.7 PERIODIC ETCHING DEPOSITION (PED) GLOW DISCHARGE

As a consequence of structural disorder, a large density of weak bonds exists in a-Si:H and related alloys. These weak bonds are detrimental to the electrical properties. Enhanced etching during film growth may remove some of the weak and strained bonds. It is well known that high hydrogen dilution of silane during deposition results in microcrystalline films at low deposition temperatures [Tsai et al. 1988]. This indicates that atomic hydrogen preferentially etches weak and distorted bonds during deposition, making it seem that the surface diffusion coefficient is higher than it actually is. Etching during deposition may also be done by fluorine-containing radicals. A normal glow discharge deposition process is in fact a balance of continuous deposition and etching. Recently, several groups have studied the effects of additional etching by periodically interrupting the glow discharge deposition process. The process is also called chemical annealing or hydrogen-radical annealing.

Tsuo et al. [1987b] developed a periodic etching deposition (PED)-CVD method for depositing a-Si:H:F. They used a constant flow of XeF_2 and a periodic flow of SiH_4 as the feed-gas in a constant RF glow discharge. The XeF_2 etch gas was introduced upstream to the substrate. When SiH_4 is present, XeF_2 reacts rapidly and spontaneously with SiH_4, producing SiF_4, HF, and Xe. During the SiH_4-on cycle, most of the XeF_2 reacts with SiH_4 before reaching the sample surface, resulting in a glow discharge deposition of a-Si:H:F. Because the XeF_2 gas spontaneously etches crystalline and a-Si, the deposited films were etched during the SiH_4-off cycle. This deposition and etching cycle is repeated for a period of 1 to 3 min. An AM1 σ_l value of 6.3 x 10^{-5} S/cm, with σ_l/σ_d = 6.6 x 10^5 and E_g = 1.82 eV, has been obtained with an a-Si:H:F film deposited by this method.

Xu et al. [1988] studied a PED-CVD process for a-Si:H using an RF hydrogen plasma for the etching cycles and a hydrogen-diluted silane plasma for the deposition cycles. They achieved a σ_l of 4.3 x 10^{-5} S/cm and a σ_l/σ_d ratio of 1.8 x 10^6 for an a-Si:H film with an E_g value of 1.81 eV. The 820 nm-thick film was deposited in 13 deposition and etching cycles with a substrate temperature of 255° C, a chamber pressure of 5 torr, and a periodic SiH_4 flow of 10 standard cubic centimeter per minute (sccm) for two minutes and off for 1 minute in each cycle. The H_2 flow rate was kept constant at 40 sccm. The RF power was kept constant at 0.12 W/cm^2. The deposition rate, averaged over both the deposition and etching cycles, was 0.34 nm/s. It is interesting to note that this averaged deposition rate for PED-CVD is about twice the deposition rate of a conventional glow discharge deposition in the same system using pure silane and similar deposition conditions.

Ishimura et al. [1988] studied an intermittent deposition method in which two kinds of process gases are exchanged at short intervals: one kind (SiH_4 or SiH_3F) is used for growing silicon film and the other kind (H_2 or SiF_4) is used for modifying the growing surface. They produced an a-Si:H:F film with an AM1 σ_l of 4 x 10^{-4} S/cm and a σ_l/σ_d ratio of 10^6.

In a modified form, atomic hydrogen is introduced during the etching period. Reduction in the density of valence band tail states and enhancement in drift mobility to 0.2 cm^2/vs at 300 K for holes was observed [Shirai, H., et al, 1990]. The improvements are attributed to the large sticking probability of the hydrogen radicals to the a-Si surface, leading to an increased concentration of hydrogen at the surface. Consequently, the diffusion of hydrogen is increased, improving the quality of the a-Si film by terminating dangling bonds [Uchida, Y., et al, 1990].

Asano et al. [1989&1990] deposited undoped a-SiC:H, a-SiH, and µc-Si films using a two-plasma-zone glow discharge apparatus. A SiH_4, C_2H_2, and H_2 gas mixture was used in one of the plasma zones for a-SiC:H deposition, and a pure H_2 gas was used in the other plasma zone for hydrogen passivation and etching. The substrate was rotated between these two zones. Thicknesses of 0.3, 0.6, and 0.9 nm per deposition cycle were studied. The etch rate in the hydrogen-plasma zone was 0.023 nm/s. The a-SiC:H films grown by 0.3 nm deposition cycles showed lower spin densities and tail states than those of continuously deposited films. An a-SiC:H film with a σ_l of 3 x 10^{-5} S/cm and an E_g value of 1.91 eV was prepared. For µc-Si deposition, Asano et al. [1990] found no significant etching of the deposited film. They believe this suggests that the macroscopic etching process, which reduces the film deposition rate, is not a necessary condition for the formation of microcrystallites in the deposited film. Instead of etching, hydrogen atoms stick to the growing film surface, decrease the reactivity of the surface, and increase the diffusion length of the adsorbed precursors. This leads to the construction of a dense Si-Si network structure and resulting in microcrystalline growth.

8.8 ILLUMINATED GLOW DISCHARGE

Photon illumination of the growing surface during glow discharge deposition of a-Si:H may have the benefit of increased surface mobility of atoms during reconstruction after the arrival of deposition precursors. The photon illumination may also break weak bonds during growth and thus reducing the number of weak bonds in the deposited film. Both visible light illumination [Sakata et al. 1990&1991] and ultraviolet light illumination [Vieira et al. 1988] have been studied. Sakata et al. observed reduced light-induced degradations in a-Si:H deposited by visible-light-illuminated glow discharge and attributed this effect to the possible reduction of latent sites such as Si-Si weak bonds during film growth. They also noted that, compared to conventional glow discharge-deposited films, illuminated glow discharge-deposited a-Si:H films have lower as-deposited photoconductivity, lower bonded-hydrogen content, and lower ratio of silicon dihydride bonds vs. total silicon-hydrogen bonds.

Vieira et al. [1988] found that the presence of UV light during μc-Si deposition enhances the deposition rate. It also influences the incorporation of impurity atoms and the way hydrogen is attached in the film. For example, when oxygen is present during the deposition, the UV light leads to the formation of Si=O bonds, instead of the Si-O bonds that form without UV illumination.

9
REMOTE-PLASMA-ASSISTED CHEMICAL VAPOR DEPOSITION METHODS

In a remote-plasma-assisted chemical vapor deposition, gas species of a single type, which can be either a chemical component of the final film or metastable species that function as an energy transfer vehicle, are remotely excited in an excitation chamber separate from the deposition chamber. The excited species are subsequently allowed to engage in specific downstream chemical reactions with feedstock gas molecules in the deposition chamber, where they decompose the feed gas to produce film-forming precursors that are in turn responsible for the film deposition. Thus, in *remote-plasma* processes, the formation of deposition precursors occurs outside of the plasma region [Parsons et al. 1988a&b]. With proper care, a single-component precursor can be generated (selective fragmentation) to ensure that a single, serial deposition pathway controls the chemical composition of the resulting film. Thereby, control over the hydrogen bonding configurations can be achieved. In a remote plasma process, the back-streaming of the downstream injected reactant gases into the upstream plasma excitation region is prevented by a combination of process pressure and gas flow rate. Because the plasma is only minimally in contact with the growing film, effects due to charge-particle bombardment and vacuum-ultraviolet (UV) light exposure can be eliminated [Johnson et al. 1989]. A disadvantage of many of the remote-plasma deposition processes is the low deposition rate. For example, Parsons et al. [1988] reported a deposition rate of 0.013 nm/s. Remote-plasma processes have also been used in etching and SiO_2 deposition [Cook 1987, Jackson, R., et al. 1987]. Remote-plasma processes vary by the method of plasma generation and the excitation gas used; examples of remote plasma-enhanced CVD techniques for a-Si:H deposition are glow discharge remote-plasma CVD, electron cyclotron resonance (ECR) microwave plasma CVD, remote microwave plasma glow discharge, and microwave hydrogen-radical-enhanced CVD.

9.1 GLOW DISCHARGE REMOTE-PLASMA CVD

Bonded hydrogen at any substrate temperature is lower for remote-plasma-enhanced CVD than for radio-frequency glow discharge, and amounts to about half of that for glow discharge at 240°C [Lucovsky, 1989]. In glow discharge, remote-plasma-enhanced CVD, the remote plasma is excited by a RF or DC power source. Parsons et al. [1988] used an RF-excited helium plasma to provide excited helium atoms that were then used to react with the SiH_4 + Ar process gas mixture introduced downstream. Intrinsic a-Si:H with defect state density, determined from sub-bandgap absorption, as low as 5×10^{15} $cm^{-3}eV^{-1}$ was deposited by this process at 235°C substrate temperature [Parsons et al. 1988b]. It was also found that a-Si:H films deposited at low substrate temperatures (125-200°C) had lower polyhydride bonds than conventional glow discharge-deposited films deposited in the same temperature range [Lucovsky et al. 1989]. Phosphorus-doped a-Si:H films deposited by this method at 235° and 100°C substrate temperatures showed doping efficiencies and electrical properties comparable to glow discharge-deposited a-Si:H doped with the same ratio of phosphine to silane at a deposition temperature (T_s) of 240°C [Parsons et al. 1988b].

The $Ar(^3P_2)$-induced CVD method reported by Toyoshima et al. [1985] is a DC glow discharge, remote-plasma technique that used metastable $Ar(^3P_2)$ for silane decomposition. Intrinsic a-Si:H films deposited by this method have predominantly monohydride bonds in the substrate temperature range of 25° to 300°C. A photosensitivity of 10^5 was achieved for a-Si:H films deposited by this method at a substrate temperature of 100°C.

9.2 ELECTRON-CYCLOTRON-RESONANCE MICROWAVE REMOTE-PLASMA CVD

In the electron cyclotron resonance (ECR)-CVD method, a plasma is generally formed in an excitation chamber from an excitation gas such as hydrogen, helium, or argon. A divergent magnetic field generated by coils or permanent magnets surrounding the plasma excitation chamber extracts charged species of the plasma from the excitation chamber into the deposition chamber. The feedstock gas is introduced in the deposition chamber where the extracted energetic gas species decomposes the feed gas (indirect feed-gas decomposition). The excitation chamber operates as a microwave cavity resonator. The magnetic flux density in the excitation chamber and tuning of the microwave cavity is controlled to achieve the helical motion of electrons around the magnetic field lines in a region inside the excitation chamber [Konuma 1992]. The frequency of this cyclotron motion in the absence of the electric field is $\omega = eB/m$, where e and m are the charge and the mass of the electron and B is the magnetic field strength, respectively. Resonance of the microwave with the electron cyclotron motion, called Electron Cyclotron Resonance (ECR), is satisfied by matching the frequencies by adjusting the field strength. The plasma production efficiency is enhanced by satisfying the ECR condition [Matsuo and Kiuchi 1983, Trevor et al. 1990]. More than 70% of the microwave power is absorbed by the plasma [Sakamoto 1977]. The microwave power required to sustain a stable plasma is reduced by one order of magnitude with respect to non-magnetically confined plasmas. Microwave plasmas operating under ECR conditions can increase the dissociation rate of molecular gases by the presence of high energy electrons, while the gas temperature remains low because of inefficient energy transfer between electrons and heavier particles. The frequency of the microwave energy source is usually 2.45 GHz, which gives a resonance for a 874-Gauss magnetic flux density. (See **Fig. 9.2-1**).

A negative static electric field (approximately -15 V) from the excitation chamber to the specimen table accelerates ions and decelerates electrons; it is generated by the divergent magnetic field along the plasma stream. Ions are often extracted into the deposition chamber by ion extraction grids positioned across the excitation chamber exit. The dominant ions bombarding the growing film surface are H_2^+, SiH_2^+, and SiH_3^+. These ions with low kinetic energy were found to have the beneficial effects of increasing the deposition rate, lowing the substrate temperature needed for high quality film deposition, and decreasing dangling bonds and polyhydride bonds in the film [Hayama et al. 1990]. The electron density and the electron temperature for ECR-CVD are high compared to those of a conventional RF glow discharge. The minimum gas pressure allowing the generation of a plasma is much lower under ECR conditions than for conventional RF discharge. This is because the threshold pressure strongly decreases (by almost an order of magnitude) as the magnetic field approaches the ECR condition [Fujita et al. 1987]. Typically, a gas pressure of 10^{-5}-10^{-3} torr is used.

The advantages of ECR microwave remote plasma CVD

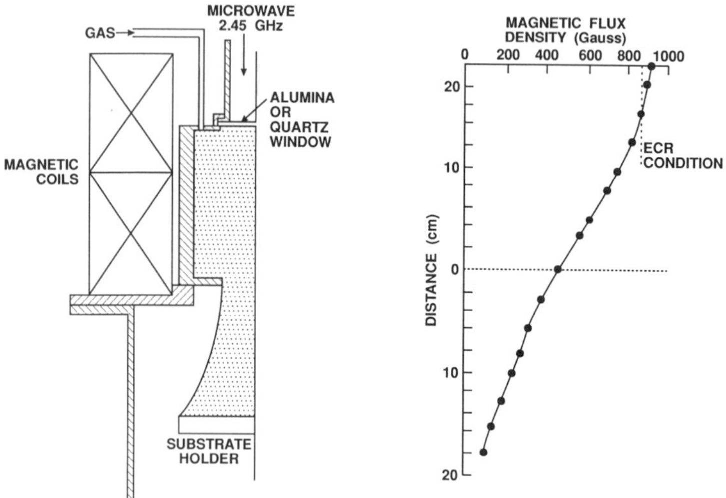

Figure 9.2-1 Electron Cyclotron Resonance plasma source [Matsuo and Kiuchi 1983].

compared to conventional glow discharge are (1) efficient energy transfer from the microwave field to the plasma [Sakamoto 1977]; (2) control of the ion energy (typically 10-50 eV) by the gas pressure to avoid high energy particle bombardment of the growing film thereby producing less stress in the film and less damage to the growing surface [Kitagawa et al. 1987b and Shirai, K., et al. 1989]; (3) high utilization of feedstock gases [Mejia et al. 1986]; (4) better control of the dissociation of the deposition gas; (5) lower operating pressures, which may lead to cleaner processing; (6) lower substrate temperature [Kitagawa et al. 1987b and 1990, Kobayashi et al. 1987], (7) reduced powered electrode effects, such as contamination, self-biasing, and hot electron generation [Shufflebotham et al. 1988], (8) high ionization ratio (10^{-2}), and (9) high plasma density ($10^{11}/cm^3$) [Shing et al. 1989].

The disadvantage of ECR microwave remote plasma CVD is that the method probably cannot be easily scaled up to large-area solar module production because of film uniformity problems and the mechanical complexity of the deposition system. It is probably only good for material studies and small-area device fabrication. However, variations of ECR-CVD involving plasma production using a Lisitano coil in a discharge tube have produced large-diameter (14 cm) [Kawai and Sakamoto 1982] plasmas with high temperature and high electron density. An electron temperature of 10 eV has been achieved using a microwave launcher inserted in a discharge tube under ECR conditions [Fujita et al. 1987].

Deposition rates up to 2 nm/s for a-Si:H have been reported [Kobayashi et al. 1987], but the material properties were poor: photosensitivity = 1×10^3 and spin density = 2×10^{16} cm^{-3}. Shing et al. [1989] achieved a photosensitivity of 1×10^5 at a deposition rate of 2.5 nm/s. For a-SiGe:H with 1.5 eV E_g, deposited at a rate of 1 nm/s, a photosensitivity of 1×10^4 has been achieved. Aoki et al. [1989] deposited a-Ge:H films in a coaxial-type ECR-CVD system and obtained films with about 0.9 eV E_g and a photosensitivity of 5 or less. So far, the most successful application of ECR-CVD in a-Si:H solar cells is probably the deposition of highly conductive p-type amorphous and microcrystalline SiC:H films. Thick (> 1 μm) films of p-type material have been deposited with an E_g value of 2.25 eV and which exhibit σ_d values as high as 10 S/cm. Using an ECR-CVD-deposited a-SiC:H:B buffer layer, a microcrystalline SiC:H p-layer, and conventional glow discharge-deposited a-Si:H i- and n-layers, a solar cell with an 11.8% conversion efficiency was fabricated [Hattori et al. 1987].

Using the principle of electron cyclotron wave resonance, Kausche et al. [1989] invented a wave resonance plasma deposition (WRPD) method that uses a lower magnetic field and gas pressure than those of ECR-CVD. In addition, the WRPD method uses a conventional RF power source, instead of a microwave source, for the wave resonance plasma. These factors make the WRPD method more suitable for large-area deposition. Using hydrogen as the plasma gas and silane or germane as the process gas, Kausche et al. [1989] deposited p-type microcrystalline silicon and a-Ge:H films with high dark and photoconductivities.

9.3 MICROWAVE-EXCITED REMOTE-PLASMA CVD

The microwave-excited remote-plasma CVD method is similar to ECR microwave remote-plasma CVD, but without the resonance. Often, the system is operated at a magnetic flux density twice as high as that required for ECR conditions, in the so called whistler mode [Watanabe et al. 1986]. The advantages of microwave-excited remote-plasma CVD over conventional glow discharge deposition are (1) low deposition pressure (10^{-3} torr), resulting in short power transients of 0.1 seconds or less, and easy maintenance; (2) higher feed-gas utilization than RF glow discharge (20% instead of 10%); and (3) lower equipment costs. The advantage over ECR-CVD is that the method can be scaled to large-area deposition. For a-Si:H films, photosensitivities of 2 x 10^6 have been achieved [Watanabe et al. 1986]. For a-SiGe:H films deposited from silane at 1.3 nm/s deposition rate, photosensitivities of 1 x 10^4 at E_g = 1.5 eV have been recorded [Watanabe et al. 1987b].

Watanabe et al. [1986] deposited a-Si:H films from a microwave-excited argon plasma (2.45 GHz at 10^{-4} to 10^{-2} torr) that decomposed silane. The films were porous when produced at low pressure (10^{-4} torr), with $\sigma_d = 5 \times 10^{-10}$ S/cm, $\sigma_l = 4 \times 10^{-4}$ S/cm, and photosensitivity = 8 x 10^5. E_g decreased from 1.95 to 1.85 eV as the substrate temperature was increased from 140° to 200° C. At the same time, the refractive index increased from 3.5 to 3.7. Watanabe et al. [1987a,b] deposited a-SiGe:H films from silane and germane decomposed by argon or hydrogen plasma or by direct 2.45 GHz microwave excitation (at 10^{-3} torr gas pressure) at high deposition rates, with a σ_l value of 10^{-5} S/cm and a photosensitivity of 10^4 with E_g = 1.5 eV. Deposition rates up to 1.3 nm/s were achieved. Watanabe et al. [1987a] deposited a-SiGe:H films from SiH_4, GeH_4, and H_2 at 1 mtorr using a substrate temperature of 200° C. A reduction in microwave power caused the preferential decomposition of germane and also affected the decomposition of silane by lowering the electron energy in the plasma. This is because the ionization potential of germane is lower than that of silane. Photoconductivity decreased as SiH_2 chain structures from low energy electrons increased with higher pressure and decreased microwave power. At E_g = 1.49 eV, they obtained values of $\sigma_l = 1 \times 10^{-5}$ S/cm, $\sigma_d = 1.8 \times 10^{-9}$ S/cm, and photosensitivity = 5.6 x 10^3. Optical bandgap decreased with chamber pressure, as did the dark and photoconductivities, whereas the refractive index increased from 3.6 to 3.9.

Johnson et al. [1989] used a remote hydrogen plasma sustained within a microwave cavity for a-Si:H deposition. The remote plasma generated monatomic hydrogen that was mixed downstream with silane molecules to produce SiH_3 radicals. B- and P-doped a-Si:H films were deposited by this method with conductivities similar to those of conventional glow discharge-deposited films.

Hourd et al. [1991] used a microwave plasma to deposit amorphous and microcrystalline silicon from hydrogen diluted SiF_4 to explore the effect of hydrogen content on transport properties. SiF_4 allowed reducing the hydrogen to lower levels than was possible with silane as a source gas.

9.4 HYDROGEN-RADICAL-ENHANCED CVD

The objective of the hydrogen-radical-enhanced (HR)-CVD method is to increase the control of CVD a-Si:H deposition processes with the aid of atomic hydrogen generated independent of the regular CVD processes. The HR-CVD method, used by Shibata el al. [1986a,b] and Hanna et al. [1988], consists of two basic processes: (1) the generation of silicon-containing fragments, SiF_n (n = 0, 1, 2), and (2) the generation of atomic hydrogen. Each of these long-lifetime radicals is prepared and controlled independently. The generation of SiF_n radicals can be done by RF or microwave glow discharge of SiF_4 or by the thermal decomposition of fluorinated silicon compounds (e.g., by passing SiF_4 gas through polycrystalline silicon grains at 1150°C or by passing it over a hot tungsten wire). Atomic hydrogen can be obtained by microwave glow discharge of H_2, by catalytic decomposition of H_2 on a hot filament, or by mercury photosensitized decomposition of H_2.

Generating H and SiF_n radicals by a 2.45 GHz microwave glow discharge from SiF_4 and H_2 has an advantage over catalytic thermal decomposition in that the hydrogen gas is free of contamination from the tungsten heating wires, and a high density of H atoms can be produced [Oda et al. 1986 and Shibata et al. 1987a]. SiF_n and atomic hydrogen react reductively, generating the precursors responsible for deposition. The chemically active species SiF_2H and SiH_2F were found to be those related to the growth of the film. The precursors condense on the substrate into an a-Si:H film, liberating HF. The gas-phase reaction is thought to scavenge those Si radicals with high sticking coefficients, resulting in a reduction in the dangling bonds in the film. The surface reactions are believed to be promoted by a chemical interaction between hydrogen and fluorine, which leads to a reduction of defects in the film and also regulates the orientation of Si-Si bonds. By controlling the hydrogen flow, the concentration of atomic hydrogen in the gas phase can be controlled, thereby controlling the chemical processes both in the gas phase and on the surface of the growing film [Hanna et al. 1987]. For example, the photoconductivity of films can be optimized by controlling the flow rates of H radicals without affecting the deposition rates.

According to Hanna et al. [1988], the advantages of microwave HR-CVD over RF glow discharge of silane are (1) a greater controllability of the silicon network structure, as the optimal conditions in terms of both properties and deposition rates can be sought independently, (2) high quality a-Si:H and its related alloys at deposition rates of more than 1 nm/s, (3) a plasma-free environment, avoiding ion bombardment of the growing film, (4) extraction of impurities such as carbon and oxygen on the growing surface because of the strong chemical activity of hydrogen atoms, and (5) dangling bonds that can be more effectively passivated by H atoms. Hanna et al. [1988] also believe that the advantages of using fluoride gases as feedstock, as compared to silane or disilane, are that (1) the gases are non-flammable, (2) the intermediate species SiF_n has a long lifetime, (3) the strong chemical activity of SiF_n radicals with H atoms permits higher deposition rates at lower temperatures without sacrificing the properties of the films, and (4) small amounts of F atoms incorporated in the films can improve its properties, particularly in the states near the valence band.

a-Si:H:F and a-SiGe:H:F films produced by microwave HR-CVD containing 8-9 at. % hydrogen display a non-dispersive hole transport above the temperature of 286 K [Oda et al. 1986 and Shibata et al. 1987c]. The a-Si:H:F films have a hole drift mobility of 2.6×10^{-1} cm^2/Vs [Hanna et al. 1988] and a hole mobility-lifetime product of 1.2×10^{-8} cm^2/V. The mobility-lifetime product and the drift mobility increase with decreasing hydrogen content corresponding to an increasing packing density of the Si network. The a-SiGe:H:F films have low sub-bandgap absorption [Oda et al. 1986]. The improved hole transport characteristics are ascribed to a reduction in the density of the tail states in the midgap and near the valence band [Hanna et al. 1987 and Shibata et al. 1987c]. A-SiGe:H films with an E_g value of 1.4 eV prepared from fluorides (SiF_4 and GeF_4) in the presence of atomic hydrogen had a high $\eta\mu\tau$ product at moderate deposition rates (5×10^{-7} cm^2/V at 0.4 nm/s, 2×10^{-8} cm^2/V at 0.7 nm/s) [Shibata et al. 1986].

10
PHOTOCHEMICAL VAPOR DEPOSITION

In photo-CVD, ultraviolet (UV) or vacuum UV (VUV) photons are used as the primary energy source to dissociate the reactant gas. Because the photons usually do not have sufficient energy to ionize or dissociate the process gas, and because there are no electrodes and no applied electric fields, ion bombardment damage of the resultant film is avoided. Compared to glow discharge, the cross contamination during deposition may also be reduced for photo-CVD because of the absence of ion bombardment on reactor walls and fixtures. Photo-CVD can be divided into *sensitized* and *direct* processes, the latter being achieved with either low pressure mercury lamps or lasers as the energy source. The UV photons are normally generated by vapor lamps (Hg, Xe, Ar, or Kr) external to the reactor chamber. Wavelengths are 184.9 and 253.7 nm for Hg, 147.0 nm for Xe, 104.8 and 106.7 nm for Ar, and 123.6 nm for Kr. Mercury lamps are the most commonly used. Source radiation from mercury lamps at 184.9 and 253.7 nm yields almost exclusively SiH_3 from silane in the primary reaction [Itozaki and Fujita 1987]. A pulsed excimer laser operated with ArF (193 nm) has also been used as the UV light source for photo-CVD of a-Si:H [Eres et al. 1989]. Energy transfer from the photons to the feed-gas in a photo-CVD process can be enhanced using sensitizers, such as mercury in the mercury-sensitized photo-CVD process [Konagai 1986].

A serious technical difficulty for the photo-CVD method is a-Si:H film deposition on the window through which UV light is introduced into the reactor. Several techniques have been proposed for keeping the windows clean, each with particular strengths and limitations (see, for example, Tsuo and Langford [1989]). Approaches to reduce film deposition on the window or to remove films deposited on the UV window during deposition include: purging the window with an inert gas during deposition [Jasinski et al. 1987a], coating the window with perfluoropolyether [Konagai 1987], a continuous liquid coating and spreading (with wiper blades) [Scapple et al. 1986], installing a mobile UV transparent window, such as a teflon film, behind the reactor window [Peters et al. 1981, Rocheleau et al. 1987], and introducing an etch gas near the window [Langford et al. 1987 and Tsuo and Langford 1989]. Another novel approach is to operate, *inside* the reactor, a UV lamp

[Kamisako et al. 1984, Walsh and Bottka 1984, Yoshida, A., et al. 1990] such as a well-confined hydrogen plasma of disk shape that serves both as a vacuum UV lamp (121.5 nm wavelength) and source for atomic hydrogen radicals [Robertson, P., et al. 1987, Sheng et al. 1988]. Yoshida et al. obtained a photosensitivity of about 10^7 in a-Si:H films prepared by the direct photolysis of disilane using windowless hydrogen discharge.

Normally, photo-CVD of a-Si:H is conducted at a pressure of 1-5 torr. However, high quality a-Si:H films have been deposited using atmospheric-pressure photo-CVD [Kim et al. 1988]. One of the advantages of photo-CVD over glow discharge is that it avoids damages caused by high energy charged particles when restarting the glow discharge plasma at interfaces (for example, the p/i interface of a p-i-n solar cell) [Konagai 1987]. The characteristics of a-Si-based materials produced by photo-CVD are very similar to those produced by glow discharge deposition. The best a-Si:H material produced by photo-CVD has the following properties: $\sigma_l = 5 \times 10^{-5}$ S/cm [Hegedus et al. 1987a and Konagai et al. 1985], $\eta\mu\tau = 9 \times 10^{-7}$ cm^2/V [Baron et al. 1988], $\sigma_d = 1 \times 10^{-10}$ S/cm [Hegedus et al. 1987a], photosensitivity = 5×10^6 [Konagai 1987], Urbach parameter = 30 meV [Nakano et al. 1989], ESR spin density = 5×10^{15} cm^{-3}, midgap density of states = 2×10^{15} cm^{-3} eV^{-1} with 2×10^{19} cm^{-3} carbon and 5×10^{19} cm^{-3} oxygen impurity levels [Rocheleau et al. 1987], and cell efficiency = 11.2% [Konagai 1987].

Tomikawa et al. [1987], using mercury sensitized photo-CVD, obtained improved properties (photosensitivity = 1×10^4 at 1.45 eV) by decreasing deposition pressure and light intensity, and increasing gas flow rate for a-SiGe:H cells. They achieved an 8.9% efficiency for a triple-junction cell. The best a-SiGe:H or a-SiGe:H:F material produced by photo-CVD with a 1.5 eV E_g has the following properties: $\sigma_l = 2 \times 10^{-4}$ S/cm [Konagai 1987], $\eta\mu\tau$ at 1.4 eV = 8×10^{-8} cm^2/V [Baron et al. 1988], $\sigma_d = 2 \times 10^{-10}$ S/cm [Sichanugrist et al. 1985 and Konagai et al. 1985], photosensitivity = 2×10^4 [Konagai 1986], Urbach parameter = 40-45 meV [Baron et al. 1988] (46 meV at 1.42 eV [Hegedus et al. 1988]). Hydrogen dilution was found to give better $\eta\mu\tau$ for a-SiGe:H than did helium dilution of silane/germane (6×10^{-8} vs. 9×10^{-9} cm^2/V at 1.4 eV) [Rocheleau et al. 1988].

Itozaki and Fujita [1987] reported mercury-sensitized photo-CVD of a-SiGe:H using a low-pressure mercury lamp (184.9 and 253.7 nm) and a $SiH_4/Si_2H_6/GeH_4/H_2$ gas mixture at 0.01-10 torr with T_s = 200° C. As the hydrogen dilution level was increased, SiH_2 bonds decreased; at 50% hydrogen dilution, SiH bonds dominated. Whereas hydrogen dilution increased the σ_l by 2 orders of magnitude, σ_d increased by 4 orders of magnitude, thus giving lower photosensitivity. Furthermore, at higher hydrogen dilution levels, a columnar structure was observed. Low pressure (0.1 torr) prevented secondary gas reactions and resulted in SiH bonds because of fewer radical interactions. The photoconductivity decreased steadily by more than a factor of 20 as the pressure was increased from 0.1 to 10 torr and σ_d increased by about a factor of 2 to 5 over the same pressure range, thus reducing the photosensitivity by a factor of 40-100. Dangling-bond density decreased by one-third as pressure decreased from 1 to 0.01 torr.

Doped films deposited by photo-CVD have both high conductivity and wide optical bandgap [Konagai 1986]. N-type, hydrogenated amorphous-microcrystalline mixed-phase Si films (σ_d = 13 S/cm, E_a = 1.98 eV) were deposited by mercury-sensitized photo-CVD using a H_2/Si_2H_6 dilution ratio of 70. P-type a-SiC:H films ($\sigma_d = 2.2 \times 10^{-5}$ S/cm, E_a = 0.29 eV, and E_g = 2.0 eV) were deposited by direct photo-CVD using a gas mixture of $B_2H_6 + Si_2H_6 + C_2H_2$.

The most important merit of photo-CVD compared to glow discharge seems to be a reduction in interfacial mixing and cross contamination for a-Si devices. This advantage is mainly applicable to a single-chamber reactor, because glow discharge has start-up surges whenever a plasma is started for new deposition; continuously operating, multichamber glow discharge reactors should not have such problems. High-energy charged particles in a glow discharge plasma may damage the p/i interface of a p-i-n solar cell and an intrinsic a-SiGe:H:F film prepared from GeF_4 by glow discharge, whereas photo-CVD avoids such damages [Konagai 1986, 1987]. However, ion bombardment can be avoided in glow discharge by a triode geometry or proper substrate biasing. a-Si:H films deposited by direct decomposition by photo-CVD of Si_2H_6 have lower H-content (~6 at.% at 275°C) than material by glow discharge (~12 at.% at 275°C) as measured both by IR and effusion. The defect state density for the photo-CVD process is ~3×10^{15} cm^{-3} at 275°C in spite of the lower H-content [Matsunami et al. 1990].

11
THERMALLY-INDUCED CHEMICAL VAPOR DEPOSITION

The thermally-induced chemical vapor deposition techniques discussed here include techniques that thermally decompose a gaseous silicon compound, usually SiH_4 or Si_2H_6, for a-Si:H deposition. Both thermal energy and catalytic surface effects contribute to the dissociation of gas molecules. The high temperature surface needed for process gas dissociation can be in the form of (1) the substrate surface as in low pressure and atmospheric-pressure thermal CVD techniques, (2) the reactor walls as in the homogeneous (HOMO) CVD techniques, (3) hot filaments as in the hot-wire CVD techniques, or (4) laser heating as in the photothermal CVD techniques.

11.1 THERMAL CHEMICAL VAPOR DEPOSITION

Thermal Chemical Vapor Deposition is normally referred to simply as CVD. Chemical vapor deposition techniques are widely used for growing epitaxial silicon thin films [Grove 1967]. In the absence of air, SiH_4 decomposes at 400°C and Si_2H_6 decomposes at 250° C. When a silane molecule strikes a substrate surface, it can either be reflected or adsorbed. If it is adsorbed, and the substrate temperature is higher than its decomposition temperature, it may decompose into Si and H_2 with the latter going back into the gas phase. When the substrate temperature is lower than 600° C, the film deposited is amorphous. The deposition rate decreases with decreasing substrate temperature, whereas the hydrogen content of the deposited a-Si:H increases with decreasing substrate temperature. Studies of gas-phase kinetics of the pyrolysis of silane and disilane and its relationship to silicon CVD have been reviewed by Jasinski et al. [1987b] and Giunta et al. [1990]. A comprehensive review of the fundamentals of thermal CVD, thermal CVD of dielectrics and semiconductors for microelectronics, and commercial CVD systems for film production can be found in the book by Sherman [1987].

A CVD process can be extremely simple and low-cost, such as the atmospheric-pressure (AP) CVD process studied by Ellis et al. [1984]. They used magnesium silicide and dilute acid as starting materials to prepare the silane-polysilane gas mixture, which was then mixed with a hydrogen carrier gas at 1 atm total pressure for the CVD process. The substrate temperature was between 420 and 530° C. The more widely used low-pressure (LP) CVD technique has the advantage of being able to perform uniform deposition on closely stacked substrates because of the surface-reaction limited nature of the deposition process. However, because silane pyrolysis requires high substrate temperatures to achieve useful deposition rates, intrinsic a-Si:H films deposited by silane CVD processes have poor electronic properties due to low hydrogen contents and high dangling bond densities. Amato et al. [1992] obtained up to 8 at.% hydrogen with LPCVD, but with considerable Si-H_2 in the a-Si film and a relatively high Urbach energy (>60 meV). For LPCVD there is no direct correspondence between the presence of the 2100 cm^{-1} vibrational peak and the amount of voids in the material [Amato et al., 1991].

Many of the reported studies on CVD-deposited a-Si:H were done before 1985. Hirose [1984b] and Kaplan [1984] reviewed these CVD-deposited a-Si:H deposition studies. Recent studies on CVD-deposited a-Si:H are more concentrated on using higher-order silanes [Chu et al. 1985, McCurdy and Gordon 1988] and doped films [Chik et al. 1989, Du et al. 1989]. Some of the reported properties of intrinsic a-Si:H films deposited by LPCVD using disilane [Rocheleau et al. 1985] include the following: $\sigma_l < 5 \times 10^{-6}$ S/cm, $\sigma_d < 5 \times 10^{-11}$ S/cm with E_a = 0.7 to 0.8 eV, diffusion length ~ 0.1 μm, Urbach energy = 48-55 meV, and mid-gap density of states > 5 x 10^{16} $cm^{-3}eV^{-1}$. Phosphorus- and boron-doped CVD-deposited a-Si:H films have high conductivity suitable for application as electrodes. Hegedus et al. [1984] fabricated a 3.2%-efficient, all-CVD, transparent conducting oxide (TCO)/p-i-n/metal a-Si:H solar cell using disilane as the process gas.

In a conventional CVD reactor, the gas and substrate temperatures are equal and silane decomposes on the substrate surface to produce a-Si:H. The problem with this approach is that the substrate temperature must be high (> 500°C for SiH_4 and > 350°C for Si_2H_6) to achieve reasonable deposition rates. To reduce the substrate temperature required for high-rate deposition of high quality films in a CVD process, the deposition radicals must be generated away from the substrate surface, similar to the gas-phase generation of deposition radicals in glow discharge and photochemical CVD techniques. These new, thermally-induced CVD techniques include homogeneous, hot-wire, and photothermal CVD.

According to a study by Chu et al. [1986a], in the pyrolytic decomposition of disilane the deposition rate is twice as high with helium dilution than with hydrogen dilution. Chu et al. [1986b] also state that, in the chemical vapor deposition by pyrolytic decomposition of 6% disilane, helium dilution produces better films ($E_a = 0.75$ eV, photoconductivity = 3-6 x 10^{-5} S/cm, dark conductivity = 2-4 x 10^{-10} S/cm) with lower gap-state densities (L_p = 0.57 μm) and better photostability than does hydrogen dilution.

11.2 HOMOGENEOUS CHEMICAL VAPOR DEPOSITION (HOMOCVD)

In a HOMOCVD approach, the substrate is cooled to below 350°C while the gas temperature is maintained at 650-700°C using a hot-wall (~700°C) reactor. In this way, the reactive film-forming intermediates are produced homogeneously in the gas phase, similar to glow discharge. A review of HOMOCVD depositions of a-Si:H has been done by Scott [1984]. A maximum σ_l of ~10^{-4} S/cm was obtained with an a-Si:H film deposited with a substrate temperature of 250°C. However, the σ_d value is high (> 10^{-9} S/cm). Silane loss at the hot reactor walls and low deposition rates are also problems with the HOMOCVD technique. Reimer et al. [1985] reported HOMOCVD-deposited a-Ge:H films with low spin densities but no sub-bandgap photoluminescence and very small photosensitivity values (< 10). Qian et al. [1990] showed that homogeneous gas phase nucleation and/or local nucleation set the real limit for growing high quality HOMOCVD a-Si:H with a high growth rate. The optical bandgap of the film can be tuned from 2.6 to 1.6 eV as the substrate temperature changes from 20 to 280°C. They believe, even at a low substrate temperature and with some polymerization, in the HOMOCVD process the precursors still have enough energy to participate in the surface kinetics.

11.3 HOT-WIRE CHEMICAL VAPOR DEPOSITION

Hot-wire CVD [Mahan et al. 1991b] is also known as catalytic thermal CVD [Matsumura 1988&1989] and evaporative surface decomposition [Doyle et al. 1988]. This method was first studied by Wiesmann et al. [1979], using a hot tungsten or carbon foil, for the deposition of a-Si:H. In more recent studies, the silane feed gas is decomposed by pyrolytic and catalytic reactions on a heated filament at a temperature of 1300-1600°C. The silicon atoms deposited on the filament are then thermally evaporated onto the substrate located only a few centimeters away from the filament. Hydrogen atoms are also generated at the filament surface by thermal and catalytic dissociations of silane [Doyle et al. 1988] and hydrogen molecules [Jansen et al. 1989]. The substrate temperature, at 200-300°C, is similar to that used by conventional glow discharge deposition. The actual film deposition precursors are dependent on the gas pressure. When the pressure is low enough that no gas-phase reactions can occur, the film precursors are the Si and H radicals generated at the filament surface. When the pressure is high, the film growth is dominated by the products of gas-phase reactions between the evaporated atoms and the ambient. Doyle et al. [1988] and Mahan et al. [1991b] used low and intermediate pressures of 4-30 mtorr for their studies. Matsumura [1988, 1989] used a relatively high pressure of around 4 torr. The deposition rate is usually between 0.2 to 0.7 nm/s [Mahan et al. 1991b]. Matsumura achieved deposition rates of up to 25 nm/s at high filament temperatures (1500°C or higher).

Problems of the hot-wire CVD method include short filament life caused by silicide formation and film contamination by the hot filament. To reduce impurity contaminations, it is necessary to mask the substrate surface during the initial filament heating and introduction of silane [Doyle et al. 1988]. Good quality hot-wire CVD films have properties close to glow discharge-deposited a-Si:H. However, a large number of hot-wire CVD intrinsic a-Si:H films reported in the literature have high σ_d (> 10^{-10} S/cm), indicating impurity contamination.

Hot-wire CVD can also be used for remote deposition using hydrogen-radical-enhanced CVD from SiF_4 and H_2 as described in Section 9.4. Using SiF_4 and hydrogen, Matsumura and Tachibana [1985b] obtained a-Si:H photosensitivity values above 10^6 and spin densities as low as 1.5 x 10^{16} cm^{-3} for deposition rates of several tenths of nm/s. Using a silane and hydrogen gas mixture, Matsumura [1988] achieved for a-Si:H (at E_g = 1.7 eV) a photosensitivity of about 1 x 10^5.

The optical bandgap can be controlled with no apparent degradation of the properties by adding germane gas to the deposition gas in a hot-wire CVD process. For a-SiGe:H (with E_g values of 1.4-1.45 eV) deposited at 1 nm/s, a photosensitivity of about 1 x 10^4 was obtained

[Matsumura 1988].

a-SiGe:H of 1.45 eV optical gap by Catalytic CVD show Urbach energy, ESR spin density, and σ_l to be about equal to those of glow discharge a-Si:H. At 1.5 eV: $\sigma_l = 10^{-5}$ S/cm, $\sigma_l / \sigma_d = 10^4$, E_o=53 meV, spin density=10^{16} cm^{-3} [Matsumura et al. 1990].

With the hot-wire technique, good quality a-Si:H material can be deposited at very low hydrogen contents. Down to 0.2 at.% hydrogen the Urbach energy remains in the 50-60 meV range, whereas for glow discharge material the Urbach energy increases with hydrogen content below about 10 at.%. (see **Figure 11.3-1**). In Figure 11.3-1, the Stutzmann data refer to a compilation of literature data.

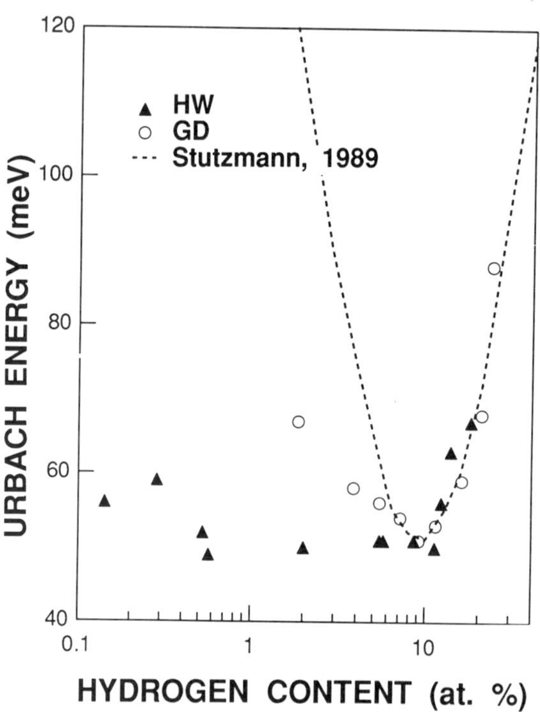

Figure 11.3-1 Urbach energy (measured by PDS) for hot-wire deposited a-Si:H films as compared with that for glow discharge deposited films [Mahan 1991b].

The hot wire technique allows a deposition rate of 1 nm/s and gives a material as good as that by glow discharge at a rate of 0.3-0.5 nm/s. The hot wire method is the only one to give low hydrogen content material (1-3 at.%). The hydrogen content for glow discharge material is typically in the 8%-10% range. Low hydrogen content is believed essential for stable material. The hot wire technique has consistently produced material with the lowest saturated defect density of $2\text{-}3\times10^{16}$ cm^{-3} of any deposition method. Glow discharge films possess typically a saturated defect density of 8×10^{16}. The hot wire technique gives the highest initial ambipolar diffusion length 220 nm as compared to 130-140 nm for optimized glow discharge material, and the highest stabilized diffusion length (150 nm vs <100 for GD). a-SiGe:H with 1.5 eV optical bandgap has given an ambipolar diffusion length of 96 nm as compared to 70 nm for glow discharge material.

11.4 PHOTOTHERMAL CHEMICAL VAPOR DEPOSITION

There are three basic mechanisms of using laser or high-intensity light in thin-film deposition: photolysis, evaporation, and pyrolysis [Solanki et al. 1985, Mayo 1986]. The photo-CVD processes discussed earlier use the photolysis mechanism. Processes using a laser for target evaporation will be discussed in the next chapter. In this section, we discuss the photothermal CVD processes that use the pyrolysis mechanism for a-Si:H film deposition. The light beam, usually a 10.6-μm-wavelength CO_2 laser beam, can either impinge upon the substrate, thereby heating it, or it can impinge on a gas mixture parallel to the substrate. In the latter case, which is also known as laser-induced (LI) CVD, infrared (10.6 μm wavelength) excitation of silane is accomplished via the absorption of infrared photons by resonant coupling to a vibration mode of SiH bonds [Hanabusa et al. 1979]. These excited silane molecules collisionally relax between absorptions to a thermal equilibrium state where the laser energy is divided between all the molecular degrees of freedom, raising the gas temperature. Consequently, the film growth in LICVD is controlled by gas-phase homogeneous thermal decomposition [Meunier et al. 1987a]. The LICVD process is similar to HOMO-CVD in using the gas-phase homogeneous thermal decomposition of SiH_4 but without the problem of reactor wall deposition. However, it seems unlikely that the LICVD process can be used for *large-area* solar-cell deposition.

Disilane gas is usually used in substrate-heating photothermal CVD to reduce the substrate temperature. The deposition rate is quite high and increases with laser power, substrate temperature, and Si_2H_6 pressure [Iwanaga and Hanabusa 1984]. An intrinsic a-Si:H film deposited by this method at a rate of 0.5 nm/s (σ_l = 5 x 10^{-5} S/cm, σ_d = 7 x 10^{-12}, E_g = 1.77 eV) was reported by Fukuda et al. [1986].

Intrinsic a-Si:H films deposited by the gas-heating LICVD method have poor photoelectric properties. The hydrogen is incorporated primarily in the SiH_2 configuration, and for substrate temperature less than 300°C the films contain some polysilane $(SiH_2)_n$ regions [Meunier et al. 1987b, González et al. 1989, Hesch et al. 1989]. For a film deposited at 400°C with a E_g value of 1.5 eV, the σ_l is about 1 x 10^{-6} S/cm and the σ_d is about 1 x 10^{-11} S/cm. The σ_l value decreases by over four orders of magnitude as the substrate temperature is decreased to 100°C and E_g is increased to 2.5 eV [Meunier et al. 1987b]. Small-area solar cells of both p-i-n and n-i-p structures with photovoltaic conversion efficiencies up to 1.1% were fabricated using this method [Branz et al. 1987].

Gas-phase homogeneous thermal decomposition of silane, used by LICVD and HOMOCVD methods, produces SiH_2 and higher-order polysilanes (Si_2H_6, Si_3H_8, etc.) [Scott 1984 and Meunier et al. 1987]. SiH_2 and possibly higher diradicals then diffuse to the substrate and serve as the film deposition precursors. However, it is now widely believed that the SiH_2 deposition precursor, with its high sticking coefficient, produces poor quality films [Luft and Tsuo 1988c]. In conventional glow discharge, high quality a-Si:H films are produced when the deposition condition favors SiH_3 as the deposition precursor because of its low sticking coefficient. Using SiH_2 radicals as the dominant deposition precursor may be responsible for poor film properties obtained by the LICVD and HOMOCVD methods.

11.5 SPONTANEOUS CHEMICAL VAPOR DEPOSITION

The purely chemical process of fluoro-oxidation of SiH_4 by F_2 can be used to deposit a-Si:H films [Hanna 1988 and Hanna et al. 1986, 1989]. Because this gas-phase reaction is a spontaneous process, it is called spontaneous CVD by Hanna et al. The depositions were done in a vacuum system with He-diluted SiH_4 and F_2 gases fed into the system through nozzles. a-Si:H films are deposited on heated substrates from some of the SiH_mF_n intermediates produced by the reaction between SiH_4 and F_2. The deposition pressure is about 0.5 torr. The deposition process can be controlled by parameters such as SiH_4 vs. F_2 flow ratio, pressure, or gas temperature. Deposition rates of 1 nm/s have been achieved. Films deposited at T_s above 200°C contain less than 0.2 at. % fluorine.

The film characteristics measured include E_g = 1.6-1.7 eV, σ_l = 5-7 x 10^{-6} S/cm, σ_d = 10^{-9}-10^{-10} S/cm, and $\eta\mu\tau$ = 2 x 10^{-7} cm^2/V [Hanna et al. 1989].

12 PHYSICAL VAPOR DEPOSITION METHODS

Physical vapor deposition and chemical vapor deposition are the two main techniques used to apply thin-film materials to substrates. However, PVD-deposited a-Si:H has not achieved as high quality as CVD-deposited a-Si:H has. A possible reason is the higher level of impurity contamination that is usually associated with PVD-deposited films. To our knowledge, the highest efficiency of a PVD-deposited a-Si:H solar cell is 5.5% [Moustakas et al. 1985]. Some of the PVD methods investigated for a-Si:H deposition are reviewed in this section. The source material in a PVD process can be either *evaporated* or *sputtered* into vapor phase before being transported onto the substrate. In a thermal evaporation technique, the method of heating the source include resistance, electron beam (e-beam), RF induction, and laser heating. Of these techniques, e-beam heating is most often used for a-Si deposition. In a sputtering technique, the methods of generating the high energy ions needed for sputtering the source include RF and DC plasmas, magnetron in various configurations [Thornton and Penfold 1978], and ion-beam sputtering. Most of these sputtering techniques have been used for a-Si deposition. Hydrogen incorporation in PVD-deposited a-Si:H alloys is usually achieved by reacting the vapor with a hydrogen-containing atmosphere [Hauser 1976]. Because molecular hydrogen is relatively stable, an atomic hydrogen source is usually needed to sufficiently hydrogenate the evaporated a-Si during deposition.

12.1 EVAPORATION

The deposition rate of an evaporative PVD process is usually much higher than that of a sputtering PVD process. Among the evaporative processes, the e-beam heating process is usually favored because the e-beam is directed to the source material and thus reduces the heating and cross-contamination from the crucible. Several types of atomic hydrogen sources have been used to hydrogenate the a-Si film during e-beam evaporation. They include passing molecular hydrogen gas through a heated tungsten tube [Miller et al. 1978, Ghosh et al. 1979] or high voltage electrodes [Dellafera et al. 1981], using a hydrogen ion gun [Grasso et al. 1982, Martin et al. 1983, Shindo et al. 1984, Christou et al. 1986, Viturro and Weiser 1986], or using an RF-generated hydrogen plasma [Anderson and Biswas 1985 and Shimizu et al. 1986]. e-beam-evaporated a-Si:H generally has poor photosensitivity, and solar cells made using e-beam-evaporated material have low efficiencies [Fang et al. 1982]. A laser beam directed to an evaporation target in a hydrogen atmosphere should have similar advantages as the e-beam evaporation, with potentially much higher deposition rates. Hanabusa and Suzuki [1981] used a Q-switched Nd:YAG laser to deposit a-Si:H films with resistivities up to 10^9 Ω·cm.

A novel combination of evaporation and ion-based techniques is the ionized-cluster beam deposition [Takagi et al. 1979 and Yamada 1984]. In this technique, e-beam-evaporated atoms were ionized and then accelerated toward the substrate. Clusters of approximately 10^3 atoms loosely coupled together are used for deposition after ionization. Both intrinsic and doped a-Si:H films have been produced by this technique [Yamada et al. 1983]. An undoped a-Si:H film was reported to have a σ_l value of 1×10^{-5} S/cm and a σ_d value of 5×10^{-10} S/cm.

12.2 SPUTTERING

RF and DC plasma sputter deposition of a-Si and a-Si:H in planar diode systems has been investigated since the early 1970s [Hauser 1976, Paul et al. 1976]. Paul and Anderson [1981], Moustakas [1984], and Thompson [1984] have reviewed the deposition methods and properties of sputter-deposited a-Si:H films. The most successful effort in solar cell device fabrication seems to be that by Moustakas et al. [1985]. An RF diode sputtering system using a sputtering gas mixture of hydrogen and argon was used by Moustakas and coworkers to deposit the microcrystalline p- and n-layers and the a-Si:H i-layer of a 5.5%-efficient solar cell on a stainless steel substrate. Recent reports of diode sputtering studies concentrate mostly on the fabrication of narrow- and wide-bandgap amorphous alloys. Girginoudi et al. [1987] reported the fabrication of a-$(SiC)_xGe_{1-x}$:H films with E_g varying from 1.0 to 2.25 eV. Girginoudi et al. [1989] reported on the structural, electrical, and optical properties of

RF-sputtered a-$Si_{1-x}Sn_x$:H films in the composition range of $0 < x < 0.51$.

In magnetron sputtering, the plasma is confined to a narrow zone near the target. Configurations of various magnetron sputtering sources were discussed by Thornton and Penfold [1978]. Pinarbasi et al. [1989 & 1990] have been able to prepare good quality a-Si:H films by DC magnetron reactive sputtering in an Ar + H_2 atmosphere. The density of deep-level states for these films (with hydrogen content between 10 and 28 at. %) is between 1 x 10^{15} and 1 x 10^{16} /cm^3. The AM1 σ_l is in the range 0.8-3.5 x 10^{-5} S/cm, and σ_d decreases from 1 x 10^{-9} to 1 x 10^{-12} S/cm as the hydrogen content increases. The hydrogen content of these films was controlled independently by the hydrogen partial pressure in the discharge. Abelson et al. [1992] found that, for the conditions which produce electronic quality a-Si:H by reactive magnetron sputtering, the total H flux arriving at the substrate varies between 0.5 and 2 times the depositing Si flux; about half of the H flux reflects. The growth surface has excess H varying between 0.5 and 2 x 10^{15}/cm^2, and this surface H coverage is linearly related to the bulk H incorporation. They also found evidence that film density varies with the energy of the arriving sputtered Si atoms.

RF diode planar magnetron sputtering has been used to deposit a-Si:H [Abdel-Rahman et al. 1989], a-SiN:H [Morimoto et al. 1983], and a-SiGe:H [Saito et al. 1984] alloys. Rudder et al. [1984 and 1985] deposited a-SiGe:H films by dual magnetron sputtering in a UHV chamber. Films were deposited with E_g near 1.4 eV, activation energies (E_a) near 0.63 eV, σ_l around 10^{-7} S/cm, and σ_d around 10^{-8} S/cm. Hydrogenated amorphous germanium deposited by RF sputtering of a Ge target in a Ar + H_2 atmosphere has a temperature-activated conductivity down to below 200 K with activation energies larger than 0.5 eV and a photosensitivity of 2.0 under AM1 illumination [de O. Graeff et al. 1990]. This is one of the highest reported photosensitivities for a-Ge:H.

Reactive ion-beam sputtering has been used to deposit a-Si:H at a rate of 0.08 nm/s [Martin et al. 1983, Yamada and Torii 1987] . Coluzza et al. [1983] used a double ion-beam sputtering technique to deposit a-Si:H films. They used an argon ion beam to sputter silicon and a hydrogen beam to hydrogenate the growing film. The films produced have E_g values around 1.7 eV and σ_d values in the range 10^{-9}-10^{-11} S/cm. However, the σ_l values are low, on the order of 1 x 10^{-9} S/cm. Unhydrogenated a-SiGe films with E_g = 1.1 eV and σ_d = 3 x 10^{-5} S/cm have been deposited by ion-beam sputtering [Bhan et al 1991].

12.3 POSTHYDROGENATION METHODS

As we know, the hydrogen content and the way hydrogen atoms are bonded in the a-Si:H material strongly affect its optical and electrical properties. All the a-Si:H deposition techniques discussed in this review so far introduce hydrogen atoms into a-Si:H material during deposition. This ensures that hydrogen atoms are distributed uniformly in the material, and the reaction chemistry during deposition determines the hydrogen content and bonding configuration. However, the parameter ranges for high quality a-Si:H film growth are usually very limited, making it difficult to incorporate hydrogen in a well-controlled manner. Methods of introducing hydrogen into a-Si after deposition (i.e., posthydrogenation) and methods of reintroducing hydrogen into a-Si:H after the originally as-deposited hydrogen atoms have been fully or partially driven out by heating (i.e., rehydrogenation) are likely to be more controllable. Because, as far as we know, all reported a-Si:H with hydrogen introduced during deposition suffers light-induced degradation, it makes it even more interesting to investigate the stability properties of posthydrogenated and rehydrogenated a-Si:H. Several reports have been published claiming better photostability for posthydrogenated a-Si:H than for as-deposited a-Si:H [Chik et al. 1983, Nakashita et al. 1984a,b, Thomas and Flachet 1985, Ohagi et al. 1987, and Chen 1988]. Tsuo et al. [1987a and 1988a,b] also observed reduced light-induced degradation of ion-beam-rehydrogenated a-Si:H.

Pankove et al. [1978] first reported that dangling bonds created by the dehydrogenation of a-Si:H can be rehydrogenated by exposure to atomic hydrogen, but not to undissociated molecular hydrogen. They used a hydrogen RF glow discharge plasma as the source of atomic hydrogen for the rehydrogenation experiments. Photoluminescence measurements [Pankove 1978] showed that the RF plasma rehydrogenation eliminated the non-radiative recombination centers. Tsuo et al. [1987a and 1988a,b] studied a rehydrogenation process using a low energy (less than 2000 eV) ion beam as the source of atomic hydrogen. AM1 σ_l/σ_d ratios were achieved as high as 9.5×10^6 (with $\sigma_l = 8.6 \times 10^{-6}$ S/cm) with a rehydrogenated a-Si:H film with an E_g value of 1.85 eV [Tsuo et al. 1988a]. The hydrogen atoms introduced by this low energy ion implantation method bonded predominantly as SiH.

There have been many reports on the posthydrogenation of a-Si since Le Comber et al. [1972] first described their posthydrogenation work. The RF hydrogen plasma has been the most widely used method. Le Comber et al. [1974] reported that an RF hydrogen plasma posthydrogenation treatment of evaporated a-Si resulted in a reduction, of more than two orders of magnitude, of the ESR signal and a reduction, by a factor of 2, in dark conductivity. PVD-deposited a-Si often contains high levels of impurities. To reduce the impurity levels, Kaplan et al. [1978] studied the RF plasma posthydrogenation of a-Si evaporated by an e-beam under UHV. The hydrogen content of their posthydrogenated films is only about 3.5 at.%. The photoconductivity is 6 x 10^{-7} S/cm, and the σ_l/σ_d ratio is only 6. De Chelle et al. [1984] studied the optical properties of RF posthydrogenation of a-Si:H deposited by RF reactive sputtering and found that posthydrogenation resulted in a decrease of the residual disorder. Pratt and Weil [1987] investigated the RF plasma posthydrogenation of evaporated a-Si, and they produced a-Si:H films with less than 1 at.% hydrogen and a σ_l/σ_d ratio of only 2.1.

RF posthydrogenation of both doped and undoped a-Si produced by thermal CVD have been studied by many research groups. Hasegawa et al. [1983] reported a σ_l value of 5 x 10^{-6} S/cm and a σ_l/σ_d ratio of about 50 for undoped, posthydrogenated, CVD-deposited a-Si:H. Nakashita et al. [1983 and 1984a] found that their posthydrogenated, undoped, CVD-deposited a-Si:H had a low activation energy—only 0.47 eV. By slightly doping their a-Si with boron during deposition, they were able to move the Fermi level of posthydrogenated a-Si:H toward the midgap. For a film with a boron gas doping ratio of 4 x 10^{-6} and an E_g value of 1.66 eV, they obtained a σ_l value of 3 x 10^{-6} S/cm and a σ_l/σ_d ratio of nearly 10^5 [Nakashita et al. 1984b].

Many hydrogenation methods other than RF plasma have also been studied. Thomas and Flachet [1985] studied the DC plasma posthydrogenation of UHV-deposited a-Si. They obtained a low σ_d value of 3 x 10^{-9} S/cm. Microwave-excited plasma has been used for posthydrogenation of CVD-deposited a-Si materials and devices; σ_d in the 10^{-8}-S/cm range for undoped a-Si:H have been obtained [Magarino et al. 1982 and Donnadieu et al. 1983]; Schottky and p-i-n diodes have been fabricated, and an AM1 photoconversion efficiency of 2.5% has been obtained for a Schottky diode [Szydlo et al. 1982]. Lee and Neudeck [1983] posthydrogenated their e-beam-evaporated a-Si by a 12.5 keV hydrogen ion beam and reduced the density of localized states near the Fermi level to 1 x 10^{19} $cm^{-3}eV^{-1}$. A thermal annealing process after the film deposition further reduced the density of states to 4 x 10^{17} $cm^{-3}eV^{-1}$ [Neudeck and Lee 1983]. One of the most interesting posthydrogenation techniques is theta-pinch plasma hydrogenation. A theta-pinch plasma ion source is capable of extracting a high-current ion beam with ion energy in the 10-20 keV range. Current densities can be as high as 200

A/cm^2 at the focus of the ion beam [Dembinski et al. 1979]. Tong et al. [1981 and 1983] hydrogenated e-beam-evaporated a-Si using a focused theta-pinch ion source with a total bombardment time of only about 1 min. The resulting a-Si:H films showed a σ_l value of about 1 x 10^{-5} S/cm and a σ_l/σ_d ratio of about 100. Tsuo et al. [1988a] posthydrogenated a-Si samples glow discharge-deposited at 480° C, using a low energy hydrogen ion beam. They obtained a σ_l value of 6.3 x 10^{-6} S/cm and a σ_l/σ_d ratio of 1.1 x 10^5 for a posthydrogenated a-Si:H sample with an E_g value of 1.82 eV. Using the same ion source, Deng et al. [1989] posthydrogenated a-Si, a-SiB, and a-SiGe films deposited by RF sputtering. They obtained a σ_l value of 7.6 x 10^{-5} S/cm and a σ_l/σ_d ratio of 1.5 x 10^5 for posthydrogenated a-Si:H (E_g = 1.72 eV), a σ_d value of 7.6 x 10^{-2} S/cm for a posthydrogenated a-SiB:H (E_g = 1.51 eV), and a σ_l value of 6.6 x 10^{-7} S/cm and a σ_l/σ_d ratio of 3.7 x 10^3 for a posthydrogenated a-Si:H (E_g = 1.63 eV). Uchida et al. [1992] posthydrogenated a-Si produced by low-pressure CVD and obtained a photosensitivity greater than 4.5 x 10^4 and the hydrogen was bonded mainly as Si-H with a penetration of about 300 nm.

As we mentioned earlier, there have been quite a number of reports of improved photostability of rehydrogenated and posthydrogenated a-Si:H materials. However, all of the improved-photostability materials reported have rather poor initial photoconductivity (less than 1 x 10^{-5} S/cm). Smith, E., et al. [1989] compared the photostability properties of six different types of materials that include glow discharge-deposited a-Si:H, glow discharge-deposited a-Si:H:F, ion-beam-rehydrogenated a-Si:H, RF-hydrogen-plasma-posthydrogenated a-Si:H, PED-CVD-deposited a-Si:H, and PED-CVD-deposited a-Si:H:F. All of these films were deposited in the same reactor. The load-locked, single-chamber reactor has a base vacuum of 1 x 10^{-8} torr or less and a leak rate of 2 x 10^{-5} S/cm or less. These films had initial (annealed) σ_l values ranging from 10^{-6} to 10^{-4} S/cm. They found that the light-induced degradation properties follow the same dependence on the initial photoconductivity values regardless of the sample preparation method. They also found, indirectly through photoconductivity, that the absolute and relative degradations in photoconductivity seems to decrease with increasing initial defect-state densities. The relative change in photoconductivity is drastically reduced when the initial photoconductivity is less than about 3 x 10^{-5} S/cm. If the initial σ_l value is below 1 x 10^{-5} S/cm, the relative degradation in σ_l is usually so small that it appears as if there was no Staebler-Wronski effect. The relative decrease in σ_l saturates at about a factor of 10 for films with initial σ_l values of about 6 x 10^{-5} S/cm or higher. This seems to indicate that many of the reported observations of reductions of the Staebler-Wronski effect for rehydrogenated and posthydrogenated a-Si:H films are caused by poor material qualities rather than actual improvements of the photostability property.

Although the quality and photostability of posthydrogenated and rehydrogenated a-Si:H materials may not be better than the state-of-the-art glow discharge-deposited a-Si:H, there may still be applications for such techniques. For example, the work by Galloni et al. [1989 and 1990] has shown that a hydrogen ion source can be used to neutralize the damage caused by ion-implantation of dopants into a-Si:H. The capability of using an ion beam to selectively hydrogenate areas of a-Si to increase the σ_l/σ_d ratio from zero to almost 10^7 [Tsuo et al. 1988a] may be useful in fabricating image-detection devices. Yet another potential application of the posthydrogenation technique is to posthydrogenate amorphous materials deposited at high temperature with better structural properties [Jones, S. J., et al. 1989].

13
ETCHING PROPERTIES OF AMORPHOUS SILICON-BASED ALLOYS

Etching is a material removal process which is the reverse of deposition. Etching is used extensively in semiconductor processings, especially in the fabrications of integrated circuits [Wolf and Tauber 1987]. For thin-film a-Si:H photovoltaic module fabrications, etching may be used in place of laser scribing. Etching is also used in fabricating a-Si:H thin-film transistor circuits. The ability to do selective etching is important in lithography operations to fabricate integrated circuits. In addition, because etching of weak bonds during chemical vapor deposition is important in achieving high quality a-Si:H, better understanding of the etching properties of a-Si:H may lead to better deposition techniques.

Etching properties of a-Si:H alloys, such as a-Si:H, a-$Si_{1-x}Ge_x$:H, and a-$Si_{1-x}C_x$:H, are more complicated than those for single crystalline silicon. Factors that may affect the etching properties of a-Si:H alloys include alloy composition, hydrogen content and bonding configurations, optical bandgap, and the Fermi level position which depends on doping level and defect density. The microstructure of a-Si:H alloys, such as voids, columnar growth structure, microcrystallinity, etc., may also affect the etching properties. Because a-Si:H alloys are usually deposited as thin films, we need also consider the effects of etchants on the substrates.

For most of the etching properties reported in this chapter, a-Si:H films used are typically around 500 nm thick deposited on Corning code 7059 glass or n-type crystalline silicon substrates [Tsuo et al. 1987b & 1991]. The RF diode glow discharge deposition conditions are 22 mW/cm^2 RF power density, 700 mtorr SiH_4 pressure, and 70 sccm SiH_4 flow rate. The substrate temperature is varied from room temperature to 500°C to obtain films of different hydrogen content.

13.1 WET CHEMICAL ETCHING

For fast etch removal of a-Si:H alloy films, such as removing films deposited on the inside walls of a stainless steel deposition chamber, commercially available $HNO_3/CH_3COOH/HF$ (3:1:2) etchants are normally used. For slow, more controlled etching of a-Si:H alloy films on glass substrates, one method is to use a KOH-based solution. Haller et al. [1988] studied the etch rates of undoped and phosphorous-doped a-Si:H deposited at 275°C and 125°C in KOH-based etchants. They found that the etch rate is faster for the 125°C films which have higher hydrogen content than the 275°C films. Tsuo et al. [1991] found that a KOH-based etchant with 23.4 wt.% KOH, 13.3 wt.% isopropyl alcohol (IPA), and 63.3 wt.% H_2O having a pH value of 13.75 etches an 1.75-eV undoped a-Si:H at a rate of 12 nm/minute.

The wet chemical etch rate of undoped a-Si:H in a KOH-based etchant depends very strongly on both the KOH concentration in the solution and on the hydrogen content of the film. For example, the wet

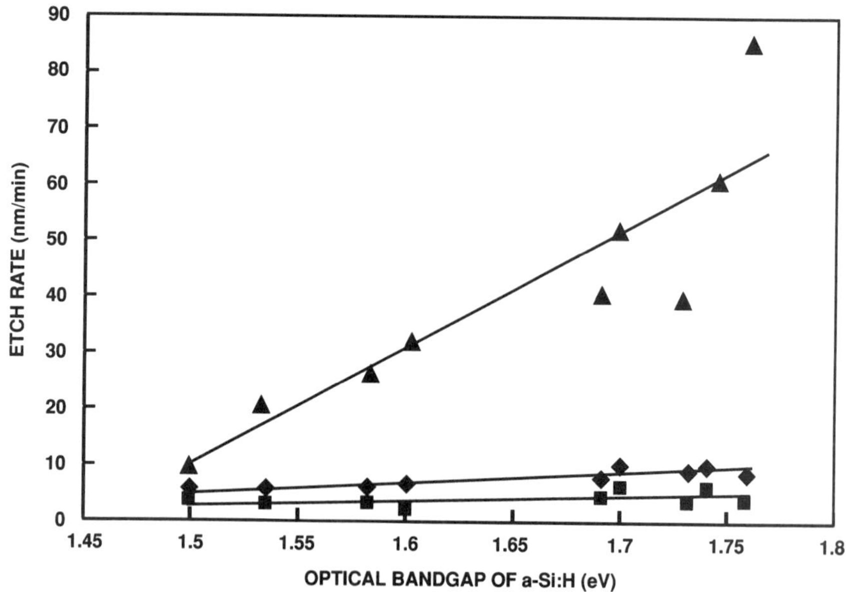

Figure 13-1. Etch rate vs. optical bandgap for a-Si:H films in KOH-IPA-H_2O etchants with 36 wt.% (top curve), 24 wt.% (middle curve), and 12 wt.% (bottom curve) KOH contents.

chemical etch rate of undoped a-Si:H in a KOH-IPA-H_2O solution with 36 wt.% KOH can be controlled from 10 nm/min to 90 nm/min by varying the hydrogen content of the film, as shown by the top curve in **Fig. 13-1** [Tsuo et al. 1991]. The 1.5 eV optical bandgap (E_g) films in Fig. 13-1 correspond to a-Si films with no hydrogen content. The 1.76 eV E_g films correspond to a-Si:H films with 9 at.% bonded hydrogen content. The middle and bottom curves in Fig. 13-1 show the E_g dependence of etch rates for solutions with 24 wt.% KOH and 12 wt.% KOH, respectively. The weight ratio of H_2O/IPA is kept at 4.74. Obviously, a-Si:H films etched faster in etchants with higher KOH concentrations. The etch rate also increased with increasing optical bandgap which corresponds to increasing hydrogen content.

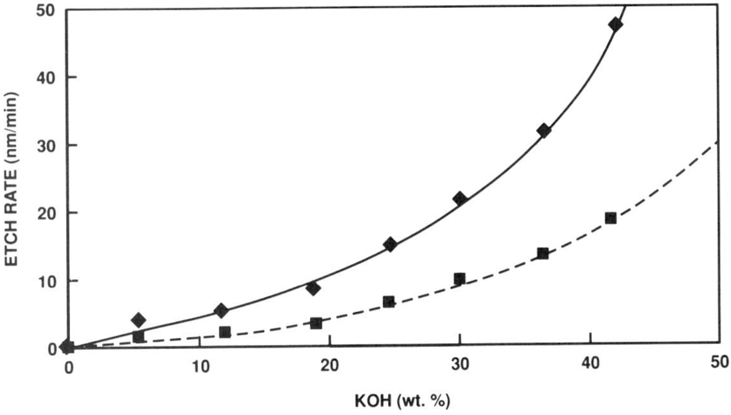

Figure 13-2 Etch rate vs. KOH wt.% for 1.50 eV E_g a-Si films with no hydrogen (upper curve) and 1.70 eV E_g a-Si:H with 6.5 at.% hydrogen.

Fig. 13-2 shows the etch rate as a function of KOH content in the KOH-based etchant discussed in the previous paragraph for two types of films. The upper curve is for a-Si:H films with 6.5 at.% hydrogen and 1.70 eV E_g. The lower curve is for a-Si films with no hydrogen.

Staebler [1979] showed that the etch rate of a-Si:H films in basic solutions depends on the optical bandgap after laser-beam annealing. The basic solutions used by Staebler include a pH 11.00 solution of one part Decontam (Kern Chemical Corp.) in four parts water and a NaOH buffer solution of the same pH. Hundhausen et al. [1987] found that the etch rate for a-Si:H in a CP6 solution (a mixture of HF, HNO_3, and CH_3COO-H) is 10 μm/min, and the etch rate for a-$SiN_{1.3}$:H in CP6 is 0.1 μm/min.

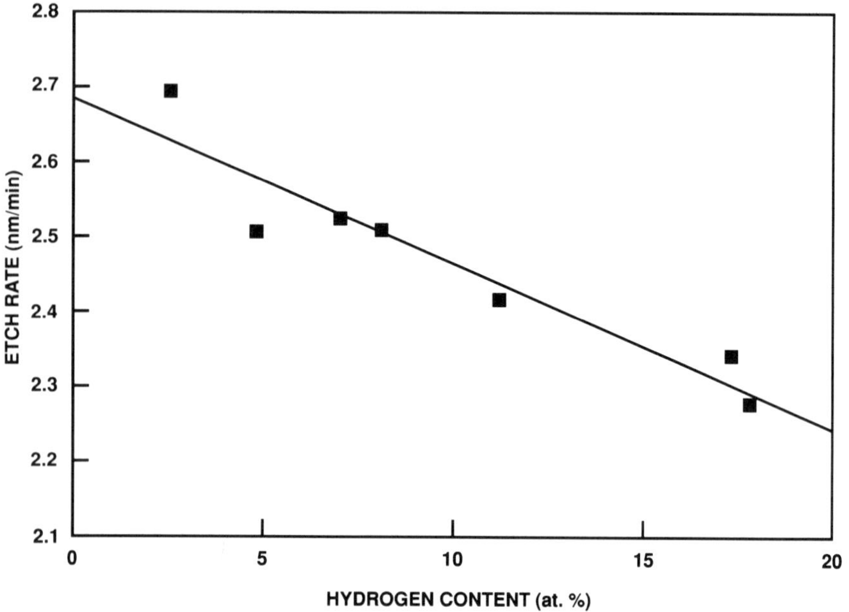

Figure 13-3 Etch rate vs. hydrogen content for undoped a-Si:H films in a RF-generated hydrogen plasma.

13.2 PLASMA ETCHING

Atoms and radicals in a plasma are far more chemically reactive than most molecules. Plasma etching processes done in vacuum chambers are also cleaner than wet-chemical etching processes. For a-Si:H alloys, plasmas generated from fluorine- and chlorine-containing gases, such as CF_4-O_2 [Haller et al. 1988], CCl_4 [Clarke et al. 1990], and NF_3 [Barkanic et al. 1989], are effective etchants. NF_3 and SF_6 can also become etchants of a-Si:H when photolyzed with uv light [Langford et al. 1987]. CF_4-O_2 plasmas are often used to etch clean a-Si:H deposition chambers. Haller et al. [1988] observed that the etch rate of a-Si:H in a CF_4-O_2 plasma increases with film hydrogen content. RF-generated hydrogen plasma is also highly reactive. It has been used to flush the a-Si:H deposition chamber and improve the p/i interface of p-i-n a-Si:H solar cells [Tsuo et al. 1989]. Tsuo et al. [1991b] found that the etch rate of a-Si:H in an RF hydrogen plasma decreases slightly with increasing bonded hydrogen content in the a-Si:H as shown in **Fig. 13-3.** To

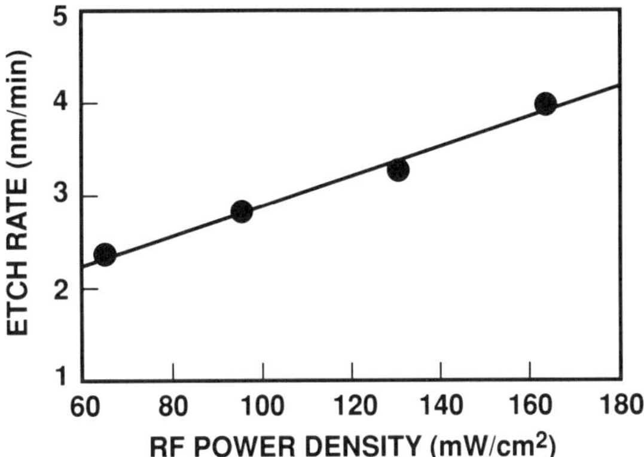

Figure 13-4 Etch rate vs. RF hydrogen plasma power density for undoped a-Si:H films with 1.81 eV optical bandgap. Hydrogen pressure was 4 torr, and sample temperature was 250°C.

generate the RF hydrogen plasma they used a hydrogen gas pressure of 4 torr, a RF power density of 64 mW/cm^2, a hydrogen flow rate of 30 sccm, and a sample temperature of 250°C. The etch rate increases with the RF power density. The hydrogen plasma etch rate of an undoped 1.81 eV E_g a-Si:H vs. the RF power density is shown in **Fig. 13-4** [Tsuo et al. 1991b]. The posthydrogenation effect [Tsuo et al. 1987a,b] by the RF hydrogen plasma is not obvious. No systematic change in the bandgap and conductivity of a-Si:H was observed after RF hydrogen plasma treatments. The etch rate dependence on the chamber pressure is very small; the etch rate increases about 10% from 2 torr to 6 torr.

Tsuo et al. [1991b] also found that the RF hydrogen plasma etching of a-Si:H is very sensitive to the existence of native oxides on the surface of a-Si:H. Etching experiments reported in the previous paragraph were carried out with a-Si:H films never exposed to air before hydrogen plasma etching. Usually, two identical films were deposited for each measurement: one film was taken out of the vacuum chamber for thickness and hydrogen content measurements, another film remained in the vacuum chamber for hydrogen plasma etching. Exposure of a-Si:H surface to air at room temperature and under room light for five hours or more completely prevents the a-Si:H from been etched by a RF hydrogen plasma. The surface oxidation rate of a-Si:H is observed to depend on illumination conditions: faster in light and slower in the dark. When a sample is kept in the dark, it takes about 20 hours for enough surface oxide to develop to prevent the sample from being etched by a RF hydrogen plasma.

The etching of a-Si:H in glow discharges of SiF_4 and H_2 were studied by Okada and Wagner [1990]. They observed that the etch rate of a-Si:H in a SiF_4 plasma (6 nm/min) is about an order of magnitude higher than that of crystalline silicon. The etch rate of a-Si:H by a hydrogen plasma observed by Okada and Wagner is lower than those observed by van Oort et al. [1987] and by Tsuo et al. [1991b].

Tsuo et al. [1991b] found that H-plasma either does not etch or only very weakly etches a-Ge:H. This may be one of the reasons why glow discharge deposited a-SiGe:H and a-Ge:H alloys have poor electronic properties.

13.3 VAPOR ETCHING

It has been reported that vapors of xenon difluoride (XeF_2) spontaneously etch crystalline Si with rates as large as 700 nm/min without requiring the application of heat, light, a plasma, or ion bombardment [Winters and Coburn 1979, Ibbotson et al. 1984]. XeF_2 is a white powder with a vapor pressure of 3.8 torr at 25°C and a melting point of 140°C. There is no observable XeF_2 etching of some Si-containing compounds such as SiO_2, Si_3N_4, or SiC. In the etching experiments reported by Tsuo et al. [1991b], XeF_2 powder weighing 10 g or less was placed in a 400 cm^3 stainless steel container that was connected to a vacuum chamber through a quarter-inch (O.D.) stainless steel tube, a toggle valve, and a micrometer valve. They found that XeF_2 vapors at a pressure of 0.1 torr spontaneously etch 1.75 eV bandgap a-Si:H films at a fast rate of 180 nm/min with sample at room temperature. This etch rate increases or decreases linearly with the pressure of XeF_2. XeF_2 vapor was used as the etch gas in a periodic etching-and-deposition CVD method for a-Si:H:F [Tsuo et al. 1987b]. XeF_2 vapor was also used to etch away the a-Si:H films deposited on the window in a photo-CVD process [Tsuo and Langford 1989b].

13.4 ION BEAM PROCESSING

Hydrogen ion beam etching is less sensitive to surface oxides than plasma etching [Ishii et al. 1991]. The etch rate of a-Si:H depends more strongly on the ion beam current density than on energy [Deng et al. 1990]. The material-removal rate of undoped a-Si:H by a hydrogen ion beam increases slowly with the ion beam energy and peaks at about 1700 eV. This rate increases linearly with the ion beam current density within the range studied (up to 1 mA/cm^2).

Ion beam doping or hydrogenation can change the etching properties of a-Si:H. Selective etching after ion beam doping or hydrogenation make it possible to use a-Si:H as an inorganic resist for ion beam lithography [La Marche and Levi-Setti 1984, Tsuo and Deb 1990].

13.5 DOPING EFFECTS

It is well known that the etch rate of certain semiconductors depends on the Fermi level position [Palik et al. 1982, Lee and Chen 1986]. N-type and undoped semiconductors have more filled states in the bandgap and thus are chemically more reactive and have a higher etch rate than do p-type semiconductors. Staebler [1979] found that phosphorus-doped a-Si:H has lower etch rates than boron-doped and undoped samples in basic solutions (§ 13.1) Haller et al. [1988] found that the etch rates of undoped a-Si:H are lower than n-type a-Si:H etch rates but higher than p-type a-Si:H etch rates during CF_4-O_2 plasma etching. Haller et al. also noted that in their studies of KOH etching of a-Si:H samples, n-type and intrinsic a-Si:H samples have similar etch rates. Tsuo et al. [1989] found that for a KOH-based wet chemical etching of a-Si:H, phosphorus-doped a-Si:H also has similar etching properties as undoped a-Si:H. However, boron-doped a-Si:H and a-SiC:H films were found to have much smaller wet-chemical and hydrogen plasma etch rates (etch rate is less than 0.1 nm/min for both wet chemical and H-plasma etching) than that of undoped and n-type a-Si:H films. La Marche and Levi-Setti [1984] showed that gallium-implanted a-Si:H has lower etch rate than as-deposited a-Si:H in 40°C aqueous NaOH (50 wt.%) solutions.

13.6 OXIDATION

The existence of high quality oxides and Si/SiO_2 interfaces is one of the main reasons for the wide applications of crystalline silicon. Similar to the oxides of crystalline silicon, a-SiO_x passivates the surface of a-Si:H. Kragler et al. [1989] found that the density of interface states between a-SiO_x and a-Si:H decreases with increasing oxide thickness from 1×10^{13} $eV^{-1}cm^{-2}$ with no oxide to 8×10^{12} $eV^{-1}cm^{-2}$ at 1.5 nm oxide thickness. The oxide has a bandgap of 8 to 9 eV, it can be removed from the surface of a-Si:H by dipping the sample in a buffered HF solution, such as 2%HF for 20 s. Yokota et al. [1985] found that room-temperature oxide growth relieves stress for a-Si:H films deposited on <100>-oriented p-type crystalline Si substrates.

Ponpon and Bourdon [1982] and Yokota et al. [1983 & 1985] found the thickness of oxides on a-Si:H grown at room temperature depends on (exposure time)^n, with n = 0.42 and 0.50, respectively. This is slower than the exponential growth rate of room-temperature oxide on crystalline Si, and indicates that the room-temperature oxide growth on a-Si:H is limited by the diffusion of oxygen through the growing oxide layer. However, Lu et al. [1988] found that the room-temperature oxidations of a-Si:H and crystalline Si are both exponential and rate-limited by surface reactions.

14
COMPARISON OF ALTERNATIVE DEPOSITION METHODS

We have reviewed most of the chemical vapor deposition, physical vapor deposition, and posthydrogenation processes that have been studied for the deposition and hydrogenation of a-Si:H and a-SiGe:H. CVD processes differ in the gas dissociation methods, the type and density of deposition precursors, the density of atomic hydrogen or fluorine for etching during deposition, the degree of ion or electron bombardments, and/or the surface mobility of film-producing radicals. Some of the alternative deposition methods show promise for improving the deposition rate of a-Si:H or the electrical properties of doped a-Si:H or a-SiGe:H films. However, the electrical properties of *intrinsic* a-Si:H, including defect density, stability, and solar-cell efficiency, have not been improved by any of the new deposition methods. For comparison, **Table 14-1** lists the best published characteristics for a-Si:H and a-Si:H:F for conventional glow discharge and some of the alternative deposition processes discussed in this book: separated plasma triode glow discharge (SPT-GD), photo-CVD, controlled plasma magnetron (CPM) glow discharge, electron cyclotron resonance (ECR) microwave plasma CVD, microwave plasma glow discharge, microwave hydrogen-radical-enhanced (HR)-CVD, catalytic thermal (hot-wire) CVD, and spontaneous CVD. The corresponding data for a-SiGe:H (mostly for E_g = 1.5 eV) are shown in **Table 14-2**. When the optical bandgap differs from 1.5 eV, it is indicated in parenthesis in Table 14-2 after the value affected.

Next to glow discharge, photo-CVD is the most studied a-Si:H deposition method. Although deposition on reactor window is still a problem in most cases, high quality films can be deposited by photo-CVD. Comparing photo-CVD to glow discharge, we find for a-Si:H films that about equivalent photosensitivity has been achieved, along with somewhat higher mid-gap density of states (2×10^{15} vs. 1×10^{15} $eV^{-1}cm^{-3}$) and Urbach parameter (45 vs. 42 meV), and the same ESR spin density (5×10^{15} cm^{-3}). Similarly, for a-SiGe:H alloys of about 1.5 eV E_g, the photosensitivity is about the same for photo-CVD and glow discharge ($1.8\text{-}2 \times 10^4$) [Luft 1988a], as is the Urbach parameter. Except for possibly better doped film properties, photo-CVD appears to have no significant advantage over glow discharge.

For a-Si:H devices deposited by glow discharge at rates of 0.5-2 nm/s, it appears to make no difference in device efficiency whether silane or disilane is used as a feed-gas [Luft 1988b]. However, there is some evidence that fluorine-containing gases (such as SiF_4) in the presence of atomic hydrogen will result in HF that might provide beneficial etching of strained or weak bonds [Shibata et al. 1987a]. HF can also attack SiO_2 formation from residual oxygen in vacuum system. Enhanced atomic hydrogen etching during deposition may also have beneficial effects on the film properties.

For a-SiGe:H or a-SiGe:H:F films of about 1.5 eV E_g and deposited by glow discharge or photo-CVD, none of the following gas mixtures made any difference in the photosensitivity of the resulting films: silane/germane, disilane/germane, and silane/germanium tetrafluoride. However, hydrogen (or inert gas) dilution is essential for achieving high photosensitivity of a-SiGe:H (which is not necessarily the case for a-Si:H). Hydrogen dilution gives a-SiGe:H photosensitivity 2 to 20 times higher than do undiluted gas mixtures [Konagai 1986 and Luft 1988a]. Approximately the same photosensitivity of 1×10^4 at 1.5 eV bandgap has been achieved by several of the new methods discussed. But glow discharge and photo-CVD have given about twice that photosensitivity (2×10^4) at the same bandgap. Aside from photosensitivity, there is still not enough data to permit comparisons of a-SiGe:H material quality among the various deposition methods.

The absence of feed-gas impurities is important to a-Si:H material properties. Equally important is the absence of reactor chamber leaks and contaminations. Better feed-gas quality, low base-vacuum pressure, and low reactor leak rate probably contributed to most of the improvements in glow discharge-deposited a-Si:H film properties in recent years. Reducing impurity concentrations of O, N, and C in 1.7 eV a-Si:H results in enhanced photosensitivity (1×10^7), low Urbach energy (42 meV), low midgap density of states (1×10^{15} $eV^{-1}cm^{-3}$), low ESR spin density (2×10^{15} cm^{-3}), and long minority carrier diffusion length (1 μm). Deposited by glow discharge from disilane, at a deposition rate of 0.3 nm/s, in a high-vacuum system with a 10^{-8} torr background pressure, p-i-n a-Si:H cells had an 11.7% efficiency for an area of 0.09 cm^2 and a 10.7% efficiency for an area of 1 cm^2 [Miyachi et al. 1987].

The photo-induced degradation of intrinsic a-Si:H (the Staebler-Wronski effect) is most easily observed by measuring the reductions of σ_l and σ_d after prolonged illumination. Very small σ_l and σ_d reductions from illumination were often observed for *poor quality* a-Si:H films deposited by a new method. This led many researchers to believe that the new method actually produced films with less or no Staebler-Wronski effect. Often, after the quality of films deposited by the new method was gradually improved, it became clear that the new material also degraded under illumination. Smith, E. B., et al. [1989] studied the photostabilities of a-Si:H and a-Si:H:F films deposited by six different methods in the same reactor. They found that the absolute and relative degradations in σ_l after illumination decreased rapidly with the initial σ_l value regardless of the deposition method (i.e., the smaller the initial σ_l value, the smaller the observed photo-induced reduction in σ_l). For example, if the initial σ_l value was below 1 x 10^{-5} S/cm, the relative degradation in σ_l was usually so small that it appeared as if there were no Staebler-Wronski effect. These results indicate that the defect concentrations in poor quality a-Si:H films are so large that they mask the effects produced by light-induced defects, or that the defect density and position in the bandgap are such that they short out the recombination processes that induce the Staebler-Wronski defect. For high quality material, σ_l and σ_l/σ_d may not be significant quality indicators.

The time constants in stretched exponential fitting the degradation of device-quality materials as well as the ratio of initial to degraded performance differs among the spin density, photoconductivity and ambipolar diffusion length and from that of corresponding p-i-n devices [von Roedern 1992].

Defect state densities in a-Si:H are not entirely defined at the time of film deposition [Street et al. 1986 and McMahon and Tsu 1987] but reach thermal equilibrium in the temperature range 100°-300° C. The thermal equilibrium defect density increases with temperature [Street and Winer 1989]. At temperatures below the thermal equilibration temperature, the defects are frozen in because of the long equilibration time, and the electronic properties of a-Si:H depend on the thermal history of the sample [Hack and Street 1988]. This can be understood by analogy to the glass transition temperature in various glassy solids. This glass transition temperature is about 200°C for undoped a-Si:H, 130°C for n-type a-Si:H, and 90°C for p-type a-Si:H. In principle, the defects in a-Si:H can be reduced by (1) decreasing T_s and thus reducing the defects present at film growth (but also increasing the hydrogen content), (2) reducing the number of sites at which defects can form, which is related to the valence-band-tail width [Smith and Wagner 1987], or (3) increasing the speed of defect equilibration at low temperature. Smith and Wagner [1987] have studied the effects of reducing T_s on defect density. They found that the neutral dangling-bond density measured at room temperature decreases with T_s from 600°C to about 250°C. When T_s is reduced to below 250° C, the low substrate temperature reduces the surface mobility of film growth constituents and causes a strongly inhomogeneous film structure and poor electrical properties. One possible method of overcoming this is to use high hydrogen dilution, because the enhanced etching of hydrogen creates the effects of a higher substrate temperature without raising the actual deposition temperature. It would also be interesting to investigate whether the defect equilibration speed can be increased by some of the postdeposition treatment techniques, such as posthydrogenation.

Although none of the alternative deposition processes has proved to be superior overall to conventional glow discharge in depositing a-Si:H-based electronic materials and solar cells, we believe continued investigations are still needed for at least three reasons.

1. The poor film qualities achieved so far may simply be because the available parameter spaces have not been fully investigated for many of the new methods. Even for the conventional glow discharge, the gas-phase chemistry and film growth kinetics for high quality a-Si:H film deposition are not yet well understood and may not have been fully optimized.

2. New deposition methods may have special advantages in depositing (1) films with high deposition rates, (2) highly transparent and highly conductive doped materials, (3) high quality narrow- or wide-bandgap alloys, or (4) films with less photo-induced degradation.

3. The investigations of some of the alternative deposition methods can contribute to our understanding of the basic mechanisms of conventional glow discharge deposition and the relationship between discharge parameters and film properties.

Table 14-1. Comparison of the best quality parameters for a-Si:H or a-Si:H:F (optical bandgap = 1.7-1.8 eV) achieved by various deposition methods.[Tsuo and Luft 1990]

Characteristic	Unit	GD	SPT-GD	CPM	ECR-CVD	MP-GD	MHRE-CVD	Photo-CVD	CT-CVD	S-CVD
Photoconductivity, σ_ℓ	S/cm	1E-4	1E-4	---	1.5E-7	9E-4	---	5E-5	1E-3	7E-6
Dark Conductivity, σ_d	S/cm	1.7E-12	2E-10	---	3E-10	4E-10	---	1E-10	---	1E-10
σ_ℓ/σ_d Ratio	---	1E7	---	5E5	5E2	2E6	---	5E6	1E6	---
Urbach Energy	meV	42	---	---	---	---	42-45	45	---	---
Density of States at the Fermi Level	$eV^{-1}cm^{-3}$	1E15	---	---	---	---	---	2E15	---	---
ESR Spin Density Ns	cm^{-3}	2E15	7E15	---	2E16	---	---	5E15	1.5E16	---
Space Charge Density	cm^{-3}	5E14	---	7E14	---	---	---	---	---	---
Hole Mobility, μh	cm^2/Vs	---	---	---	---	---	2.6E-1	---	---	---
$\eta\mu\tau$	cm^2/V	8E-6	---	---	---	1E-5	1.2E-8	---	---	2E-7
Index of Refraction	---	3.44	---	---	---	3.7	---	---	---	---
Deposition Rate	nm/s	2	1	1.5	2	0.3	1.2	1.7	25	---

GD = Glow Discharge; SPT-GD = Separated Plasma Triode Glow Discharge; Photo-CVD = Photochemical Vapor Deposition; CPM = Controlled Plasma Magnetron Glow Discharge; ECR-CVD = Electron-Cyclotron-Resonance Microwave Plasma Chemical Vapor Deposition; MP-GD = Microwave Plasma Glow Discharge; MHRE-CVD = Microwave Hydrogen-Radical-Enhanced CVD; CT-CVD = Catalytic Thermal Chemical Vapor Deposition; S-CVD = Spontaneous Chemical Vapor Deposition.

Table 14-2. Comparison of the best quality parameters for a-SiGe:H or a-SiGe:H:F (optical bandgap = 1.5 eV) achieved by various deposition methods. When the optical bandgap is different from 1.5 eV it is indicated in parentheses. [Tsuo and Luft 1990]

Characteristic	Unit	GD	CPM	ECR-CVD	MP-GD	MHRE-CVD	Photo-CVD	CT-CVD
Photoconductivity, σ_ℓ	S/cm	3E-4	---	1E-5	1E-5	2E-5(1.3)	2E-4(1.5)	1E-4(1.4)
Dark Conductivity, σ_d	S/cm	1.4E-10	---	---	1.8E-9	1E-8(1.4)	2E-10	1E-8(1.4)
σ_ℓ/σ_d Ratio	---	1.8E4	---	1E4	1E4	1E4	2E4	1E4(1.4)
Urbach Energy	meV	46	---	---	---	---	46(1.52)	---
Density of States at the Fermi Level	$eV^{-1}cm^{-3}$	3E16	---	---	---	---	---	---
ESR Spin Density Ns	cm^{-3}	6.5E15	---	---	---	---	---	---
Space Charge Density	cm^{-3}	6.4E15	---	---	---	---	---	---
$\eta\mu\tau$	cm^2/V	1E-6	---	---	---	5E-7(1.4)	6E-8(1.4)	---
Index of Refraction	---	3.77	---	---	3.9(1.49)	---	---	---
Deposition Rate	nm/s	---	---	1	1,3	---	0.05	1

GD = Glow Discharge; SPT-GD = Separated Plasma Glow Discharge; Photo-CVD = Photochemical Vapor Deposition; CPM = Controlled Plasma Magnetron Glow Discharge; ECR-CVD = Electron-Cyclotron-Resonance Microwave Plasma Chemical Vapor Deposition; MP-GD = Microwave Plasma Glow Discharge; MHRE-CVD = Microwave Hydrogen-Radical-Enhanced CVD; CT-CVD = Catalytic Thermal Chemical Vapor Deposition; S-CVD = Spontaneous Chemical Vapor Deposition.

15
MICROCRYSTALLINE SILICON AND SILICON CARBIDE

This chapter discusses the depositions and properties of microcrystalline silicon and silicon carbide (μc-Si and μc-SiC) films. These are polycrystalline films with grain size ranging from about 5 nm to several micrometers. Microcrystalline films also include amorphous material in which the volume fraction of crystalline inclusions is greater than a threshold value at which the onset of substantial changes in certain key parameters such as electrical conductivity, bandgap, and absorption constant occurs. Compared to a-Si:H, μc-Si has low optical absorption, high electrical conductivity and mobility, and high doping efficiency. Microcrystallinity has been shown to improve the photoelectric properties of doped a-Si alloys such as a-SiGe:H and a-SiC:H [Johnson et al. 1988a,b]. Hattori et al. [1987a] reported for doped μc-SiC with 16% carbon content a bandgap of 2.25 eV, and a dark conductivity of 10 S/cm. In contrast, the dark conductivity for a-SiC:H:B of 2.0 eV bandgap is in the order of 10^{-8} S/cm. The hydrogen content of μc-Si depends on the deposition condition. Johnson et al. [1988] showed that a large fraction (30 to 40%) of the hydrogen content of μc-Si is in the form of H_2.

Most μc-Si films with crystalline grain sizes between 10 and 30 nm are diphasic material with mixtures of amorphous and microcrystalline phases (3-phase if we also include the grain boundaries). The mixed phase with small grain sizes makes it difficult to model electrical and optical properties. Charge carrier transport may be limited by thermionic emission over potential barriers between different materials and grains. The exact hydrogen distribution in the material, which is sometimes called μc-Si:H, is also difficult to determine. However, in general, the *optical* properties of a microcrystalline film are mostly determined by those of the amorphous phase, while the *electrical* properties are mostly determined by those of the crystalline phase [Hattori et al. 1987b]. Tsu et al. [1982] determined the critical microcrystalline volume fraction for a phosphorous-doped glow discharge Si:H:F alloy to be 0.18. This value coincides with the theoretical percolation limit and serves to explain the conduction process in this two-phase material. Matsuo [1988] suggests a 2-phase model for the microcrystalline and amorphous matrix that at high hydrogen dilution (100:1) during deposition consists of a grain-like zone of microcrystalline silicon (average grain size 5 nm) and amorphous silicon embedded in a more disordered grain-boundary-like zone containing $(SiH_2)_n$ chains and SiH_2 clusters.

The microcrystallinity (the volume fraction of crystals) can be determined by Raman spectroscopy, which shows peaks at 520 cm^{-1} for μc-Si and 740 cm^{-1} [Hattori et al. 1988b] or ~800 cm^{-1} [Chayahara et al. 1986, Feng, Z. C., et al. 1988] for μc-SiC. It is unclear which of these two wavenumbers is a proper indicator of microcrystalline SiC. The bandgap determined by Tauc plots for microcrystalline films can be meaningless because of the mixed phases. The initial nucleation and film growth influence strongly the electrical properties of Si and SiC thin films (< 50 nm thick) [Carlson et al. 1988]. The properties of thick (> 50 nm) microcrystalline p- and n-layer films are independent of deposition method and whether the feed-gas is fluorinated. Other methods of determining crystallinity include x-ray diffraction, Raman scattering, high-resolution microscopy, mass density, conductivity, and Hall mobility measurements.

Microcrystalline Si and SiC is used in solar cells as window layers and as tunnel junctions. Microcrystalline SiC is also used in light emitting diodes, field-effect transistors, bipolar transistors, and other thin-film solid state devices that have to withstand high temperatures, high frequencies, and high power loads [Ganguly et al. 1991]. In the quest for higher sunlight conversion efficiencies for a-Si:H photovoltaic solar cells, a number of approaches appear promising: 1) the use of multiple junction devices, 2) the use of highly conductive and transparent p- and n-layers, and 3) the reduction of interface recombination losses. Microcrystalline Si with weaker absorption and higher conduction are useful in all these approaches.

Microcrystalline doped layers have much higher doping efficiency than amorphous doped layers. Their use in single- or multi-junction p-i-n a-Si:H devices will increase the magnitude of the built-in electric field across the i-layer, thereby improving both the fill factor and the V_{oc}. In addition, the improved conductivity of the p- and n-layers will decrease the series resistance, further improving the fill factor and the efficiency of the solar cell. V_{oc} in the range of 0.95 to 0.99 V have been obtained using high conductivity (1 to 20 S/cm), low activation energy (0.02 to 0.05 eV), microcrystalline p^+-layers of Si:H:F [Guha et al. 1986a,b]. Microcrystalline n-layers with σ_d in the range of 1 to 80 S/cm and E_a of 0.01 to 0.05 eV can improve the red response significantly. Tunnel junctions formed at the n_1/p_2 interface in $p_1i_1n_1/p_2i_2n_2$ multijunction cells where the n- and p-layers are microcrystalline, have low series resistance and non-rectifying characteristics [Sasaki 1986].

The V_{oc} in typical p-i-n photovoltaic devices is dominated by recombination of carriers at the p/i interface [Hegedus 1988b, Yoshida et al. 1988a]. Recombination of light-generated carriers at the p/i interface can be reduced by adding a compositionally graded buffer layer between the p- and i-layers [Catalano et al. 1987b, Hegedus 1988b, Yoshida et al. 1988a]. The increase of the bandgap in the buffer layer results in the reduction of the prefactor in the diode equation and thus results in an increase in the V_{oc} [Yoshida et al. 1988a]. Compositionally graded SiC (either amorphous or microcrystalline) is a good candidate material for such a buffer layer. Fill factors up to 0.77 have been measured for devices with graded buffer layers at the p/i interface [Arya 1988].

As we have shown, µc-Si, a-SiC and µc-SiC films have received considerable attention for use as wide bandgap i-layers, doped layers, and interface buffer layers in p-i-n amorphous silicon devices. The applications of these materials and their advantages over amorphous silicon are summarized in **Table 15-1**.

15.1 MICROCRYSTALLINE FILM DEPOSITION

Most methods of µc-Si and µc-SiC film depositions are similar to a-Si:H deposition methods, except using higher substrate temperature and/or enhanced etching. Higher deposition temperature is often not a preferred option due to increased impurity contaminations. Increased etching, which removes weak or strained Si-Si bonds, during deposition by using higher power density (typically 200 to 300 mW/cm^2 for RF glow discharge deposition) and/or high hydrogen dilution (>95%) has similar effects as increasing the substrate temperature but without the disadvantage of increased impurity contaminations. **Fig. 15.1-1** shows the transition from a-Si:H to µc-Si RF glow discharge deposition, as observed by Raman backscattering (of 514.5 nm-wavelength excitation), when the hydrogen dilution ratio is increased for a silane and phosphine (1%) gas mixture with a total flow rate of 400 sccm [Tsai 1988a]. Other deposition parameters are 200°C substrate temperature, 10 W RF power, and 0.4 torr gas pressure. In this case the maximum degree of crystallinity occurs at about 98% hydrogen dilution. A disadvantage of the additional etching during deposition is a reduced deposition rate; for example, Nguyen et al. [1992] reported a 1.6 nm/min deposition rate for a glow discharge µc-Si deposition using a silane and hydrogen gas mixture with a flow ratio of 1:80.

Table 15-1 Applications for Amorphous and Microcrystalline Si and SiC Layers

Material	Application	Advantage	Representative References
µc-Si	p-layer	better conductivity	Guha 1986
	n-layer	better conductivity	Carlson 1988
a-SiC	i-layer	wider bandgap	Catalano 1989
	p^+-layer	better conductivity & blue response	Kuwano 1988
	buffer layer	higher V_{oc} & FF, and better blue response	Carlson 1988
µc-SiC	p^+-layer	better conductivity	Kuwano 1988
	n-layer	higher voltage, better red response	Carlson 1988
	buffer layer	higher V_{oc} and FF	Catalano 1987b

Several methods have been used to fabricate microcrystalline silicon alloy layers, such as electron-cyclotron resonance plasma-enhanced CVD [Hattori et al. 1987a], photo-CVD [Konagai et al. 1985], plasma magnetron CVD [Kuwanu & Tsuda 1988], plasma-enhanced CVD [Guha & Ovshinsky 1988, Dusane 1992], remote plasma-enhanced CVD [Lucovsky et al. 1992], hydrogen-radical-enhanced CVD [Shimizu et al. 1990], spontaneous CVD [Komiya et al. 1990], and reactive sputtering [Tonouchi et al. 1990, Feng et al. 1992]. Hydrogen dilution during glow discharge deposition from silane [Rajeswaran et al. 1983], disilane, or

SiF_4 can induce a change in the deposited film from a purely amorphous phase to a mixed phase of amorphous and crystalline materials. Microcrystalline Si and SiC are typically grown with high hydrogen dilution (hydrogen to feed gas ratio of 100 to 10,000) and depleted feed gas mixtures [Hollingsworth et al. 1987b], at high power densities, or by a combination of both. Microcrystallinity is facilitated by low pressure and high substrate temperature. A crystalline phase can be detected (by Raman spectroscopy) at dilution ratios (H_2/SiH_4) of 20:1 or greater at normal RF power levels [Johnson et al. 1988a]. The hydrogen

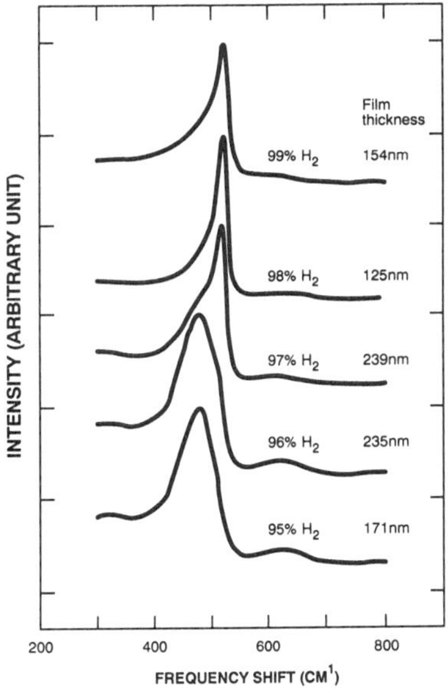

Fig. 15.1-1 Raman spectra as a function of H_2 concentration in 1% PH_3/SiH_4 showing the transition from a-Si:H to μc-Si [C.C. Tsai et al. 1988a]

concentration in the films increases with dilution ratios of 10:1 to 100:1 (Figure **15.1-2**) [Johnson et al. 1988a,b]. As the hydrogen dilution is increased, the SiH bonding configuration gradually changes into dihydride (SiH_2) bonding, the ESR spin density increases (by about one order of magnitude for a change of 10:1 to 100:1 dilution), and the films become microcrystalline [Matsuo et al. 1988]. At high power densities crystallinity is seen even for films deposited from undiluted silane [Matsuda 1983]. Matsuda [1983] also showed that the grain size of a μc-Si can be controlled by changing the substrate bias and thus the ion bombardment of the growing film in a triode glow discharge system.

Adding certain impurities or alloying elements, such as N, C, O, and B, can make the fabrication of microcrystalline silicon alloys more difficult [Hiraki et al. 1983, Guha & Ovshinsky 1988]. For instance, the microcrystallinity is suppressed for greater than 6% carbon content in glow discharge deposition [Carlson et al. 1988]. On the other hand, phosphorous doping enhances microcrystalline formation [Kaya et al. 1984, Matsuda et al. 1980a]. At high growth rates, amorphous silicon, rather than microcrystalline silicon, is deposited. The deposition and etching reactions compete with each other and a careful balance between growth and etching is required to achieve microcrystalline films [Guha & Ovshinsky 1988].

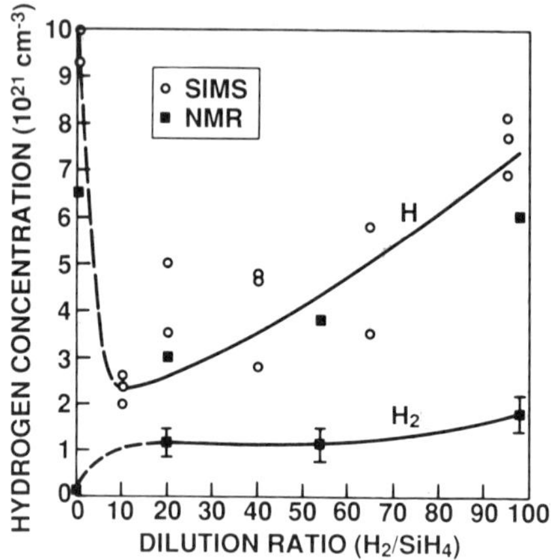

Figure 15.1-2 Hydrogen content and molecular hydrogen vs. hydrogen dilution of feed-gas for μc-Si:H [Johnson et al. 1988b]

Generally, it requires a certain minimum film thickness (>30 nm) to achieve microcrystallinity. It is exceedingly difficult to achieve 10 nm or thinner microcrystalline silicon films. Transmittance electron microscopy, infrared spectroscopy, and Raman analyses indicate that even under the proper conditions of hydrogen dilution and high power density, supposedly microcrystalline films of Si or SiC have a very small microcrystalline phase fraction at a small film thickness, and that the microcrystalline fraction only gradually increases as the film thickness is increased [Hiraki 1983, Imamura et al. 1984]. Thus, thin layers often remain amorphous. Yang, L., et al., [1992] have been able to deposit 10 nm-thick microcrystalline n-layers on 10 nm-thick a-Si:H after appropriate surface treatment. For doped *microcrystalline* layers, the dark conductivity drops rapidly as the thickness decreases, for instance, from 10 S/cm at 200-nm thickness to 10^{-9} S/cm at 20 nm thickness for μc-Si:H:B or from 1 x 10^{-1} S/cm at 200 nm to 10^{-10} S/cm at 50 nm for μc-SiC:H:B. This is illustrated in Figure **15.1-3.** The conductivity for a-Si:H:B also decreases rapidly below 30 nm, but less rapidly than for microcrystalline films: from 10^{-4} S/cm at 30 nm to 10^{-7} S/cm at 10 nm [Carlson et al. 1988]. The variations in conductivities for the three microcrystalline films in Figure 15.1-3 at the 10 to 50 nm thickness range may have resulted from differences in the deposition parameters for these films. Fluorinated gases may help nucleation for thin (less than 20 nm) microcrystalline films. It is possible that fluorinated microcrystalline p-layers retain their crystalline properties in a preferential manner to non-fluorinated films [Guha et al. 1986b]. Faraji et al. [1992] deposited μc-Si:O:H by RF glow discharge with a mixture of silane, hydrogen, and oxygen. The maximum Hall mobility observed for this material is 35 $cm^2V^{-1}s^{-1}$.

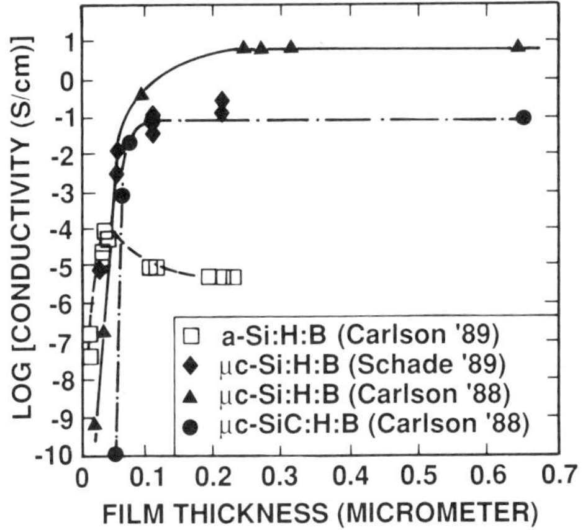

Figure 15.1-3 Dark Conductivity vs. Film Thickness for μc-Si:H:B, μc-SiC:H:B, and a-Si:H:B

Microcrystalline growth depends also on the type of substrate; for example, it is easier to grow thin microcrystalline Si or SiC (doped or undoped) on a-Si:H by RF glow discharge than on tin oxide; e.g., 20 nm thin μc-SiC:H films were achieved on a-Si:H but not on tin oxide [Carlson et al. 1988]. When p-i-n devices are deposited on glass (as they frequently are), the p-layer is deposited onto the transparent conductive layer, such as tin oxide, that is coated onto the glass. To overcome the problem of thin silicon carbon layer remaining amorphous when deposited on tin oxide, a thin amorphous silicon layer can be deposited on the tin oxide, before laying down the microcrystalline SiC layer. Another problem with forming μc-Si:H:B on tin-oxide by RF glow discharge is the reduction of the SnO_2 surface by the high hydrogen content in the plasma, perhaps forming elemental Sn (which increases the absorption of light) and subsequently forming SiSn alloys. ZnO appears to be more resistant to the hydrogen plasma than SnO_2 [Hollingsworth et al. 1987b], which is one of the reasons why ZnO is being investigated as a conductive transparent coating to either protect the underlaying tin oxide or to totally replace tin oxide. For devices deposited in the n-i-p sequence on an opaque substrate such as stainless steel, the n-layer is deposited first on the substrate. Since it is easier to achieve microcrystallinity in n-type material than in p-type, this n-type material can be microcrystalline, even in thin layers. The p-layer is then deposited last, often at a much lower temperature. As it is deposited on a-Si instead of on, say, tin oxide, it can also more easily achieve microcrystallinity in thin layers, thus giving good conductivity and low absorption of blue light.

15.2 DOPED MICROCRYSTALLINE FILMS

Doped films of Si or SiC are used as p- or n-layers in p-i-n devices. The doped layers should be highly transparent to minimize optical absorption and have high conductivity to reduce parasitic series resistance of the devices. As the thickness of p-type window layers is increased, the optical absorption also increases, especially at short wavelengths. To ensure high transparency, the doped layers must be thin, normally no more than 10 to 20 nm-thick [Guha et al. 1986b]. On the other hand, as we have seen, the conductivity decreases rapidly as the thickness decreases (Figure 15.1-2). Thus, for maximum solar cell conversion efficiency, a compromise must be made between high conductivity and low absorption in the p-layer.

One way to reduce the absorption losses in doped layers is to increase the doping efficiency. For example, in p-layers (amorphous and microcrystalline, Si and SiC) doped with B_2H_6, only 1% of B-atoms contribute to conduction (1% doping efficiency), whereas 99% of B-atoms contribute to absorption. Increasing the doping efficiency reduces the total number of boron atoms in the film for a given electrical conductivity and reduces the optical absorption. Better doping efficiency than with B_2H_6 is obtained by using BF_3 [Guha and Kulman 1986a] or $B(CH_3)_3$ [Kuwano & Tsuda 1988]. $B(CH_3)_3$ also results in a ~0.1 eV wider bandgap than with B_2H_6 [Carlson et al. 1988]. N-layers are normally doped with PH_3, but PF_5 has also been used and gives a slightly higher doping efficiency [Matsuda et al. 1980a].

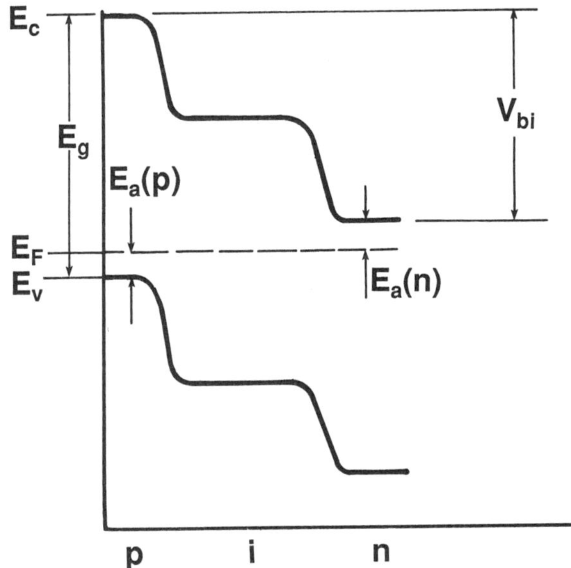

Figure 15.2-1 Schematic bandgap picture for p-i-n device showing relation between E_g, E_a, and V_{bi}

The V_{oc} of solar cells is limited by recombination of carriers at interfaces and in the bulk material. The reduction of carrier recombination at the interfaces will be discussed in Section 15.3. The recombination in the bulk materials can be reduced by using thinner intrinsic layers and by increasing the built-in field of the p-i-n junction. The built-in voltage (diffusion potential) is equal to the optical bandgap minus the activation energies for the n- and p-layers ($qV_{bi} = E_g - E_a(p) - E_a(n)$) where q = is the electron charge [Carlson 1984, Hegedus 1988]. The activation energy (E_a) is E_c-E_f for electron conduction and E_f-E_v for hole conduction. Figure **15.2-1** illustrates schematically the relationship between these quantities. Thus, a high E_g together with low p- and n-layer E_a will result in a high built-in voltage, and that in turn can result in a high V_{oc}. The V_{oc} follows the built-in voltage until bulk recombination in the intrinsic layer dominates [Hollingsworth et al. 1987b]. As the E_a increases, the built-in potential decreases.

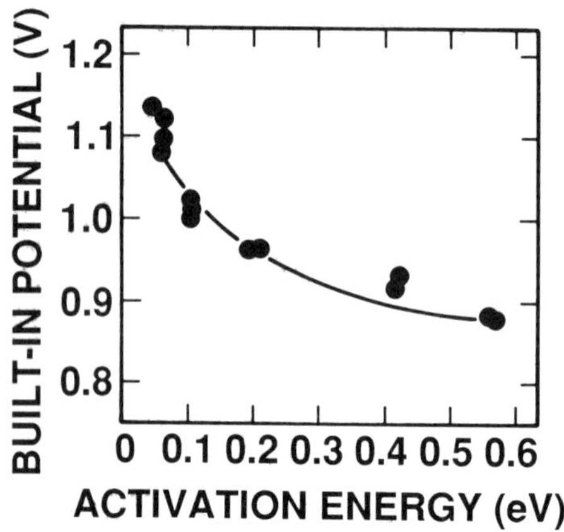

Figure 15.2-2 Built-in Potential vs. Activation Energy for p-Type a-SiC:H:B [Hattori et al. 1987b]

Figure **15.2-2** shows the measured built-in voltage for a p(a-SiC:H:B)/i(a-Si:H) heterojunction cell as a function of the activation energy of the p-layer; the built-in potential decreases with increasing E_a; from 1.1 V at 0.05 eV activation energy to 0.88 V at 0.6 eV [Hattori et al. 1987b]. The relationship determined experimentally [Tawada 1982, Hamakawa 1989] and by calculation [Guha et al. 1986] between the V_{oc} and built-in diffusion potential for various heterojunction solar cell is shown in Figure **15.2-3**. The computed relationship is for a 1.72 eV E_g a-Si alloy solar cell for varying p^+-layer activation energies and zero activation energy for the n^+-layer.

The reason that high conductivity p- and n-layers are sought is that high conductivity layers have low activation energy and, consequently result in a greater built-in potential (for a given bandgap) and a smaller contact resistance [Carlson 1984]. Microcrystalline doped layers have a lower activation energy than amorphous layers, e.g. for p-type Si:H, it is 0.05 eV vs. 0.3 eV and, for n-type Si:H, it is 0.05 eV vs. 0.2 eV [Madan & Shaw 1988]. Therefore microcrystalline doped layers give devices with a higher built-in potential. We will discuss the considerations and data for p- and n-layers in turn.

15.2.1 Microcrystalline P-layers

Higher built-in potential in a p-i-n solar cell can be achieved by using higher optical bandgap materials in the p- and i-layers. The higher built-in potential can in turn result in a higher open-circuit voltage **(Fig. 15.2-3).** The higher optical bandgap for the p-layer of SiC compared to that of Si will increase the built-in potential; thus heterojunction p(SiC)/i(Si)/n(Si) devices have a higher built-in field than homojunction p(Si)/i(Si)/n(Si) devices.

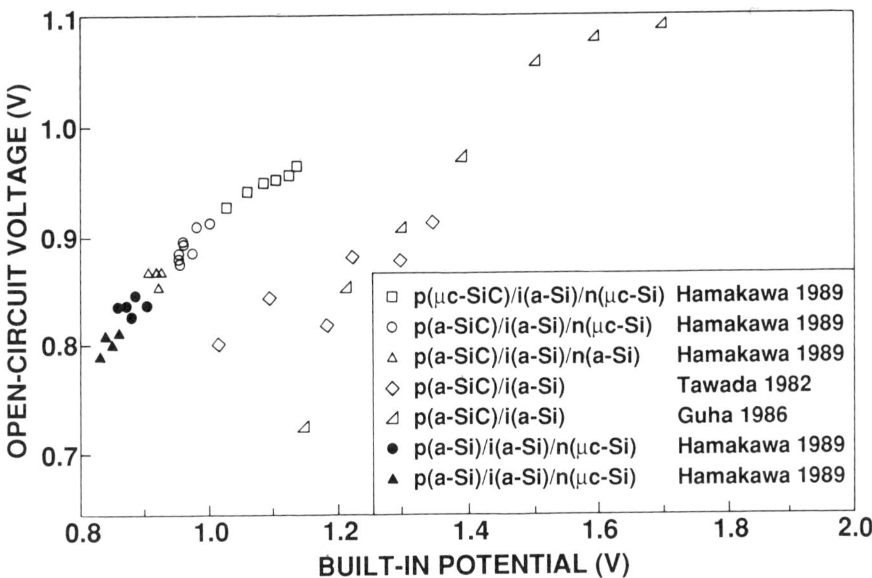

Figure 15.2-3 Open-circuit Voltage dependance on the Built-in Potential

Experimental data of the best open-circuit voltages in high-efficiency single-junction a-Si:H p-i-n devices achieved with microcrystalline Si, microcrystalline SiC, and amorphous SiC p-layers are shown in **Table 15.2-1**. The built-in field is higher for devices with microcrystalline p-layers or a-SiC p-layers than with a-Si.

The highest open-circuit voltage achieved is 0.967 V for a 3.3-mm^2-area solar cell with the structure: glass/SnO_2/ZnO/p a-SiC:H (buffer)/p μc-SiC:H/i a-Si:H/n μc-Si:H/Ag. The thickness of each semiconductor layer was 2 nm, 20 nm, 500 nm, and 30 nm, respectively. The SiC layers were formed by ECR CVD, while the other layers were formed by conventional RF plasma CVD. The other characteristics for the cell were efficiency = 12.0%, J_{sc} = 17.7 mA/cm^2, FF = 0.703. The optical band gap for the μc-SiC:H was 2.25 eV and the dark conductivity >1 S/cm [Hamakawa 1989].

Table 15.2-1 Characteristics for High-Efficiency Single-Junction Cells with p-Layers of various materials (Measured values, unless indicated otherwise). [Luft 1988-1990]

P^+-Layer	Best V_{oc}(a) (V)	E_a (eV)	Efficiency %	Reference
a-Si:H:B	0.76	0.33	-	Hollingsworth 1987b
μc-Si:H:B	1.0 (calc)	-	-	Yamanaka 1988
	0.96	0.04	10.7	Guha 1986
a-SiC:H:B	0.91	-	11.5	Shibata 1988
	0.917	0.40	11.4	Catalano 1989b
	0.895	-	12.0	Shimada 1990
μc-SiC:H:B	0.95 (d)	-	(c)	Guha 1989a
	0.965 (b)	0.05	11.8	Hattori 1987a&c
	0.967	-	12.0	Hamakawa 1989

(a) at room temperature, (b) a-SiC/μc-SiC combination, (c) quantum efficiency at 400 nm = 0.66, (d) μc-Si in an a-SiC matrix.

The structure for the Hattori [1987b] device in the above table was glass/SnO_2/ZnO/ECR,p(a-SiC:H)/ECR,p(μc-SiC:H:B)/GD,i(a-Si:H)/GD,n(μc-Si:H)/Ag. The μc-SiC:H:B had E_a = 0.05 eV, and E_g = 2.25 eV, and >1 S/cm conductivity. The a-SiC:H layer was used for better nucleation of the μc-SiC:H layer and may have totally been etched away in the subsequent microcrystalline layer deposition. The microcrystallinity of the p-layer was confirmed by Raman spectroscopy. The device had no separate buffer layer at the p/i interface. The low E_a of 0.05 eV should have resulted in a built-in voltage of about 1.1 to 1.15 eV per Figure 15.2-2.

The next highest V_{oc} is 0.96 V (J_{sc} = 16 mA/cm^2, FF = 0.70, efficiency = 10.7%) achieved with μc-Si:H:B by Guha [1986]. The V_{oc} of 0.91 V (J_{sc} = 18.2 mA/cm^2, FF = 0.693) for a-SiC:H:B by Shibata, A., et al. [1988] in Table 15.2-1 was achieved with a p-layer structure of 0.1 to 0.5 atomic layers of boron between 1 nm- and 4 nm-thick layers of undoped a-SiC:H; the so called delta-doped p-layer. This structure shows enhanced carrier concentration.

The 0.917 V by Catalano et al. [1989b] is for a-SiC:H:B and was achieved with a p-layer doped with $B(CH_3)_3$ (J_{sc} = 16.9 mA/cm^2, FF = 0.736, efficiency = 11.4%). The p-layer activation energy was 0.40 eV. For a-SiC:H:B doped with diborane, the E_a is typically 0.45 eV [Catalano 1989a]. The electrical conductivity for a-SiC:H:B films doped with $B(CH_3)_3$ is about one order of magnitude higher than for such films doped with diborane, giving higher E_g (+0.1 eV), V_{oc} (+0.07 V), and FF (+0.05) [Catalano et al. 1989b]

15.2.1.1 *Microcrystalline Si:H:B P-Layers*

According to theoretical investigation by Yamanaka et al. [1988], devices with μc-Si:H:B p-layer can reach higher efficiency than devices with a-SiC:H:B p-layer (by about 1 or 1.25 percentage points), mainly because of significantly higher potential V_{oc} (1.0 V). The potential increase in V_{oc} is caused by the potential increase in the built-in voltage due to the higher conductivity (0.4-20 S/cm) of microcrystalline silicon in the p-layer as compared to amorphous silicon p-layer (10^{-3} S/cm) or as compared to a-SiC:H:B (10^{-5} to 10^{-9} S/cm). In fact, an open-circuit voltage of 0.96 V has been achieved with fluorinated p^+-type microcrystalline Si:H:F:B; (J_{sc} = 16 mA/cm^2, FF = 0.70, Efficiency = 10.7%) [Guha et al. 1986b]

The dark conductivities of *thick* (>1 μm) μc-Si p-layers with 1.9 to 2.5 eV bandgap range from 0.4 to 20 S/cm for films deposited by RF (13.6 MHz) glow discharge or photo-CVD. Conductivities as high as 1000 S/cm have been achieved for μc-Si:H:B films with the solid state crystallization method [Matsuyama et al. 1990]. In contrast, heavily boron-doped a-Si:H:B has an optical bandgap of 1.5 to 1.6 eV and a conductivity of about 10^{-3} S/cm [Guha et al. 1989b]. The activation energy for μc-Si:H:F:B is 0.02 to 0.05 eV as compared to 0.3 to 0.4 eV for a-Si:H:B [Guha et al. 1986b]. The dark conductivity of μc-Si:H:B is a function of the hydrogen dilution in the gas phase (it increases by 2 orders of magnitude with an increase in dilution ratio from 150 to 300). The optical bandgap increases from 1.95 to 2.2 eV over the same dilution range [Nishida et al. 1985]. Grain sizes of 8 to 12 nm and crystallite fractions over 80% have been observed [Guha et al. 1986b]. The film properties are generally independent of deposition method or the use of fluorinated gases. However, the dark conductivity drops rapidly as the thickness of the layer is reduced as discussed in Section 2.3.2 and 15.1.

Amorphous silicon solar cells with a μc-Si:H:B p-layer exhibit a lower Staebler-Wronski degradation as compared to cells with a-SiC:H p-layer [Carlson et al. 1988].

15.2.1.2 *Microcrystalline SiC:H:B(:F) P-Layers*

Microcrystalline SiC:H:B (10 to 20 nm thick) is used for p-layers of a-Si:H pin solar cells because of its better conductivity compared to a-SiC:H:B, increasing V_{oc} by 10% to 0.965 V [Hattori et al. 1987a]. E_a as low as 0.05 eV were measured for a-SiC:H:B for E_g of 2.0 to 2.9 eV [Hattori et al. 1987a]. For up to 6 at.% carbon, the conductivity and activation energy are improved compared to a-SiC:H:B [Carlson et al. 1988]. In many cases, what is believed to be μc-SiC is actually μc-Si crystals in an a-SiC matrix; the material having a Raman peak at 520 cm^{-1} but not at 740 cm^{-1} [Catalano et al. 1989].

Microcrystalline p^+-type SiC:H:B can be made using methane and silane mixtures, but more carbon can be incorporated into the microcrystalline film before the microcrystallinity is suppressed when using disilylmethane and silane. For example, relatively high-conductivity, partial microcrystalline films containing up to 6 at.% carbon can be made with disilylmethane, but such microcrystalline films can only be made with 3 to 4 at.% carbon when methane (CH_4) is used [Goldstein et al. 1988, Catalano et al. 1989]. A new feedstock for μc-SiC:H:B p-layers is SiH_2F_2. Phosphorous counter-doping to improve micro-crystallinity of SiC:H:B p-layers does not give improved device performance [Carlson et al. 1988].

The large improvement in dark conductivity of μc-SiC:H:B over a-SiC:H:B can be seen in Figure **15.2-4**. The σ_d for μc-SiC:H:B p-layers decreases with increasing carbon content (and increasing E_g) but less rapidly than for a-SiC:H:B. Similar dark conductivities have been obtained for films deposited by the controlled plasma magnetron (CPM) method [Kuwano & Tsuda 1988] and for films by ECR-CVD [Hattori et al. 1988b]. The microcrystallinity is suppressed for greater than 6 at.% carbon content for glow discharge deposition [Carlson et al. 1988]. The dark conductivity for thin layers can be improved by several orders of

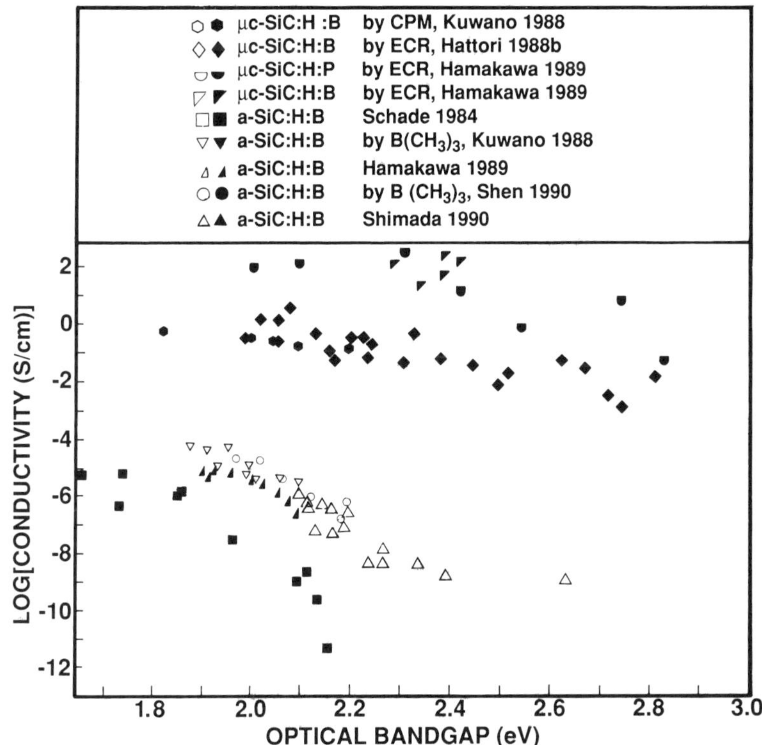

Figure 15.2-4. Photo- and Dark Conductivities vs. Optical Bandgap for p- and n-Type Microcrystalline and Amorphous SiC:H. Open symbols for photoconductivity, filled symbols for dark conductivity.

magnitude by modifying the deposition conditions [Carlson et al. 1988]. Guha & Ovshinsky [1988] have achieved a σ_d of 0.5 S/cm, E_a of 0.05 eV, E_g of 2.1 eV, an absorption coefficient at 500 nm of 3 x 10^4 cm^{-1}, with 50 nm thick SiC layers, having 2 at.% F, resulting in 6 to 10 nm size microcrystals of 60% volume fraction. Thick μc-Si:H:B and μc-SiC:H:B p-layers of 8 to 12 nm grain size with σ_d in the range of 0.1 to 20 S/cm and with E_a of 0.02 to 0.05 eV have been deposited by DC proximity glow discharge, RF glow discharge, photo-CVD, and ECR-CVD.

Only two research groups [Energy Conversion Devices, which uses stainless steel substrates, and Hamakawa's group in Osaka University, which uses ECR-CVD] have demonstrated improvements in cell efficiency using microcrystalline p-layers. With a μc-SiC:H:B p-layer, Hattori et al. of the Osaka University group [1987b] achieved a conversion efficiency of 12% (V_{oc}=0.967 V, J_{sc}=17.7 mA/cm^2, FF=0.703) for a 0.033 cm^2 area cell.

The problems of depositing μc-SiC:H p-layers on tin oxide-coated glass by conventional glow discharge may be caused by the difficulty to achieve microcrystallinity for thin p-layers. Catalano et al. [1989] believes that the combination of necessary buffer layer and thicker-than-usual microcrystalline p$^+$-layer may significantly reduce the short-circuit current density despite the wide bandgap.

15.2.2 Microcrystalline N-Layers

15.2.2.1 *Microcrystalline Si:H:P N-Layers*

Microcrystalline Si:H:P n$^+$-layers in a-Si:H cells improve the red response at 700 nm from 0.28 to 0.40 because of lesser absorption in the n-layer [Carlson et al. 1988].

Conductivities of **thick** (>1 μm) microcrystalline Si:H:P layers range from 1 to 80 S/cm with E_a of 0.01 to 0.05 eV. The σ_d for microcrystalline n-type Si:H:P increases with increasing hydrogen dilution [Nishida et al. 1985]. Crystallite sizes are about 20 nm [Catalano et al. 1989b]. The film properties are generally independent of the deposition method whether the film is deposited by RF (13.6 and 40 MHz) glow discharge, RF glow discharge with a magnetic field, or by photo-CVD. The use of fluorinated gases is not essential for good properties, since results achieved with such gases have been duplicated with hydrogen-diluted gases at higher discharge powers. The electrical properties of μc-Si:H:P are determined mainly by the phosphorous content, not by the microcrystalline fraction; the crystallite grain size plays a minor role [Spear et al. 1981, Nakatani et al. 1983, Tsai et al. 1989].

The use of μc-Si:H:P n-layers has been reported to improve the thermal stability in some cells, but these cells had 200 nm n-layer thickness instead of 20 nm as used for a-Si:H:P n-layers [Carlson et al. 1988b].

15.2.2.2 *Microcrystalline SiC:H:P N-Layers*

The degree of microcrystallinity of SiC-layers is considerably better for n^+-layers than for p^+-layers [Catalano et al. 1989]. Solar cells with μc-SiC:H:P n-layer (just as cells with μc-Si:H:P n-layer) have better red quantum efficiency than cells with a-SiC:H:P n-layer and have 50 to 100 mV higher V_{oc} [Carlson et al. 1988]. Similarly as for p-layers, there is some question of whether the microcrystals are Si or SiC in an a-SiC matrix. Some researchers have only confirmed the existence of μc-Si [Catalano et al. 1989b].

15.3 COMPOSITIONALLY GRADED BUFFER LAYERS

The V_{oc} in typical a-Si:H solar cells is dominated by recombination at the p/i interface [Hegedus 1988b]. The effective interface recombination velocity is 10^3 to 10^4 cm/s [Rothwarf 1988]. The p/i interface recombination of carriers is much higher for cells with an a-Si:H:B p^+ layer on intrinsic a-Si:H than with an a-SiC:H:B p^+ layer [Hollingsworth et al. 1987b]. A 10 nm- to 15 nm-thick compositionally graded SiC:H buffer layer between the p-layer and the i-layer reduces carrier recombination and results in increased V_{oc}, I_{sc} (by enhanced blue

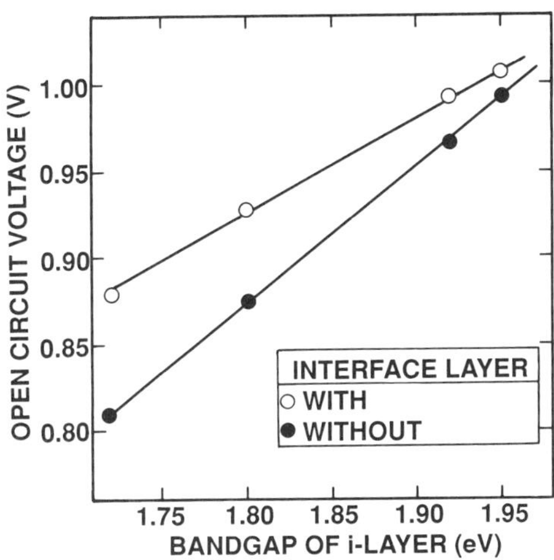

Figure 15.3-1. Open-circuit voltage as a function of the intrinsic layer bandgap with and without an a-SiC interface layer between the p- and i-layers for p(a-SiC)/i(a-SiC)/n(a-Si) device [Yoshida et al. 1988b].

response), and FF (but does not change the built-in voltage) [Hegedus 1988b]. The buffer layer can be either amorphous or microcrystalline. Figure **15.3-1** shows the relation between the optical bandgap of the a-SiC:H i-layer for a p(a-SiC)/i(a-SiC)/n(a-Si) cell and the V_{oc} with and without a graded a-SiC:H interface layer. The improvement in V_{oc} in this case ranges from 0.07 V to 0.02 V depending on the i-layer bandgap. Other researchers have observed even greater improvements in the V_{oc}: 0.09 V [Kim et al. 1987b] and 0.10 V [Hegedus et al. 1987]. 5 to 7% improvement in FF [Carlson et al. 1988, Yamamoto et al. 1987b] to values as high as 0.77 [Arya 1986], and 2 mA/cm^2 in short-circuit current density [Kim et al. 1987b] have been recorded. Increasing the bandgap of the intrinsic layer and the interface buffer layer increases the V_{oc}, but decreases the FF because of deterioration in the film quality [Yoshida et al. 1988b]. Hegedus et al. [1987] proposed that the buffer layer reduces interface recombination by maintaining a spacial separation of photogenerated electrons in the i-layer from holes in the doped p-layer and that the improvements observed in the V_{oc} are due to the reduction in interface recombination losses. He found that the thickness of the graded layer has the most dominant effect. However, no improvements were observed unless the carbon and oxygen impurities in the p- and i-layers were below the 10^{19} cm^{-3} level.

15.4 POST-DEPOSITION CRYSTALLIZATION METHODS

Typical glow discharge-deposited μc-Si films have grain sizes on the order of 10 nm. Such grain sizes are too small for μc-Si to be useful as active semiconductor materials in devices like solar cells and thin-film transistors. Crystallization methods, such as laser annealing, rapid thermal annealing (RTA), and furnace annealing (solid-phase crystallization, SPC) have been used to enlarge grain sizes to the micrometer range at low substrate temperatures. The initial film for crystallization is usually a-Si:H deposited at low temperature using glow discharge or low-pressure CVD from silane or disilane. A comparison of these crystallization methods for TFT applications is shown in **Table 15.4-1** [Little et al. 1991]. Because of recent advances discussed later in this section, we changed the mobility of SPC films to "excellent" from the "good" originally listed by Little et al. Other annealing methods, such as incoherent-light sources, electron beams, strip heaters, and RF induction, have also been studied but with poor film quality or low throughput [Givargizov 1990, Hayama and Saito 1992]. The onset of crystallization during the transition from a-Si:H to μc-Si may be detected *in situ* using optical methods such as UV reflectivity, IR absorption, Raman scattering, and x-ray diffraction.

Solid-phase crystallization (SPC) of a-Si:H films can be achieved by annealing the film at temperatures above 500°C. The annealing temperature is usually below 600°C to avoid thermal damage to the glass substrate (e.g., Corning 7059) that the a-Si:H is deposited on. It is possible to increase this upper temperature limit to 800°C by using special glass substrates, like Corning Code 1729 glass [Troxell 1987]. The grain size of a SPC μc-Si film depends on the annealing temperature and duration. It also depends on the hydrogen content and microstructure of the initial a-Si:H film. Grain size increases as film thickness increases [Kuriyama et al. 1991 and Little et al. 1991]. SPC generally produces the largest grain sizes. However, the throughput is low with treatment time as long as hundreds of hours.

Table 15.4-1 Comparison of microcrystalline silicon fabrication methods [Little et al. 1991]

μc-Si Material	Mobility	Large area capability	Uniformity	Throughput
As-deposited	fair	good	good	good
Furnace anneal (SPC)	excellent	good	good	fair
Rapid thermal anneal (RTA)	good	poor	poor	fair
Laser anneal	excellent	poor	poor	fair

SPC μc-Si films with large grain sizes have shown very high mobility. Kuwano et al. [1992] achieved the mobility of 623 $cm^{-2}V^{-1}s^{-1}$ in SPC (550°C, 8 hours) μc-Si with a n-type carrier concentration of 3.0 x 10^{15} cm^{-3} on textured quartz substrates. This mobility value is about 50% of that of single crystalline Si at the same carrier concentration. The grain size of these films ranges from 4 to 6 μm. To increase the grain size of μc-Si, the number of nucleation sites during SPC needs to be reduced. Kuwano et al. achieved this by substrate texturing and by doping a surface layer of the initial a-Si with phosphorus (SPC of Si occurs at a lower temperature with phosphorus doping than without doping.). Another method is to reduce the nucleation rate by increasing the structural disorder of the as-deposited film by depositing the film at a low substrate temperature [Nakazawa and Tanaka 1990]. However, Tsu et al. [1986] reported that they obtained maximum grain size when the substrate temperature is in the range where the deposited film changes from the amorphous to the crystalline state. They believe that at low deposition temperatures, the grain size of the SPC μc-Si or Ge is limited by the trapping of residual gases in the voids, whereas at high deposition temperatures, the nucleation limits the grain growth.

Modern rapid thermal annealing (RTA) equipment can treat sample areas up to 30 cm x 40 cm at power densities of ~3000 W/cm^2 [Fair 1992]. RTA can achieve high crystalline fractions in thermal treatments of a second or less. The peak temperature required for complete crystallization increases as the crystallization time for RTA is decreased—about 900°C for 1 second and 1150°C for 0.01 seconds. Grain size is generally maximized by reducing the annealing temperature and increasing the crystallization time—up to about 0.5 μm for 10 seconds [Fair 1992]. A dark conductivity of 100 S/cm was obtained by RTA for 20 nm thick p-type and n-type material as compared to 10^{-8} to 10^{-9} S/cm for plasma deposited microcrystalline doped materials of the same thickness [Catalano et el. 1990].

In laser annealing, a short-wavelength laser, such as an eximer laser with a 193-nm wavelength pulsed beam, is used to melt the silicon film on top of a transparent substrate. The temperature of the substrate is usually kept below 400°C during the Si film melting and crystallization process. The Si film is melted for fractions of a milliseconds before solidification. The grain size of the laser crystallized film increases with increasing substrate temperature and increasing laser energy density [Bachrach et al. 1990, Kuriyama et al. 1991]. Initial a-Si:H films with low hydrogen content (3 at.%) was found to have large grain size from eximer laser annealing because the reduced hydrogen evolution allows the use of higher laser intensity for crystallization [Shimizu et al. 1991]. Large grain size (~0.3 μm) and high field effect mobilities (230 $cm^2/V{\cdot}s$ for electrons, 140 $cm^2/V{\cdot}s$ for holes) have been obtained for eximer laser annealed μc-Si films [Kuriyama et al. 1991 and Shimizu et al. 1991]. A SiO_2 capping layer on top of the deposited a-Si:H is found to help the molten crystallization process [Hayama and Saito 1992]. Continuous-wave (cw) lasers [Thomas et al. 1981, Ivanda et al. 1991] and broad-beam light sources [Kakkad et al. 1991] have also been used to achieve μc-Si films with grain size on the order of 1 μm.

Even though crystalline Si has an indirect bandgap, it is possible to obtain visible photoluminescent and electroluminescent emissions with peak photon energy higher than 1.6 eV from Si with crystalline sizes of 3 to 30 nm [Canham 1990, Lehmann and Gösele 1992, Nishida, A., et al. 1992]. One explanation of this light emission phenomenon from crystalline Si is quantum size effects, resulting from confinement of electrons in an array of quantum wires or a network of quantum dots. It is widely observed that as the size of the quantum wires or dots is reduced, the degree of confinement increases, causing an increase in the energy of the emitted photoluminescence. Most of the light emitting Si is produced by electrochemical etching of crystalline Si (producing what is known as *porous Si*). The instability of the light emission properties of porous Si can be eliminated by hydrogenation and oxidation techniques [Xiao et al. 1992]. Bustarret et al. [1992] showed that plasma-deposited hydrogenated Si with a crystalline volume fraction of 40% can also yield visible photoluminescent and electroluminescent emissions by applying electrochemical etching and oxidation. This means that it is feasible to make light-emitting devices from plasma-deposited μc-Si films.

16
SAFETY

16.1 HAZARDS

Amorphous silicon production by plasma-assisted chemical vapor deposition involves the use of dangerous gases. The gases can be categorized as 1) toxic (Arsine, Phosphene, SiF_4), 2) flammable, (H_2, CH_4), 3) pyrophoric (Phosphene, SiH_4), or 4) compressed (N_2, Ar). Codes and regulations apply regarding their storage and use.

Other hazards that are regulated are: vacuum equipment (implosion hazard), waste disposal (process effluent, scrubber solution, pump oil), electric shock, fire, and radio-frequency (RF) radiation.

16.2 CODES AND REGULATIONS

In the United States there are congressional acts, federal, state, and municipal codes and regulations regarding the environment, safety, and health. Examples of congressional acts are the Resource Conservation and Recovery Act (RCRA), the Comprehensive Environmental Response Act, the Occupational Health and Safety Act (OSHA), the Superfund Amendment Reauthorization Act - Title III, and the Compensation and Liability Act (CERCLA). The federal government regulates the use of hazardous gases through statutory and regulatory programs administered by the Department of Labor Occupational Health and Safety Administration (DOL OSHA) and by the U.S. Environmental Protection Agency (EPA). The Occupational Health and Safety Administration regulates matters of health and safety in the work place. The Environmental Protection Agency protects the environment and public health by regulating the discharge of materials into the environment. State regulation vary from state to state. Municipal government often adopt regulations such as the Uniform Fire Code, the Uniform Building Code, the National Electric Code, and the National Fire Protection Association (NFPA) standards.

16.3 EQUIPMENT AND FACILITIES

Equipment for hazardous production materials covers items such as gas cabinets, gas distribution systems, distribution systems for pyrophoric liquids and solids, distribution systems for highly toxic liquids and solids, reaction vessels and/or deposition systems, effluent removal, exhaust systems, and storage facilities.

Personal protective equipment standards consider minimum requirements to ensure personal safety of employees who are involved with the handling of hazardous production materials such as for alarms, safety interlocks, toxic gas monitors, hydrogen monitoring, corrosive gas monitoring, fire-rated laboratory doors, fire extinguishers, and emergency response team stations.

16.4 OPERATION

Administrative and engineering controls of hazardous operations are required. Administrative controls include safety analysis review (including an accident risk analysis covering the probability, severity, and risk of accidents), operational readiness review, safe operating procedures, procurement policies for hazardous materials, minimum staffing requirements, training requirements, and a verification program of critical components and systems. Engineering controls are such items as redundant controls, automatic fail-safe devices, emergency system power shut-offs, interlock protection, design of the gas distribution system, reaction vessels or deposition chambers, effluent removal and exhaust systems, and storage facilities.

Because of the extremely hazardous nature of gases such as silane, it is important to eliminate exposure to even small quantities. Cross purge assemblies enable gas cylinders to be connected and disconnected from the system without exposing the rest of the system to air or allowing hazardous gases to be released into the atmosphere. Purging pyrophoric or flammable gas lines and the entire pumping system (including scrubber) with an inert gas such as nitrogen also eliminates the high risk of an explosion that results when pyrophoric gases come in contact with air or oxygen. An example of a safe a-Si:H alloy deposition system is shown in **Figure 16-1**. Further information about safety issues in a-Si:H depositions may be found in the proceedings of the 1988 Photovoltaic Safety Conference [Luft 1988].

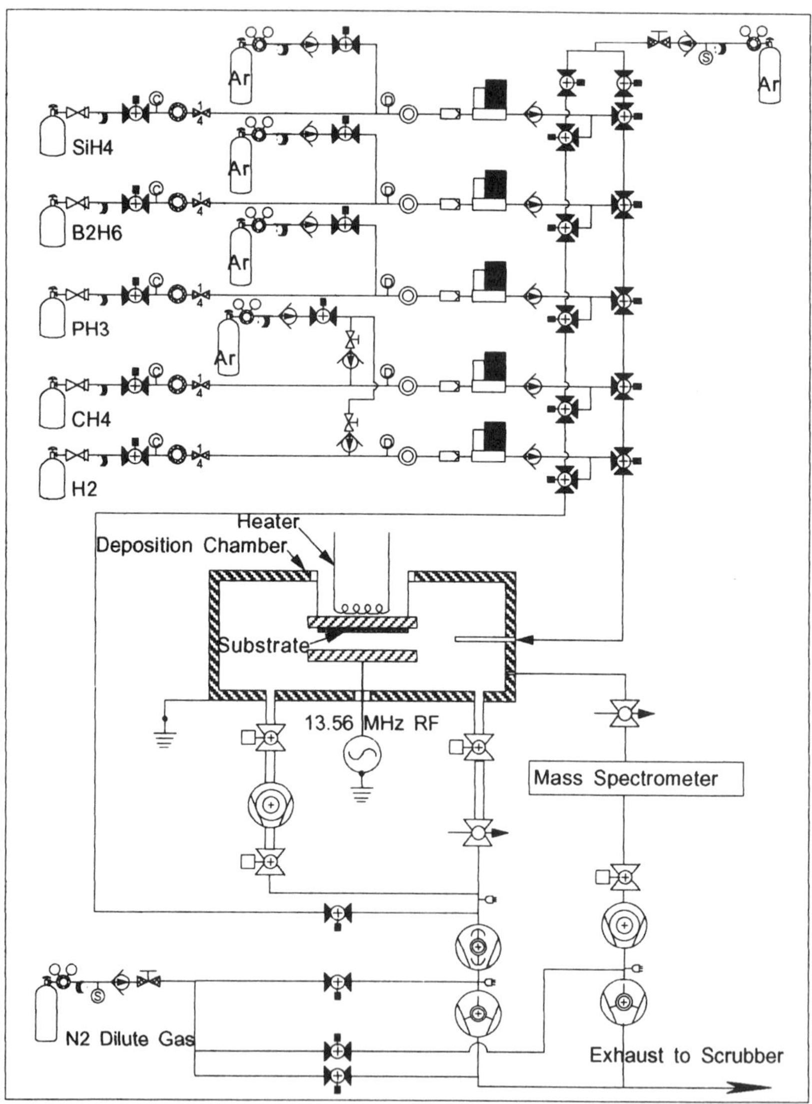

Figure 16-1 Schematic of safe gas lines for glow discharge reactor

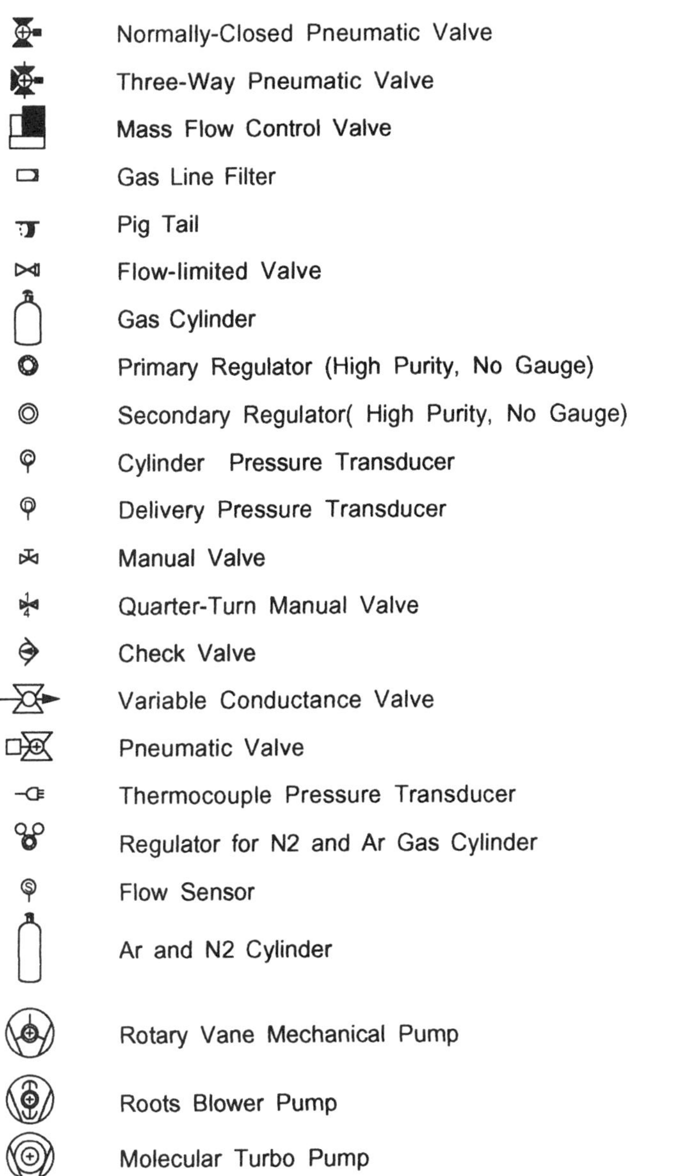

Figure 16-1 Schematic of safe gas lines for glow discharge system (continued)

REFERENCES

Abdel-Rahman, M., H. Madkour, H.H. Hassan, and S. El-Desouki, "Photoelectronic Properties of Hydrogenated Amorphous Silicon Films Deposited by R.F. Sputtering and Glow Discharge Methods," Appl. Phys. Comm., Vol. 9, 1989, pp. 229-241.

Abeles, B., C.R. Wronski, T. Tiedje, and G. D. Cody, "Exponential Absorption Edge in a-Si:H Films," Solid State Communications, 36, 1980, pp. 537-540.

Abelson(1), J., and G. de Rosny, "The Relation between Contact Potential and Planar Conduction as a-Si:H Films undergo Gas Adsorption or Temperature Changes," J. Physique, Vol. 44, 1983, pp. 993-1003.

Abelson(2), J.R., J.R. Doyle, L. Mandrell, and N. Maley, "Reactive Magnetron Sputtering: In Situ Analyses of Particle Fluxes and Interactions with the Growth Surface," Mat. Res. Soc. Symp. Proc., Vol. 268, 1992, pp. 83-94.

Adams, A.C., In J. Mort and F. Jansen, "Plasma Deposited Thin Films," CRC Press, Inc. Boca Raton, Florida, 1986, pp. 129-159.

Adler(1), D., "Density of States in the Gap of Tetrahedrally Bonded Amorphous Semiconductors," Phys. Rev. Lett., Vol. 41, 1978, pp. 1755-1758.

Adler(2), D., "Origin of the photo-induced changes in hydrogenated amorphous silicon", Solar Cells, Vol. 9, 1983, pp. 133-148.

Adler(3), D., "Defects and Density of Localized States," Semiconductors and Semimetals, Vol. 21A, 1984, pp. 291-318.

Ahn, B.-C., K. Shimizu, T. Satoh, H. Kanoh, O. Sugiura, and M. Matsumura, "Hot-wall chemical-vapor-deposition of amorphous-silicon and its application to thin-film transistors," Jpn. J. Appl. Phys., Vol. 30, 1991, pp. 3695-3699.

Aljishi(1), S., Z E. Smith, D. Slobodin, J. Kolodzey, V. Chu, R. Schwarz, and S. Wagner, "Electronic Transport and the Density of States Distribution in a-(Si,Ge):H,F Alloys," Mat. Res. Soc. Symp. Proc., Vol. 70, 1986, pp. 269-274.

Aljishi(2), S., J.D. Cohen, S. Jin, and L. Ley, "Band Tails in Hydrogenated Amorphous Silicon and Silicon-Germanium Alloys," Phys. Rev. Lett., Vol.64, 1990, pp. 2811-2814.

Allan, D.C. and J.D. Joannopoulos, "Theory of Electronic Structure of a-Si:H," Topics in Applied Physics, Vol. 56, 1984, pp. 5-60.

Alvarez(1), F., and I. Chambouleyron, "Photoelectronic properties of amorphous silicon nitride compounds," Solar Energy Materials, Vol. 10, 1984, pp. 151-170.

Alvarez(2), F., I. Chambouleyron, A. Gobbi, C. Mendonca, and P. L. Castro, "On the Influence of an External DC Substrate Bias on Boron and Phosphorous Doping Efficiencies in a-Si:H," J. Non-Cryst. Solids, Vol. 77&78, 1985, pp. 527-530.

Amato(1), G., G.D. Mea, F. Fizzotti, C. Manfredotti, R. Marchisio, and A. Paccagnella, "Hydrogen bonding in amorphous silicon with use of the low-pressure chemical-vapor-deposition technique," Phys. Rev. B, Vol. 43, 1991, pp. 6627-6628.

Amato(2), G., F. Fizzotti, C. Manfredotti, P. Menna, G. Noble, and R. Spagnolo, "Optical and structural properties of amorphous silicon obtained by low-pressure chemical vapor deposition", Phys. Stat. Sol. B, Vol. 170, 1992, pp.119-128.

Anderson, J.C. and S. Biswas, "Activated Reactive Evaporation of Hydrogenated Amorphous Silicon," J. Non-Crystal. Solids, Vols. 77&78, 1985, pp. 817-820.

Ando, K., M. Aozasa, and R.G. Pyon, "Bias Effects on the Deposition of Hydrogenated Amorphous Silicon Film in a Glow Discharge," Appl. Phys. Lett., Vol. 44, 1984, pp. 413-415.

Antoine, A.M., B. Drevillon, and P. Roca i Cabarrocas, "In Situ Investigation of the Growth of RF Glow-Discharge Deposited Amorphous Germanium and Silicon Films," J. Appl. Phys., Vol. 61, 1987, pp. 2501-2508.

Aoki, T., S. Kato, M. Hirose, and Y. Nishikawa, "DC Bias Effects on Growth of a-Ge:H in Coaxial-Type ECR Plasma," Jpn. J. Appl. Phys., Vol. 28, 1989, pp. 849-855.

Aozasa, M., R.G. Pyon, and K. Ando, "Bias Effects on Preparation of Amorphous Silicon in a Triode Glow Discharge," Thin Solid Films, Vol. 136, 1986, pp. 263-274.

Arya(1), R.R., A. Catalano, R.S. Oswald, "Amorphous Silicon p-i-n Solar Cells with Graded Interface," Appl. Phys. Lett. Vol. 49, 1986, pp. 1089-1091.

Arya(2), R.R., "High Efficiency Amorphous Silicon Based Solar Cells: A Review," Mat. Res. Soc. Symp. Proc., Vol. 118, 1988, pp. 569-580.

Asano(1), A., T. Ichimura, and H. Sakai, "Preparation of Highly Photoconductive Hydrogenated Amorphous Silicon Carbide Films with a Multiplasma-zone

Apparatus," J. Appl. Phys., Vol. 65, 1989a, pp. 2439-2444.

Asano(2), A., and H. Sakai, "Improvement in the boron doping efficiency of hydrogenated amorphous silicon carbide films using BF_3," Appl. Phys. Lett. Vol. 54, 1989b, pp. 904-906.

Asano(3), A., "Effects of Hydrogen Atoms on the Network Structure of Hydrogenated Amorphous and Microcrystalline Silicon Thin Films," Appl. Phys. Lett., Vol. 56, 1990, pp. 533-535.

Ashida, Y., Y. Mishima, M. Hirose, Y. Osaka, and K. Kojima, "Hydrogenated Amorphous Silicon Produced by Pyrolysis of Disilane in a Hot Wall Reactor," Jpn. J. Appl. Phys., Vol. 23, 1984, pp. L129-L131.

Ast, D.G. and M.H. Brodsky, In the 14th Int'l Conf. of Physics of Semiconductors, B.H.L. Wilson, editor, Inst. of Physics Conf. Series No. 43 (Hilger, London), 1979, p.1159.

ASTM, "Standard Test Methods for Measuring Resistivity and Hall Coefficient and Determining Hall Mobility in Single-Crystal Semiconductors," Designation: F76-86, Annual Book of ASTM Standards, American Society for Testing and Materials, Buffalo, New York, 1992.

Auciello, O. and D.L. Flamm, editors, "Plasma Diagnostics," Academic Press - Harcourt Brace Jovanovich, San Diego, CA, 1988.

Azuma, K., M. Tanaka, T. Watanabe, M. Makatani, and T. Shimada, "High Conversion Efficiency Amorphous Silicon Solar Cell Formed from Disilane under High Rate Deposition Conditions," Proc. of the 19th IEEE PV Specialists Conf., 1987, pp. 558-563.

Bachrach, R.Z., K. Winer, J.B. Boyce, S.E. Ready, R.I. Johnson, and G.B. Anderson, "Low Temperature Crystallization of Amorphous Silicon Using an Eximer Laser," J. Electronic Materials, Vol. 19, 1990, pp. 241-248.

Balberg, I., A.E. Delahoy, and H.A. Weakliem, "Self-Consistency and Self-Sufficiency of the Photocarrier Grating Technique," Appl. Phys. Lett., Vol.53, 1988, pp. 992-994.

Banerjee, A. and S. Guha, "Study of Back Reflectors for Amorphous Silicon Alloy Solar Cell Application," J. Appl. Phys., Vol. 59, 1991, pp. 1030-1035.

Bar-Yam, Y., D. Adler, and J.D. Joannopoulos, "Structure and Electronic States in Disordered Systems," Phys. Rev. Lett., Vol. 57, 1986, p.467.

Barkanic, J.A., D.M. Reynolds, R.J. Jaccodine, H.G. Stenger, J. Parks, and H. Vedage, "Plasma Etching Using NF_3: A Review," Solid State Technology, 1989, pp. 109-115.

Baron(1), B.N., R.E. Rocheleau, and S.S. Hegedus, "Photochemical Vapor Deposition of Amorphous Silicon Photovoltaic Devices," SERI Subcontract Semi-Annual Report, January 1988.

Baron(2), B.N., C.M. Fortmann, S.S. Hegedus, W.A. Buchanan, D.E. Albright, N. Saxena, T.X. Zhou, & T.W.F. Russell, "Amorphous Silicon and Silicon-Germanium Thin-Film Materials and Solar Cells," First "Sunshine" Workshop on Solar Cells, February 1990.

Bauer, G.H. and G. Bilger, "Properties of Plasma-Produced Amorphous Silicon Governed by Parameters of the Production, Transport and Deposition of Si and SiH_x," Thin Solid Films, Vol. 83, 1981, pp. 223-229.

Bennett(1), M.S. and K. Rajan, "The Stability of Multijunction a-Si Solar Cells," Proc. 20th IEEE PV Specialists Conf., 1988, pp. 67-72.

Bennett(2), M.S., and J.C. Tu, "Amorphous Hydrogenated Silicon p-i-n Solar Cell Grown from Hydrogen-Dilute Silane," Mat. Res. Soc. Symp. Proc., Vol. 192, 1990, pp. 45-50.

Beyer, W., R. Hager, H. Schmidbaur, and G. Winterling, "Improvement of the Photoelectric Properties of Amorphous SiC_x:H by Using Disilylmethane as a Feeding Gas," Appl. Phys. Lett., Vol. 54, 1989, pp. 1666-1668.

Bhan, M.K., L.K. Mahotra, and S.C. Kashyap, "Electrical and Optical Properties of Ion-Beam-Sputtered Amorphous Silicon-Germanium Alloy Films," Thin Solid Films, Vol. 203, 1991, pp. 23-32.

Bhat(1), P.K., A.J. Rhodes, T.M. Searle, I.G. Austin, and J. Allison, "The 0.9 eV Defect Luminescence Band in Sputtered and Forms of Plasma-Deposited a-Si:H," Phil. Mag. B, Vol. 47, 1983, pp. L99-105.

Bhat(2), P.K., H. Chatham, A. Madan, "Preparation and Properties of High-Deposition-Rate a-Si:H Films and Solar Cells using Disilane," Annual Subcontractor Report for the period 1 May 1987 - 30 April 1988, Work performed by Glasstech Solar Inc., Wheatridge, Colorado, SERI/STR-211-3364, 1988, Solar Energy Research Inst., NTIS Accession No. DE88001191.

Blayo, N. and B. Drévillon, "Interaction between Growing Amorphous Silicon and Glass Substrate Evidence by *in situ* infrared ellipsometry," Appl. Phys. Lett.,

Vol. 57, 1990, pp. 786-788.

Bocko, P.L., C.T. Stutts, T. Hayashi, and F. Okamoto, Corning Incorporated, "Glass Substrates for Flat Panel Displays," SEMI Technical Education Program, Flat Panel Display Manufacturing Technology, Sept. 14&15, 1989.

Boeuf, J.P., Ph. Belenguer, and J. Wang, "Radiofrequency Discharge Modeling," Mat. Res. Soc. Symp. Proc., Vol. 165, 1990, pp. 17-29.

Böhm, M., A.E. Delahoy, F.B. Ellis, Jr., E. Eser, S.C. Gau, F.J. Kampas, and Z. Kiss, "Single Chamber Manufacturing Process for Amorphous Silicon Solar Cells," Proc. 18th IEEE PV Specialists Conf., 1985, pp. 888-893.

Boulitrop(1), F.,"Recombination processes in a-Si:H. A study by optically detected magnetic resonance," Phys. Rev. B, Vol. 28, 1983, p. 6192,

Boulitrop(2), F., N. Proust, J. Magarino, E. Criton, J.F. Peray, and M. Dupre, "A Study of Hydrogenated Amorphous Silicon Deposited by Hot-Wall Glow Discharge," J. Appl. Phys., Vol. 58, 1985, pp. 3494-3498.

Boyce, J.B., M. Stutzmann, and S.E. Ready, "Molecular Hydrogen in Amorphous Si:NMR Studies," J. Non-Crystal. Solids, Vol. 77 & 78, 1985, pp. 265-268.

Branz(1), H.M., L.K. Liem, C.J. Harris, S. Fan, J.H. Flint, D. Adler, and J.S. Haggerty, "Laser-Induced Chemical Vapor Deposition of Hydrogenated Amorphous Silicon: Photovoltaic Devices and Material Properties," Solar Cells, Vol. 21, 1987, pp. 177-188.

Branz(2), H.M., "Charge Trapping Model of Metastability in Doped Hydrogenated Amorphous Silicon," Phys. Rev. B, Vol. 38, 1988, 7474.

Branz(3), H.M., and M. Silver, "Potential Fluctuations Due to Inhomogeneity in Hydrogenated Amorphous Silicon and the Resulting Charged Dangling Bond Defects," Phys. Rev. B, Vol. 42, 1990, p. 7420.

Branz(4), H.M., "Comment on Excitation-energy dependence of optically induced ESR in a-Si:H," Phys. Rev. B, Vol. 41, 1990, pp. 7887-7890.

Brodsky(1), M.H., "On the Deposition of Amorphous Silicon Films from Glow Discharge Plasmas of Silane," Thin Solid Films, Vol. 40, 1977, pp. L23-L25.

Brodsky(2), M.H., M. Cardona, and J.J. Cuomo, "Infrared and Raman Spectra of the Silicon-Hydrogen Bonds in Amorphous Silicon Prepared by Glow Discharge and Sputtering," Phys. Rev. B, Vol. 16, 1977, pp. 3556-3571.

Brodsky(3), M.H., "Amorphous Semiconductors," Second Edition, Topics in Applied Physics, Vol. 36, 1985.

Brumberger, H., "Small-Angle X-Ray Scattering," Gordon and Breach Science Publishers, New York, 1967.

Burnham, N.A., A.B. Swartzlander, A.J. Nelson, and L.L. Kazmerski, "Auger Line Shape Analysis of Hydrogenated Amorphous Silicon," Solar Cells, Vol. 21, 1987, pp. 135-140.

Bustarret, E., M. Ligeon, J.C. Bruyère, F. Muller, R. Hérino, F. Gaspard, L. Ortega, M. Stutzmann, "Visible Light Emission at Room Temperature from Anodized Plasma-Deposited Silicon Thin Films," Appl. Phys. Lett., Vol. 61, 1992, pp. 1552-1554.

Canham, L.T., "Silicon Quantum Wire Array Fabrication by Electrochemical and Chemical Dissolution of Wafers," Appl. Phys. Lett., Vol. 57, 1990, pp. 1046-1048.

Canillas, A., E. Bertran, J.L. Andújar, and B. Drévillon, "In Situ Spectroellipsometric Study of the Nucleation and Growth of Amorphous Silicon," J. Appl. Phys., Vol. 68, 1990, pp. 2752-2759.

Canon, Inc., News Release on Jan. 8, 1993.

Cardona, M., "Vibrational Spectra of Hydrogen in Silicon and Germanium," Phys. Stat. Sol. B, Vol. 118, 1983, pp. 463-481.

Carius, R., "Time Resolved Electroluminescence in a-Si:H p-i-n Junctions," Mat. Res. Soc. Symp. Proc., Vol.192, 1990, pp. 101-106.

Carlson(1), D.E., and C.R. Wronski, "Amorphous Silicon Solar Cells," Appl. Phys. Lett., Vol. 28, 1976, pp. 671-673.

Carlson(2), D.E., "Solar Energy Conversion," In Topics in Applied Physics, Vol. 55, The Physics of Hydrogenated Amorphous Silicon I, (Eds. Joannopoulos & Lucovsky), Springer-Verlag, 1984a, pp. 203-244.

Carlson(3), D.E., A. Catalano, R.V. D'Aiello, C.R. Dickson, and R.S. Oswald, "The Effect of Light Soaking on a-Si:H Films Containing Impurities," Amer. Inst. of Physics Conf. Proc., No. 120, 1984b, pp. 234-241.

Carlson(4), D.E., A. Catalano, R.V. D'Aiello, C.R. Dickson, and R.S. Oswald, "Research on High-Efficiency, Single-Junction, Monolithic, Thin-Film a-Si Solar

Cells," Annual Subcontract Report for the Period 2/1/1984-1/31/1985, STR-211-2813, Solar Energy Research Inst., 1985a.

Carlson(5), D.E. and C. R. Wronski, Chapter 10 in "Amorphous Semiconductors," Topics in Applied Physics, Vol. 36, Springer-Verlag, N.Y., 1985b.

Carlson(6), D.E., "Hydrogenated Microvoids and Light-Induced Degradation of Amorphous-Silicon Solar Cells," Appl. Phys. A, Vol. 41, 1986, pp. 305-309.

Carlson(7), D.E., A. Catalano, R.V. D'Aiello, C.R. Dickson, and R.S. Oswald, "The Effects of Impurities on the Diffusion Length in Amorphous Silicon," Proc. 19th IEEE PV Specialists Conf., 1987, pp. 330-335.

Carlson(8), D.E., R.R. Arya, M.S. Bennett, A. Catalano, "Research on High Efficiency Single-Junction Monolithic Thin Film Amorphous Solar Cells," Semiannual Report for the period 2/1/1987-7/31/1987, STR-211-3375, Solar Energy Research Inst., 1988. NTIS Accession No. DE89000843.

Catalano(01), A., R.R. Arya. C. Fortmann, J. Morris, J. Newton, and J.G. O'Dowd, "High Performance, Graded Bandgap a-Si:H Solar Cells," Proc. 19th PV Specialists Conf., 1987b, pp. 1506-1507.

Catalano(02), A. and G. Wood, "Short Wavelength Response in a-Si:H p-i-n Diodes: A Simple Method to Minimize Interface Recombination," Proc. MRS, Vol. 118, 1988, pp. 581-586.

Catalano(03), A., D.E. Carlson, R.R. Arya, M.S. Bennett, "Research on High Efficiency Single-Junction Monolithic Thin Film Amorphous Solar Cells," Phase 1 Annual Subcontract Report for the period 2/1/1987-1/31/1988, STR-211-3582, Solar Energy Research Inst., 1989a.

Catalano(04), A., R.R. Arya, M. Bennett, L. Yang, J. Morris, B Goldstein, B. Fieselman, J. Newton, S. Wiederman, "Progress on High Efficincy a-SiGe Solar Cells and Submodules," Proc. of the 1989 Amorphous Silicon Subcontractors' Review Meeting, June 1989b, pp. 107-116.

Catalano(05), A.W., R.R. Arya, M.S. Bennett, "Research on Stable, High Efficiency, Large-Area, Amorphous Silicon Based Solar Cells," Phase 2 Annual Subcontract Report for the Period 1 February 1988 - 1 February 1989, STR-211-3580, Solar Energy Research Inst., 1989c. NTIS Accession No. DE89009491.

Catalano(06), A., "Research on Stable, Large-Area, Amorphous Silicon-Based Submodules," Phase 2 Semi-Annual Subcontract Report for the Period 1

February 1989 - 31 July 1989, TP-211-3805, Solar Energy Research Inst., 1989d. NTIS Accession No. DE90000339.

Catalano(07), A., "Research on High-Efficiency, Large-Area, Amorphous Silicon-Based Solar Cells," Final Subcontract Report for the period 1 February 1989 to 28 February 1990, SERI/TP-211-3906, Solar Energy Research Inst., 1990a.

Catalano(08), A., et al. "Research on Stable, High-Efficiency Amorphous Silicon Multijunction Modules," Phase I Semi-annual Technical Progress Report for the period 1 May 1990 to 31 October 1990, SERI TP-214-4271, Solar Energy Research Inst., 1990b.

Catalano(09), A., R. Arya, M. Bennett, B. Fieselmann, Y. Li, J. Morris, J. Newton, R. Podlesny, S. Wiederman, L. Yang, "Research on Stable, High-Efficiency, Amorphous Silicon Multijunction Modules," Semiannual Subcontract Report, 1 May 1990 - 31 October 1990, SERI/TP-214-4271, 1991a. Work performed by Solarex, Newtown, Pennsylvania. Golden, Co: National Renewable Energy Laboratory. NTIS Accession No. DE91002138.

Catalano(10), A., M. Bennett, L. Chen, B. Fieselmann, Y. Li, J. Newton, R. Podlesny, S. Wiederman, L. Yang, "Research on Stable, High-Efficiency, Amorphous Silicon Multijunction Modules," Annual Subcontract Report, 1 May 1990 - 30 April 1991, SERI/TP-214-4405, 1991b. Work performed by Solarex, Newtown, Pennsylvania. Golden, Co: National Renewable Energy Laboratory. NTIS Accession No. DE91015002.

Catalano(11), A., R.R. Arya, M. Bennett, L. Chen, R. D'Aiello, B. Fieselmann, Y. Li, J. Newton, R. Podlesny, S. Wiederman, L. Yang, "Research on Stable, High-Efficiency, Amorphous Silicon Multijunction Modules", Phase II Semiannual Subcontract Report, 1 May 1991 - 31 October 1991, SERI/TP-214-4720, 1992a. Work performed by Solarex, Newtown, Pennsylvania. Golden, Co: National Renewable Energy Laboratory. NTIS Accession No. DE92001213.

Chahed, L., G. Vuye, M.L. Theye, Y.M. Li, K.D. Mackenzie, and W. Paul, "Comparative Study of the Optical Absorption Spectra of a-Si:H Derived from Photothermal Deflection Spectroscopy and Photoconductivity Measurements," Proc. of the 12th Int'l Conf. on Amorphous and Liquid Semiconductors, Prague, August 1987.

Chapman, B., "Glow Discharge Processes," John Wiley & Sons, New York, 1980.

Chatham(1), R.H., and A. Gallagher, "Ion Chemistry in Silane DC Discharges," J. Appl. Phys, Vol. 58, 1985, pp. 159-169.

Chatham(2), R.H. and P.K. Bhat, "Preparation and Properties of High Deposition a-Si:H Films and Solar Cells using Disilane," Annual Subcontractor Report for the period 1 May 1988 - 30 April 1989, Work performed by Glasstech Solar Inc., SERI/STR-211-3562, 1989, Solar Energy Research Inst., Golden, Colorado, NTIS Accession Number DE89009467.

Chaudhuri, P., S. Ray, A.K. Barua, "The Effect of Mixing Hydrogen with Silane on the Electronic and Optical Properties of a-Si:H Thin Films," Thin Solid Films, Vol. 113, 1984, pp. 261-270.

Chayahara, A., A. Matsuda, T. Imura, and Y. Osaka, "Formation of Polycrystalline SiC in ECR Plasma," Jpn. J. Appl. Phys., Vol.25, 1986, pp. L564-L566.

Chen, L., J. Tauc, J.K. Lee, and E.A. Schiff, "Photomodulation spectroscopy of defects in hydrogenated amorphous silicon-germanium alloys," J. Non-Cryst. Solids, Vol. 114, 1989, pp. 585-587.

Chen, Y.-F., "Elimination of Light-Induced Effect in Hydrogenated Amorphous Silicon," Appl. Phys. Lett., Vol. 53, 1988, pp. 1277-1278.

Chen, Y.L., C. Wang, G. Lucovsky, D.M. Maher, and R.J. Nemanich, "Transmission Electron Microscopy and Vibrational Spectroscopy Studies of Undoped and Doped Si,H and Si,C:H Films," J. Vac. Sci. Technol. A, Vol. 10, 1992, pp. 874-880.

Chenevas-Paule, A., R. Bellissent, M. Roth, and J.I. Pankove, "Correlation between Staebler Wronski Effect and Medium Range Order in a-Si:H by SANS," J. Non-Crys. Solids, Vol. 77-78, 1985, pp. 373-376.

Chik(1), K.P., P.K. Lim, B.Y. Tong, P.K. John, P.K. Gogna, and S.K. Wong, "Photoconductivity of Evaporated Amorphous Silicon Films Post-Hydrogenated in a Theta-Pinch Plasma," Phys. Rev. B, Vol. 27, 1983, pp. 3562-3570.

Chik(2), K.P., P.H. Chan, K.H. Tam, B.Y. Tong, S.K. Wong, and P.K. John, "Thermoelectric Power and Electronic Transport in Thermal LPCVD Amorphous Silicon-Boron Films," Phil. Mag. B, Vol. 59, 1989, pp. 543-559.

Chittick, R.C., J.H. Alexander, and H.F. Sterling, "The Preparation and Properties of Amorphous Silicon," J. Electrochem. Soc., Vol. 116, 1969, pp. 77-81.

Christou, A., P. Tzanetkis, Z. Hatzopoulos, G. Kyriakidis, W. Tseng, and B.R. Wilkins, "Schottky Barrier Formation on Electron Beam Deposited Amorphous $Si_{1-x}Ge_x$:H Alloys and Amorphous ($Si/Si_{1-x}Ge_x$):H Modulated Structures," Appl.

Phys. Lett., Vol. 48, 1986, pp. 408-410.

Chu(1), T. L., S.S. Chu, S.T. Ang, D.H. Lo, A. Duong, C.G. Hwaung, and L. Book, "Hydrogenated Amorphous Silicon Films by the Pyrolysis of Disilane," Mat. Res. Soc. Symp. Proc., Vol. 49, 1985, pp. 21-26.

Chu(2), T. L., S.S. Chu, S.T. Ang, D.H. Lo, A. Duong, and C.G. Hwang, "Deposition and Photoconductivity of Hydrogenated Amorphous Silicon Films by the Pyrolysis of Disilane," J. Appl. Phys., Vol. 59, 1986a, pp. 1319-1322.

Chu(3), T. L., S. S. Chu, S. T. Ang, A. Duong, and Y. X. Han, "Properties of Amorphous Silicon Films Deposited at Rates Higher Than 1 μm/min.," Mat. Res. Soc. Symp. Proc., 70, 1986b, pp. 49-53.

Clarke, P.E., D. Field, and D.F. Klemperer, "Optical Spectroscopic Study of Mechanisms in CCl_4 Plasma Etching of Si," J. Appl. Phys., Vol. 67, 1990, pp. 1525-1534.

Cody(1), G.D., T. Tiedje, B. Abeles, B. Brooks, and Y. Goldstein, "Disorder and the Optical-Absorption Edge of a-Si:H," Phys. Rev. Lett., Vol. 47, 1981, pp. 1480-1483.

Cody(2), G.D., B.G. Brooks, and B. Abeles, "Optical Absorption above the Optical Gap of Amorphous Silicon Hydride," Solar Energy Materials, Vol. 8, 1982, pp. 231-240.

Collins(1), R. W., C.Y. Huang, and H. Windischmann, "Oxidation and Etching of Hydrogenated Amorphous Silicon: An In-Situ Ellipsometry Study," Solar Energy Materials, Vol. 12, 1985a, pp. 1-10.

Collins(2), R.W., H. Windischmann, J.M. Cavese, and J. Gonzalez Hernandez, "Optical Properties of Dense Thin-Film Si and Ge Prepared by Ion-Beam Sputtering," J. Appl. Phys., Vol. 58, 1985b, pp. 954-957.

Collins(3), R.W., "In-Situ Ellipsometry Studies of the Growth of Hydrogenated Amorphous Silicon by Glow Discharge," J. Vac. Sci. Technol., Vol. A4, 1986a, pp. 514-517.

Collins(4), R.W., and A. Pawlowski, "The Nucleation and Growth of Glow Discharge Hydrogenated Amorphous Silicon," J. Appl. Phys., Vol. 59, 1986b, pp. 1160-1166.

Collins(5), R.W., and J.M. Cavese, "Effect of Deposition Conditions on the Nucleation and Growth of Glow Dicharge a-Si:H," J. Appl. Phys., Vol. 61,

1987a, pp. 1869-1882.

Collins(6), R.W., and J.M. Cavese, "Surface Roughness Evolution on Glow Discharge a-Si:H," J. Appl. Phys., Vol. 61, 1987b, pp. 1662-1664.

Collins(7), R.W., "In Situ Study of P-Type Amorphous Silicon Growth from B_2H_6:SiH_4 Mixtures: Surface Reactivity and Interface Effects," Appl. Phys. Lett., Vol. 53, 1988a, pp. 1086-1088.

Collins(8), R.W. and J.M. Cavese, "Process Monitoring for a-Si:H Materials and Interfaces," Mat. Res. Soc. Symp. Proc., Vol. 118, 1988b, pp. 19-30.

Collins(9), R.W., "Ellipsometric Study of a-Si:H Nucleation, Growth and Interfaces," in Amorphous Silicon and Related Materials, edited by H. Fritsche, (World Scientific Publishing, Singapore) Vol. B, 1988c, pp. 1003-1044.

Coluzza, C., D. Della Sala, G. Fortunato, S. Scaglione, and A. Frova, "a-Si:H Produced by Double Ion-Beam Sputtering," J. Non-Crystal. Solids, Vols. 59 and 60, 1983, pp. 723-726.

Cook, J.M., "Downstream Plasma Etching and Stripping," Solid State Technology, April 1987, pp. 147-151.

Crandall(1), R.S., "Band-Tail Absorption in Hydrogenated Amorphous Silicon," Phys. Rev. Lett., Vol. 44, 1980, pp. 749-752.

Crandall(2), R.S., "Modeling of Thin Film Solar Cells: Uniform Field Approximation," J. Appl. Phys., Vol. 54, 1983, pp. 7176-7186.

Crandall(3), R.S., "Modeling of Thin-Film Solar Cells: Nonuniform Field," J. Appl. Phys., Vol. 55, 1984, pp. 4418-4425.

Crandall(4), R.S., K. Sadlon, J. Kalina, and A.E. Delahoy, "Direct Measurement of the Mobility-Lifetime Product of Holes and Electrons in an Amorphous Silicon p-i-n Cell," Mat. Res. Soc. Symp. Proc. Vol. 149, 1989, pp. 423-427.

Crandall(5), R.S. and I. Balberg, "Mobility-Lifetime Products in Hydrogenated Amorphous Silicon," Appl. Phys. Lett., Vol. 58, 1991, pp. 508-510.

Crandall(6), R.S. and I. Balberg, "Mobility-lifetime Products in Hydrogenated Amorphous Silicon," Appl. Phys. Lett., Vol. 58, , 1991, pp. 508-510.

Craven, A.J., A.J. Patterson, A.R. Long, and J.I.B. Wilson, "Small Angle Electron Scattering in a-Si and a-Si:H Films," J. Non-Cryst. Solids, Vol. 77-78,

1985, pp. 217-220.

Crowley, J.L., "Plasma Enhanced CVD for Flat Panel Displays," Solid State Technology, Feb. 1992, pp. 94-97.

Curtins(1), H., M. Favre, N. Wyrsch, M. Brechet, K. Prasad, and A.V. Shah, "High-Rate Deposition of Hydrogenated Amorphous Silicon by the VHF-GD Method," Proc. 19th IEEE PV Specialists Conf., 1987a, pp. 695-698.

Curtins(2), H., N. Wyrsch, and A.V. Shah, "High-Rate Deposition of Amorphous Hydrogenated Silicon: Effect of Plasma Excitation Frequency," Electronics Letters, Vol. 23, 1987b, pp. 228-230.

Curtins(3), H., M. Favre, Y. Ziegler, N. Wyrsch, and A.V. Shah, "Comparison of Light-Induced Degradation in Low and High-Rate Deposited VHF-GD a-Si:H: Effect of Film Inhomogeneities," Mat. Res. Soc. Symp., Vol. 118, 1988, pp. 159-166.

Czanderna(1), A. W., Editor, "Methods of Surface Analysis," Elsevier, Amsterdam, 1975.

Czanderna(2), A.W. and D.M. Hercules, Editors, "Ion Spectroscopies for Surface Analysis," Plenum Press, New York, 1991.

Dalal(1), V. and F. Alvarez, "Minority Carrier Transport in Depletion Layers of n-i-p a-Si:H Solar Cells," J. de Physique, Vol. 42-C4, 1981, pp. 491-494.

Dalal(2), V., Research on High-Efficiency, Stacked, Multijunction Amorphous Silicon Alloy Thin-Film Solar Cells, Annual Subcontract Report, 11 Oct. 1983--31 Oct. 1984, SERI/STR-211-2730, Golden, CO: Solar Energy Research Inst., 1985.

Dawson, R.M., C.R. Wronski, and M. Bennett, "Densities of States below Midgap Determined from the Space-Charge-Limited-Currents of Holes in Intrinsic Hydrogenated Amorphous Silicon," Appl. Phys. Lett., Vol. 58, 1991, pp. 272-274.

de O.Graeff, C.F., P.V. Santos, G. Marcano, and I. Chambouleyron, "Staebler-Wronski Effect in Hydrogenated Amorphous Germanium Films," Proc. 21st IEEE PV Specialists Conf., 1990, pp. 1564-1568.

De Chelle, F., J.M. Berger, A. Deneuville, J.C. Bruyere, S.P. Coulibaly, J.P. Ferraton, and A. Donnadieu, "Temperature Dependent Studies of the Optical Properties of Post-Hydrogenated Sputtered a-Si:H," J. Non-Cryst. Solids, Vol. 64,

1984, pp. 1-10.

Dellafera, P., R. Labusch, and H.H. Roscher, "An Alternative Method of Preparing Hydrogen-Doped Evaporated Amorphous Silicon Preliminary Report," Phil. Mag. B, Vol. 43, 1981, pp. 169-172.

Dembinski, M., P.K. John, and A.G. Ponomarenko, "High-Current Ion Beam from a Moving Plasma," Appl. Phys. Lett., Vol. 34, 1979, pp. 553-555.

Demichelis, F., et al., "Physical Properties and Structure of a-SiC:H Alloy Films," Nuovo Cimento D (Italy), Vol. 9D, Ser. 1, No. 4, 1987, pp. 393-408.

Deng(1), X.J., Y.S. Tsuo, and J.U. Trefny, "Ion-Beam Hydrogenation of Sputter-Deposited Amorphous Silicon," Proc. of the Industry-University Advanced Materials Conf. II, 1989, pp. 190-196.

Deng(2), X.J., Y.S. Tsuo, and J.U. Trefny, "Ion-Beam Hydrogenation of Sputter-Deposited Amorphous Silicon and Amorphous Silicon-Germanium Alloys," Proc. 21th IEEE PV Specialists Conf., 1990, pp. 1591-1594.

Dersch, H., J. Stuke, and J. Beichler, "Light-induced dangling bonds in hydrogenated amorphous silicon", Appl. Phys. Lett., Vol. 38, 1980. pp. 456-458.

Dickson, C. R., J. Pickens, and A. Wilczynski, "Amorphous Silicon Solar Cell Modules Fabricated with a Single-Chamber Load-Lock Deposition System," Solar Cells, Vol. 19, 1986, pp. 179-188.

Donnadieu, A., J.P. Ferraton, S.P. Coulibaly, C. Ance, J.M. Berger, and F. De Chelle, "Boron Doping Effect on Optical Properties of a-Si Films Prepared by CVD : Post-Hydrogenation and H-Exodiffusion Study," J. Non-Cryst. Solids, Vols. 59 and 60, 1983, pp. 305-308.

Doyle, J., R. Robertson, G.H. Lin, M.Z. He, and A. Gallagher, "Production of High-Quality Amorphous Silicon Films by Evaporative Silane Surface Decomposition," J. Appl. Phys., Vol. 64, 1988, pp. 3215-3223.

Drevillon, B., J. Perrin, J. M. Siefert, J. Huc, A. Lloret, G. deRosny, and J. P. M. Schmitt, "Growth of Hydrogenated Amorphous Silicon due to Controlled Ion Bombardment from a Pure Silane Plasma," Appl. Phys. Lett., Vol. 42, 1983, pp. 801-803.

Du, N., S. Salkalachen, J. Yao, H.R. Froelich, P.K. John, and B.Y. Tong, "Current-Voltage Characteristics of n-Amorphous Low-Pressure Chemical Vapor

Deposited Silicon Films on p-Crystalline Silicon," J. Appl. Phys., Vol. 66, 1989, pp. 5894-5900.

Dusane, S.R., "Gap States in Hydrogenated Microcrystalline Silicon Grown by Glow Discharge Technique," J. Appl. Phys., Vol. 72, 1992, pp. 2923-2926.

Dutta, R., P.K. Banerjee, and S.S. Mitra, "Amorphous Silicon-Carbon-Fluorine Alloy Films," Phys. Rev. B, Vol. 27, 1983, pp. 5032-5038.

Eggert, J. R., and W. Paul, "Midgap Injection-Induced Absorption in Amrophous Silicon," Phys. Rev. B, Vol. 35, 1987, pp. 7993-7998.

Eicke, A. and G. Bilger, "XPS and SIMS Characterization of Metal Oxide/Amorphous Silicon-Carbon Interfaces," Surface and Interface Analysis, Vol. 12, 1988, pp. 344-350.

Eliot, S.R., "Phto-induced changes in glow-discharge-deposited amorphous silicon: The Staebler-Wronski effect," Philos. Mag., Vol. B39, 1979, pp. 349-356.

Ellis, Jr., F.B., R.G. Gorden, W. Paul, and B.G. Yacobi, "Properties of Hydrogenated Amorphous Silicon Prepared by Chemical Vapor Deposition," J. Appl. Phys., Vol. 55, 1984, pp. 4309-4317.

Evans, C.E., and R.J. Blattneer, "Modern Experimental Methods for Surface and Thin-Film Chemical Analysis," Ann. Rev. Mater. Sci., Vol. 8, 1978, pp. 181-214.

Fair, J.E., "Rapid Thermal Processing for Active Matrix Devices," Solid State Technology, August 1992, pp. 47-52.

Fang, P.H., C.C. Schubert, P. Bei, and J.H. Kinnier, "Combined Microcrystal and Amorphous Silicon Cells," Appl. Phys. Lett., Vol. 41, 1982, pp. 365-366.

Fang, R.C., Y.Z. Song, M. Yang, and W.D. Jiang, "Photoluminescence of Hydrogenated a-SiN_x Films," J. Non-Cryst. Solids, Vol. 77&78, 1985, pp. 913-916.

Fang, J.C., L. Ley, H.R. Shanks, K.J. Gruntz, and M. Cardona, "Bonding of Fluorine in Amorphous Hydrogenated Silicon," Phys. Rev. B, Vol. 22, 1980, pp. 6140-6148.

Faraji, M., S. Gokhale, S.M. Choudhari, M.G. Takwale, and S.V. Ghaisas, "High Mobility Hydrogenated and Oxygenated Microcrystalline Silicon as a Photosensitive Material in Photovoltaic Applications," Appl. Phys. Lett., 1992 Vol. 60, pp. 3289-3291.

Fedders, P.A. and A.E. Carlsson, "Defect States at Floating and Dangling Bonds in Amorphous Si," Phys. Rev. B, Vol. 37, 1988. p. 8506-8508.

Feng, G.F., M. Katiyar, Y.H. Yang, J.R. Abelson, and N. Maley, "Growth and Structure of Microcrystalline Silicon by Reactive DC Magnetron Sputtering," Mat. Res. Soc. Proc., Vol. 258, 1992, pp. 179-184.

Feng, Z.C., A.J. Mascarenhas, W.J. Choyke, and J.A. Powell,"Raman Scattering Studies of Chemical-Vapor-Deposited Cubic SiC Films of (100) Si," J. Appl. Phys., Vol. 64, 1988, pp. 3176-3186.

Fieselmann(1), B., M. Milligan, A. Wilczynski, J. Pickens, and C.R. Dickson, "Doping and Alloying Amorphous Silicon using Silyl Compounds", Conf. Record 19th IEEE PV Specialists Conf., 1987, pp. 1510-1511.

Fieselmann(2), B.F. and B. Goldstein, "Preparation of p-Type Materials for a-Si Solar Cells using DC Plasma Discharge of $B(CH_3)_3$," Mat. Res. Soc. Symp. Proc., Vol. 149, 1989, pp. 441-445.

Fischer, D., R. Tscharner, Y. Ziegler, H. Keppner, and A. Shah, "Amorphous Silicon p-i-n Solar Cells Fabricated by High Deposition Rate VHF-Glow Discharge Process," Proc. 9th Euro. Comm. PV Conf., 1989, pp. 995-997.

Folkerts, L. and J.M. Gordon, "Amorphous Silicon Solar Cells: Thermodynamic Models for Realistic Performance Characteristics," Solar Cells, Vol. 23, 1988, pp. 201-215.

Fortmann(1), C.M., and J.C. Tu, "Defects in Amorphous Silicon Germanium Alloys," Proc. 20th IEEE PV Specialists Conf., 1988, pp. 139-142.

Fortmann(2), C.M., "a-SiGe:H Alloy Material Limitations and Device Considerations," 21st IEEE PV Spec. Conf. Proc., 1990, pp. 1293-1500.

Fritzsche(1), H., "Characterization of Glow-Discharge Deposited a-Si:H," Solar Energy Materials, Vol. 3, 1980, pp. 447-501.

Fritzsche(2), H. and M. Pollak, editors, "Hopping and Related Phenomena," Adv. in Disordered Semiconductors, Vol. 2, 1990, World Scientific, Singapore.

Fujita, H., H. Handa, M. Nagano, and H. Matsuo, "Characteristics of Microwave Plasma and Preparation of a-Si Thin Film," Jpn. J. Appl. Phys., Vol. 26, 1987, pp. 1112-1116.

Fukuda(1), N., S. Ogawa, K. Abe, Y. Ohashi, and S. Kobayashi, "High Rate

Deposition of a-Si:H Films from Disilane (Si_2H_6)," 1st Int'l PV Sci. & Eng. Conf., 1984, pp. 107-110.

Fukuda(2), N., K. Miyachi, H. Tanaka, T. Igarashi, and S. Yamamoto, "Control of a-Si:H Film Properties by Photo-Assisted Plasma CVD," Mat. Res. Soc. Symp. Proc., Vol. 70, 1986, pp. 25-30.

Furukawa, S., and N. Matsumoto, "Estimation Methods for Localized-State Distribution Profiles in Undoped and Phosphorous-Doped a-Si:H," Phys. Rev. B, Vol. 27, 1983, pp. 4955-4960.

Gallagher(01), A., "Surface Reactions in Discharge and CVD Deposition of Silane," Mat. Res. Soc. Symp. Proc., Vol. 70, 1986a, pp. 3-10.

Gallagher(02), A., "Amorphous Silicon Deposition Rates in Diode and Triode Discharges," J. Appl. Phys., Vol. 60, 1986b, pp. 1369-1373.

Gallagher(03), A., and J. Scott, "Gas and Surface Processes Leading to a-Si:H Films," SERI Advanced R&D Meeting, 13-16 May 1986, Denver, CO; Solar Cells, Vol. 21, 1987a, pp. 147-152.

Gallagher(04), A., "Surface Reactions in Silane Discharges," The Physics of Ionized Gases, SPIG 86 (J. Puric and D. Belic, Eds.), World Scientific, Singapore, 1987b, pp. 229-238.

Gallagher(05), A., "Diagnostics of Glow Discharges Used to Produce Hydrogenated Amorphous Silicon Films," Annual Subcontract Report, 15 April 1986-14 June 1987, STR-211-3288, Solar Energy Research Inst., 1987c.

Gallagher(06), A. "Apparatus Design for glow discharge a-Si:H Film Deposition," Int. J. Solar Energy, Vol. 5, 1988a, pp. 311-322.

Gallagher(07), A., "Neutral Radical Deposition from Silane Discharges," J. Appl. Phys., Vol. 63, 1988b, pp. 2406-2413.

Gallagher(08), A., D.A. Doughty, J.R. Doyle, "Diagnostics of Glow Discharges used to Produce a-Si:H Film," Proc. of the SERI subcontractors' review meeting, June 19-20, 1989, Golden, Colorado, pp. 47-56, NIST Accesion No. DE89009423, SERI/CP-2113514.

Gallagher(09), A., D.A. Doughty and J. Doyle, "Diagnostics of Glow Discharges used to Produce a-Si:H Film," Annual Technical Report SERI/TP-211-3747, National Renewable Energy Laboratory subcontract DB-4-04078-1, 1990, NTIS Accession No. DE90000329.

Gallagher(10), A., R. Ostrom, G. Stutzin, and D. Tanenbaum, Annual Technical Report, National Renewable Energy Lab., subcontract DD-11001-1, 1992.

Galloni(1), R., Y.S. Tsuo, and F. Zignani, "Ion Implantation and Hydrogen Passivation in Amorphous Silicon Films," Nuclear Instruments and Methods in Physics Research, Vol. B39, 1989, pp. 386-388.

Galloni(2), R., Y.S. Tsuo, D.W. Baker, and F. Zignani, "Doping and Hydrogenation by Ion Implantation of Glow Discharge Deposited Amorphous Silicon Films," Appl. Phys. Lett., Vol. 56, 1990, pp. 241-243.

Ganguly, G., S.C. De, S. Ray, and A.K. Barua, "Polycrystalline Silicon Carbide Films Deposited by Low-Power Radio-Frequency Plasma Decomposition of SiF_4-CF_4-H_2 gas mixtures," J. Appl. Phys., Vol. 69, 1991, pp. 3915-3923.

Garscadden, A., "PECVD of Discharge Models Review," Mat. Res. Soc. Symp. Proc., Vol. 165, 1990, pp. 3-15.

Ghosh, A.K., T. McMahon, E. Rock, and H. Wiesmann, "Optical and Electrical Properties of Evaporated Amorphous Silicon with Hydrogen," J. Appl. Phys., Vol. 50, 1979, pp. 3407-3413.

Girginoudi, D., A. Thanailakis, and A. Christou, "Amorphous $(SiC)_xGe_{1-x}$:H Films Prepared by RF Sputtering: Optical and Electrical Properties," J. Appl. Phys., Vol. 62, 1987, pp. 3353-3359.

Giunta, C.J., R.J. McCurdy, J.D. Chapple-Sokol, and R.G. Gordon, "Gas-Phase Kinetics in the Atmospheric Pressure Chemical Vapor Deposition of Silicon from Silane and Disilane," J. Appl. Phys., Vol. 67, 1990, pp. 1062-1075.

Givargizov, E.I., "Oriented Crystallization on Amorphous Substrates," Plenum Press, New York, 1991.

Gleason(1), K.K., J. Baum, A.N. Garroway, A. Pines, and J.A. Reimer, "Multiple Quantum NMR Study of Hydrogen Clustering in Amorphous Silicon," Mat. Res. Soc. Symp. Proc., Vol. 70, 1986, pp. 83-88.

Gleason(2), K.K., M.A. Petrich, and J.A. Reimer, "Hydrogen Microstructure in Amorphous Semiconductors," Mat. Res. Soc. Symp. Proc., Vol. 95, 1987a, pp.171-176.

Gleason(3), K.K., K.S. Wang, M.K. Chen, and J.A. Reimer, "Monte Carlo Simulations of Amorphous Hydrogenated Silicon Thin-Film Growth," J. Appl. Phys., Vol. 61, 1987c, pp. 2866-2873.

Goldfarb, S.R., "Mass Spectrometry for IC Fabrication," Semiconductor International, Oct. 1986, pp. 55-60.

Goldstein(1), B., J. Dresner, and D.J. Szostak, "The diffusion of holes in undoped a-Si:H," Phil. Mag. B, Vol. 46, 1982, pp. 63-70.

Goldstein(2), B, C.R. Dickson, I.H. Campbell, & P.M. Fauchet, "Electrical, Optical and Structural Properties of p^+ Microcrystalline Si:C:H Deposited by RF Glow Discharge," Proc. 8th Euro. Comm. PV Solar Energy Conf., 1988, pp. 969-975.

González, P., M.D. Fernández, B. León, and M. Pérez-Amor, "Hydrogenated Amorphous Silicon Films Deposited by Laser CVD," Proc. 9th Euro. Comm. PV Solar Energy Conf., 1989, pp. 1017-1020.

Gottscho, R.A., and M.L. Mandich, "Time-Resolved Optical Diagnostics of Radio Frequency Plasmas," J. Vac. Sci. Technol., Vol. A3, 1985, pp. 617-624.

Grasso, V., A.M. Mezzasalma, and F. Neri, "A New Evaporation Method for Preparing Hydrogenated Amorphous Silicon Films," Solid State Comm., Vol. 41, 1982, pp. 675-677.

Green, M.A., "Solar Cells," Prentice-Hall, Inc. Englewood Cliffs, NJ, 1982.

Griffith, R.W., "Introduction to Basic Aspects of Plasma-Deposited Amorphous Semiconductor Alloys in Photovoltaic Conversion," in Solar Material Science, L.E. Murr, ed., New York: Academic Press, 1980, pp. 665-731.

Grove, A.S., "Physics and Technology of Semiconductor Devices," 1967, John Wiley & Sons, Inc., New York.

Guha(01), S., "Light-Induced Effects in Amorphous Silicon Alloys - Design of Solar Cells with Improved Stability," J. Non-Cryst. Solids, Vols. 77 & 78, 1985, pp. 1451-1460.

Guha(02), S., and J. Kulman, U. S. Patent No. 4,600,801, July 15, 1986a.

Guha(03), S., J. Yang, P. Nath, and M. Hack, "Enhancement of Open Circuit Voltage in High Efficiency Amorphous Silicon Alloy Solar Cells," Appl. Phys. Lett., Vol. 49, 1986b, pp. 218-219.

Guha(04), S., J.S. Payson, S.C. Agarwal, and S.R. Ovshinsky, "Fluorinated Amorphous Silicon-Germanium Alloys Deposited from Disilane-Germane

Mixture," presented at the 12th Intern. Conf. on Amorphous and Liquid Semiconductors, Prague, Czechoslovakia, August 1987, J. Non-Cryst. Solids, 97&98, 1987, pp. 1455.

Guha(05), S., "Research on High-Efficiency, Multi-Gap, Multijunction, Amorphous-Silicon-Based Alloy Thin-Film Solar Cells," Phase 1 Semiannual Subcontract Report for the Period 1 March-31 August 1987, SERI/STR-211-3374, Solar Energy Research Inst., 1988a. NTIS Accession Number DE89000844.

Guha(06), S., J. Yang, A. Pawlikiewicz, T. Glatfelter, R. Ross, and S. R. Ovshinsky, "A Novel Design for Amorphous Silicon Alloy Solar Cells," Proc. 20th IEEE PV Specialists Conf., IEEE, New York, 1988b, p. 79-84.

Guha(07), S., and S.R. Ovshinsky, U. S. Patent No. 4,775,425, October 4, 1988c.

Guha(08), S., A. Bannerjee, J. Burdick, E. Chen, T. Glatfelter, "Research on High-Efficiency, Multiple-Gap, Multijunction, Amorphous Silicon-Based Alloy Thin-Film Solar Cells," Annual Subcontract Report for the Period 3/1/1988-2/28/1989, STR-211-3581, Solar Energy Research Inst., 1989a

Guha(09), S., J. Yang, A. Pawlikiewicz, R. Ross, T. Glatfelter, J. Burdick, and A. Banerjee, "Progress in High-Efficiency, Multiple-Gap, Multijunction Amorphous Silicon-Based Alloy Thin-Film Solar Cells," SERI/CP-211-3514, Solar Energy Research Inst., 1989b, pp. 119-126.

Guha(10), S., "Research on Stable, High-Efficiency Amorphous Silicon Multijunction Modules," Subcontract report 1/1/1992-6/30/1992, United Solar Systems Corporation. NREL TR-411-5063, 1992.

Guha(11), S., J. Yang, S.J. Jones, Y. Chen, and D.L. Williamson, "Effect of microvoids on initial and light-degraded efficiencies of hydrogenated amorphous silicon alloy solar cells," Appl. Phys. Lett., Vol. 61, 1992, pp. 1444-1446.

Hack, M. and R.A. Street, "Realistic Modeling of the Electronic Properties of Doped Amorphous Silicon," Appl. Phys. Lett., Vol. 53, 1988, pp. 1083-1085.

Haller, I., Y.H. Lee, J.J. Nocera, Jr., and M.A. Jaso, "Selective Wet and Dry Etching of Hydrogenated Amorphous Silicon and Related Materials," J. Electrochem. Soc.:Solid-State Sci. & Tech., Vol. 135, 1988, pp. 2042-2045.

Hamakawa(1), Y., Y. Tawada, K. Nishimura, K. Tsuge, M. Kondo, K. Fujimoto, S. Nonomura, & H. Okamoto, "Design Parameters of High Efficiency a-SiC:H/a-Si:H Heterojunction Solar Cells," Proc. 16th IEEE PV Specialists Conf., 1982,

pp. 679-8684.

Hamakawa(2), Y., Y. Matsumoto, G. Hirata, and H. Okamoto, "Optoelectronics and Photovoltaic Applications of Microcrystalline SiC," Mat. Res. Soc. Symp. Proc., Vol. 164, 1989, pp. 291-301.

Hamakawa(3), Y., "Recent Progress of Amorphous Silicon Solar Cell Technology in Japan," Proc. 22nd IEEE PV Specialists Conf., 1991, pp. 1199-1206.

Hamasaki(1), T., M. Ueda, M. Hirose, and Y. Osaka, "Growth Kinetics of Amorphous Hydrogenated Silicon Studied by Pulsed RF Discharge," J. Non-Crystal. Solids, Vol. 59/60, 1983a, pp. 679-682.

Hamasaki(2), T., M. Ueda, M. Hirose, and Y. Osaka, "High-Rate Deposition of a-Si:H using SiH_4," Extended Abstracts, 15th Conf. on Solid State Devices and Materials, Tokyo, Japan, 1983b, pp. 193-196.

Hamasaki(3), T., M. Ueda, A. Chayahara, M. Hirose, and Y. Osaka, "High-rate Deposition of Amorphous Hydrogenated Silicon from a SiH_4 Plasma," Appl. Phys. Lett., Vol. 44, 1984, pp. 600-602.

Hanabusa(1), M., A. Namiki, and K. Yoshihara, "Laser-Induced Vapor Deposition of Silicon," Appl. Phys. Lett., Vol. 35, 1979, pp. 626-627.

Hanabusa(2), M. and M. Suzuki, "Reactive Laser-Evaporation for Hydrogenated Amorphous Silicon," Appl. Phys. Lett., Vol. 39, 1981, pp. 431-432.

Hanaki, K., T. Hattori, and Y. Hamakawa, "Characterization of High-Conductive p-Type a-SiC:H Produced by Highly Hydrogen Dilution," Technical Digest of the Int'l PVSEC-3, Tokyo, Japan 1987, pp. 49-52.

Hanna(1), J., S. Oda, H. Shibata, H. Shirai, A. Miyauchi, A. Tanabe, K. Fufuda, T. Ohtoshi, O. Tokuhiro, H. Nguyen, and I. Shimizu, "Reactive Deposition of a-Silicon and Si-Based Alloys," Mat. Res. Soc. Symp. Proc., Vol. 70, 1986, pp. 11-16.

Hanna(2), J., N. Shibata, K. Fukuda, H. Ohtoshi, S. Oda, and I. Shimizu, "Preparation of a-Si and its Related Materials by Hydrogen Radical Enhanced CVD," Disordered Semiconductors, edited by Kastner, Thomas, and Ovshinsky, Plenum Publ. Co., 1987, pp. 435-446.

Hanna(3), J., A. Kamo, M. Azuma, N. Shibata, H. Shirai, and I. Shimizu, "Chemical Reactions in Propagation of Si:H(F)-Networks," Mat. Res. Soc. Symp.

Proc., Vol. 118, 1988, pp. 79-84.

Hanna(4), J., A. Kamo, T. Komiya, H. D. Nguyen, I. Shimizu, and H. Kokado, "Preparation of Si Thin Films by Spontaneous Chemical Deposition," Mat. Res. Soc. Symp. Proc., Vol. 149, 1989, pp. 11-16.

Hasegawa, S., Y. Amano, T. Inokuma, and Y. Kurata, "Relationship between the Stress and Bonding Properties of Amorphous SiN_x:H Films," J. Appl. Phys., Vol. 72, 1992, pp. 5676-5681.

Hata(1), N., S. Yamasaka, H. Oheda, A. Matsuda, H. Okushi, K. Tanaka, "A Photoluminescence Study of Amorphous-Microcrystalline Mixed-Phase Si:H Films," Japan. J. Appl. Phys., Vol. 20, 1981, pp. L793-796.

Hata(2), N. et al., Proc. Int'l Ion Engineering Congress - ISIAT'83 & IPAT'83, Inst. of Electrical Engineering of Japan, 1983a, p. 1457.

Hata(3), N., A. Matsuda, K. Tanaka, K. Kajiyama, N. Moro, and K. Sajiki, "Detection of Neutral Species in Silane Plasma using Coherent Anti-Stokes Raman Spetroscopy," Jpn. J. Appl. Phys., Vol. 22, 1983b, pp. L1-L3.

Hata(4), N., and K. Tanaka, "Silana Plasma and Surface Processes in Amorphous Silicon Deposition," J. Non-Crystal. Solids, Vol. 77 & 78, 1985, pp. 777-780.

Hata(5), N., and S. Wagner, "A comprehensive defect model for amorphous silicon," J. Appl. Phys. Vol. 72, 1992, pp. 2857-2872.

Hattori(1), Y., S. Mizuki, K. Maekawa, H. Okamoto, and Y. Hamakawa, "DC-Biased RF Plasma CVD for Hydrogenated Amorphous Silicon," Proc. of 1st Int'l PV Sci. & Eng. Conf., 1984, pp. 731-734.

Hattori(2), Y., D. Kruangam, K. Katoh, Y. Nitta, H. Okamoto, & Y. Hamakawa, "High-Conductivity Wide Band Gap p-Type a-SiC:H Prepared by ECR CVD and its Application to High Efficiency a-Si Basis Solar Cells," Proc. 19th PV Specialists Conf., 1987a, pp. 689-694.

Hattori(3), Y., D. Kruangam, T. Toyama, H. Okamoto, and Y. Hamakawa, "High Efficiency Amorphous Heterojunction Solar Cell Employing ECR-CVD Produced p-Type Microcrystalline SiC Film," Tech. Digest 3rd Int'l PV Sci. & Eng. Conf., 1987b, pp. 171-174.

Hattori(4), Y., D. Kruangam, T. Toyama, H. Okamoto, and Y. Hamakawa, "Valence Control of p-Type a-SiC:H Having the Optical Band Gap More than 2.5 eV by Electron-Cyclotron Resonance CVD (ECR CVD)," J. Non-Cryst.

Solids, Vol. 97&98, 1987c, pp. 1079-1082.

Hattori(5), Y., D. Kruangam, T. Toyama, H. Okamoto, and Y. Hamakawa, "Highly Conductive p-Type Microcrystalline SiC:H Prepared by ECR Plasma CVD," Applied Surface Science, Vol. 33/34, 1988b, pp. 1276-1284.

Hauser, J.J., "Effect of Hydrogen on Amorphous Silicon," Solid State Comm., Vol. 19, 1976, pp. 1049-1051.

Hayama, H. and T. Saito, "Temperature Rise During Silicon-on-Glass Recrystallization Produced by AC Magnetic Fields," J. Appl. Phys., Vol. 72, 1992, pp. 2817-2822.

Hayama, M., H. Murai, and K. Kobayashi, "Ion Bombardment Effects on Undoped Hydrogenated Amorphous Silicon Films Deposited by the Electron Cyclotron Resonance Plasma Chemical Vapor Deposition Method," J. Appl. Phys., Vol. 67, 1990, pp. 1356-1360.

Hebner, G.A., and M.J. Kushner, "Phase and Energy Distribution of Ions Incident on Electrodes in Radio-Frequency Discharges," J. Appl. Phys., Vol. 62, 1987, pp. 2256-2260.

Hegedus(1), S.S., R.E. Rocheleau, and B.N. Baron, "CVD Amorphous Silicon Solar Cells," Proc. 17th IEEE PV Specialists Conf., 1984, pp. 239-244.

Hegedus(2), S.S., R.E. Rocheleau, J.M. Cebulka, and B.N. Baron, "Density of Midgap States and Urbach Edge in Chemically Vapor Deposited Hydrogenated Amorphous Silicon Films," J. Appl. Phys., Vol. 60, 1986, pp. 1046-1054.

Hegedus(3), S.S., R.M. Tullman, H.S. Lin, J.M. Cebulka, W.A. Buchanan, R. Dozier, and R.E. Rochelau, "Low Bandgap Amorphous Silicon-Germanium Alloys for Thin Film Solar Cells Using a Novel Photo-CVD Reactor," Proc. 19th IEEE PV Specialists Conf., 1987a, pp. 867-871.

Hegedus(4), S. S., M. Schmidt, & N. Salzman, "Measurement of the Built-In Potential in Amorphous Silicon p-i-n Solar Cells," 19th IEEE PV Specialists Conf., 1987b, pp. 210-215.

Hegedus(5), S.S., R.E. Rocheleau, R.M. Tullman, D.E. Albright, N. Saxene, W.A. Buchanan, K.E. Schubert, and R. Dozier, "Photo-Assisted CVD of a-Si:H Solar Cells and a-SiGe:H Films," Proc. 20th IEEE PV Specialists Conf., 1988a, pp. 129-134.

Hegedus(6), S.S., "The Open Circuit Voltage of Amorphous Silicon p-i-n Solar

Cells," 20th IEEE PV Specialists Conf., 1988b, pp. 102-107.

Hentzell, H.T.G., P.A. Psaras, and K.N. Tu, "Interfacial Reaction between Amorphous Silicon and Palladium Thin Films," Mater. Lett., Vol. 3, 1985, pp. 255-260.

Herd, S.R., P. Chaudhari, and M.H. Brodsky, "Metal Contact Induced Crystallization in Films of Amorphous Silicon and Germanium," J. Non-Cryst. Solids, Vol. 7, 1972, pp. 309-327.

Hesch, K., P. Hess, H. Oetzmann, and C. Schmidt, "Precision Surface Temperature Measurement and Film Characterization for LICVD of a-Si:H from SiH_4," Applied Surface Science, Vol. 36, 1989, pp. 81-88.

Hiraki, A., Y. Fukushima, T. Sato, H. Kiyono, H. Terauchi, and T. Imura, "Transformation of Microcrystalline State of Hydrogenated Silicon to Amorphous One due to Presence of More Electronegative Impurities," J. of Non-Crystal. Solids, Vol. 59&60, 1983, pp. 791-794.

Hirose(1), M., T. Hamasaki, Y. Mishima, H. Kurata, and Y. Osaka, in Tetrahedrally Bonded Amorphous Semiconductors, R.A. Street, D.K. Biegelsen, and J.C. Knights, eds., New York: Am. Inst. Phys., 1981, p.10.

Hirose(2), M., "Preparation and properties of a-SiN:H," in JARECT Vol.6, Amorphous Semiconductor Technologies & Devices (1983), Y. Hamakawa (ed.), OHMSHA Ltd. and North-Holland Publishing Co. pp. 173-180.

Hirose(3), M., "Chemical Vapor Deposition," Semiconductors and Semimetals, Vol. 21A, 1984b, pp. 109-122.

Hishikawa(1), Y., S. Okamoto, K. Wakisaka, N. Nakamura, S. Tsuda, S. Nakano, M. Ohnishi, and Y. Kuwano, "An Approach to High-Efficiency a-Si Solar Cells by Accurate Determination and Analysis of the Optical Absorption Coefficient," Proc. 9th Euro. Comm. PV Conf., 1989, pp. 37-40.

Hishikawa(2), Y., S. Tsuge, N. Nakamura, S. Tsuda, S. Nakano, and Y. Kuwano, "Effect of Substrates and Film Thickness on the Structural, Optical, and Electrical Properties of Hydrogenated Amorphous Silicon Films," Appl. Phys. Lett., Vol. 57, 1990, pp. 771-788.

Hollingsworth(1), R.E., P.K. Bhat, and A. Madan, "Amorphous Silicon Carbide Solar Cells," Proc. 19th PV Specialists Conf., 1987a, pp. 684- 688.

Hollingsworth(2), R.E., P.K. Bhat, and A. Madan, "Microcrystalline and Wide

Band Gap p^+ Window Layers for a-Si p-i-n Solar Cells," J. Non-Cryst. Solids, Vol. 97&98, 1987b, pp. 309-312.

Hotta(1), S., Y. Tawada, H. Okamoto, and Y. Hamakawa, "Effects of the Substrate Potential on the Incorporation Manner of Hydrogen and Impurity in a-Si:H Films," J. de Physique, Col. C4, Suppl. au No. 10, Vol. 42, 1981, pp. C4-631-634.

Hotta(2), S., N. Nishimoto, Y. Tawada, H. Okamoto, and Y. Hamakawa, "Optimization of GD a-Si:H Film Property for Photovoltaic Device by Means of the Cross Field Plasma Deposition Technique," Jpn. J. Appl. Phys., 21, Suppl. 21-1, 1982, pp. 289-295.

Hourd, A.C., D.L. Melville, W.E. Spear, "Electronic Properties of Amorphous and Microcrystalline Silicon Prepared in a Microwave Plasma from SiF_4," Philosophical Mag. B, 1991, Vol. 64, pp. 533-550.

Houssaini, S., M. Vergnat, A. Bruson, G. Marchal, and C. Vettier, "Study of the Hydrogen Stability in Evaporated Amorphous $Si_{1-x}Sn_x$:H ($0 \leq x \leq 0.2$) Alloys by Neutron Scattering and Exodiffusion Measurements," J. Appl. Phys., Vol. 73, 1993, pp. 483-485.

Huang, C-Y., C. Salupo, L.F. Szabo, G.P. Ceasar, and W. Javurek, "A New Type of Stable and Sensitive UV Detector Fabricated with Amorphous Silicon Based Alloys," Mat. Res. Soc. Symp. Proc., Vol. 118, 1988, pp. 411-416.

Hudgens(1), S.J., A.G. Johncock, and S.R. Ovshinsky, "Low Pressure Microwave Discharge Process for High Deposition Rate Amorphous Silicon Alloy," J. Non-Crystal. Solids, Vol. 77&78, 1985a, pp. 809-812.

Hudgens(2), S.J. and A.G. Johncock, "High Deposition Rate Amorphous Silicon Alloy Xerographic Photoreceptor," Mat. Res. Soc. Symp. Proc., Vol. 49, 1985b, pp. 403-408.

Hundhausen, M., P. Santos, L. Ley, F. Habraken, W. Beyer, R. Primig, and G. Gorges, "Characterization of Superlattices Based on Amorphous Silicon," J. Appl. Phys., Vol. 61, 1987, pp. 556-560.

Ibaraki, N., and H. Fritzsche, "Properties of amorphous semiconducting a-Si:H/a-SiNx:H multilayer films and of a-SiNx:H alloys," Phys. Rev. B, Vol. 30, 1984, pp. 5791-5799.

Ibbotson, D.E., D.L. Flamm, J.A. Mucha, and V.M. Donnelly, "Comparison of XeF_2 and F-atom Reactions with Si and SiO_2," Appl. Phys. Lett., Vol. 44, 1984,

pp. 1129-1131.

Ichikawa(1), Y., "Recent Progess in Stability of a-Si Solar Cells," Techinical Digest of the Int'l PVSEC-5, Kyoto, Japan, 1990a, pp. 617-621.

Ichikawa(2), Y., S. Fujikake, T. Yoshida, T. Hama, and H. Sakai, "A Stable 10% Solar Cell with a-Si/a-Si Double-Junction Structure," Proc. 21st IEEE PV Specialists Conf., 1990b, pp. 1475-1480.

Ichimura(1), T., T. Ihara, T. Hata, H. Ohsawa, H. Sakai, and Y. Ushida, "Electrical and Structural Properties of a-SiGe:H Films," J. Non-Crystal. Solids, Vol. 77&78, 1985, pp. 901-904.

Ichimura(2), T., T. Hara, T. Hama, M. Ohsawa, H. Sakai, and Y. Uchida, "Properties of a-SiGe:H and a-SiGe:H:F Alloys," J. Non-Cryst. Solids, Vol.77&78, 1985, pp. 881-884.

Imamura, T., H. Kaya, H. Yerauchi, H. Kiyono, A. Hiraki, and M. Ichihara, "Structure Change of Microcrystalline Silicon Films in Deposition Process," Jpn. J. Appl. Phys., Vol. 23, 1984, pp. 179-183.

Inoue, G., and M. Suzuki, "Reactions of $SiH_2(X^1A_1)$ With H_2, CH_4, C_2H_4, SiH_4 and Si_2H_6 at 298 K," Chemical Phys. Lett., Vol. 122, 1985, pp. 361-364.

Inoue, T., M. Konagai, and K. Takahashi, "Photochemical Vapor Deposition of Undoped and n-type Amorphous Silicon Films Produced from Disilane," Appl. Phys. Lett., Vol. 43, 1983, pp. 774-776.

Ishihara(1), S., K. Masatoshi, H. Takashi, W. Kiyotaka, T. Arita, and K. Mori, "Effects of Discharge Parameters on Deposition Rate of Hydrogenated Amorphous Silicon for Solar Cells from Pure SiH_4 Plasma," J. Appl. Phys., Vol. 62, 1987a, pp. 485-491.

Ishihara(2), S., M. Kitagawa, and T. Hirao, "Anomalous Deposition Rate Dependence of a-Si:H on Substrate Temperature," J. Appl. Phys., Vol. 62, 1987b, pp. 3060-3061.

Ishihara, T., S. Terazono, H. Sasaki, K. Kawabata, T. Itagaki, H. Morikawa, M. Deguchi, K. Sato, M. Usui, K. Okaniwa, M. Aiga, M. Otsubo, and K. Fujikawa, "High Efficiency Triple-Junction Amorphous Solar Cells," Proc. 19th IEEE PV Specialists Conf., 1987, pp. 749-755.

Ishii, M., K. Nakashima, I. Tajima, and M. Yamamoto, "Properties of Silicon Surface Cleaned by Hydrogen Plasma," Appl. Phys. Lett. Vol. 58, 1991, pp.

1378-1380.

Ishikawa, A., D. Kruangam, H. Okamoto, and Y. Hamakawa, "Transparent Polymer Film Type Substrate for Amorphous Silicon Solar Cell," Proc. 18th IEEE PV Specialists Conf., 1985, pp. 487-492.

Ishimura, T., Y. Okayasu, H. Yamamoto, and K. Fukui, "A Study on the Surface Reaction in the Growth of Amorphous Silicon by Intermittent Deposition Method," Proc. 20th IEEE PV Specialists Conf., 1988, pp. 114-118.

Itabashi, N., K. Kato, N. Nishiwaki, T. Goto, C. Yamada, and E. Hirota, "Measurement of the SiH_3 Radical Density in Silane Plasma using Infrared Diode Laser Absorption Spectroscopy," Jpn. J. Appl. Phys, Vol. 27, 1988, L1565-1567.

Itozaki(1), H., N. Fujita, and H. Hitotsuyanagi, "Amorphous Silicon-Germanium Deposited by Photo-CVD," Mat. Res. Soc. Symp. Proc., Vol. 49, 1985, pp. 161-166.

Itozaki(2), H., and N. Fujita, "Properties of a-SiGe and Application to Stacked Solar Cells," Mat. Res. Soc. Symp. Proc., Vol. 95, 1987, pp. 507-515.

Ivanda, M., K. Furić, O. Gamulin, M. Peršin, and D. Gracin, "CW Laser Crystallization of Amorphous Silicon: Thermal or Athermal Process," J. Appl. Phys., Vol. 70, 1991, pp. 4637-4639.

Iwanaga, T. and M. Hanabusa, "CO_2 Laser CVD of Disilane," Jpn. J. Appl. Phys., Vol. 23, 1984, pp. L473-475.

Jackson, R. L., J.E. Spencer, J.L. McGuire, and A.M. Hoff, "Afterglow Chemical Vapor Deposition of SiO_2," Solid State Technology, April 1987, pp. 107-111.

Jackson(1), W. B., and N. M. Amer, "Piezoelectric Photoacoustic Detection: Theory and Experiment," J. Appl. Phys., Vol. 51, 1980, pp. 3343-3353.

Jackson(2), W. B., N. M. Amer, A. C. Boccara, and D. Fournier, "Photothermal Deflection Spectroscopy and Detection," Appl. Optics, Vol. 20, 1981, pp. 1333-1334.

Jackson(3), W. B., S.M. Kelso, C.C. Tsai, J.W. Allen, and S.-J. Oh, "Energy Dependence of the Optical Matrix Element in Hydrogenated Amorphous and Crystalline Silicon," Phys. Rev. B, Vol. 31, 1985, pp. 5187-5198.

Jacobson, R.L., F.R. Jeffrey, R.K. Westerberg, R.C. Williams, "Amorphous Silicon P-I-N Layers Prepared by a Continuous Deposition Process on Polyimide

Web," Proc. 19th IEEE PV Specialists Conf., 1987, pp. 588-592.

Janai, M., R. Weil, and B. Pratt, "Properties of Fluorinated Glow-Discharge Amorphous Silicon," Phys. Rev. B, Vol. 31, 1985, pp. 5311-5321.

Jansen(1), F., J. Mort, and M. Morgan, "Nature and Distribution of Radicals in RF and DC Silane Discharges; Effects on Deposition Rate and Physical Properties of a-Si:H," Canadian J. Chem., Vol. 63, 1985, pp. 217-220.

Jansen(2), F., M.A. Machonkin, N. Palmieri, and D. Kuhman, "Thermal Expansion and Elastic Properties of Plasma-Deposited Amorphous Silicon and Silicon Oxide Films," Appl. Phys. Lett., Vol. 50, 1987, pp. 1059-1061.

Jansen(3), F., I. Chen, and M.A. Machonkin, "On the Thermal Dissociation of Hydrogen," J. Appl. Phys., Vol. 66, 1989, pp. 5749-5755.

Jasinski(1), J.M., E.A. Whittaker, G.C. Bjorklund, R.W. Dreyfus, R.D. Estes, and R.E. Walkup, "Detection of SiH_2 in Silane and Disilane Glow Discharge by Frequency Modulation Absorption Spectroscopy," Appl. Phys. Lett., Vol. 44, 1984, pp. 1155-1157.

Jasinski(2), J.M., B.S. Meyerson, and T.N. Nguyen, "Exicimer Laser-Induced Deposition of Silicon Nitride Thin Films," J. Appl. Phys., Vol. 61, 1987a, pp. 431-433.

Jasinski(3), J.M., B.S. Meyerson, and B.A. Scott, "Mechanistic Studies of Chemical Vapor Deposition," Annu. Rev. Phys. Chem., Vol. 38, 1987b, pp. 109-140.

Jasinski(4), J.M., and J.O. Chu, "Absolute Rate Constants for the Reaction of Silylene with Hydrogen, Silane and Disilane," J. Chem. Phys., Vol. 88, 1988, pp. 1678-1687.

Jasinski(5), J.M., "Direct Kinetic Studies of Silicon Hydride Radicals," Mat. Res. Soc. Symp. Proc., Vol. 165, pp. 41-48, 1990.

Jia, Q.X., and W.A. Anderson, "Sputter Deposition of $YBa_2Cu_3O_{7-x}$ Films on Si at 500°C with Conducting Metallic Oxide as a Buffer Layer," Appl. Phys. Lett., Vol. 57, 1990, pp.304-307.

Joannopoulos, J.D. and G. Lucovsky, "The Physics of Hydrogenated Amorphous Silicon I & II," Topics in Applied Physics, Vol. 56, 1984.

Johnson(1), N.M., S.E. Ready, J.B. Boyce, C.D. Doland, S.H. Wolff, & J.

Walker, "Hydrogen Incorporation in Undoped Microcrystalline Silicon," Appl. Phys. Lett. Vol. 53, 1988a, pp. 1626-1628.

Johnson(2), N.M., S.H. Wolff, C.D. Doland, and J. Walker, "Dependence of Hydrogen Incorporation in Undoped a-Si:H and μc-Si:H on Hydrogen Dilution During PECVD," Mat. Res. Soc. Symp. Proc., Vol. 118, 1988b, pp. 85-90.

Johnson(3), N.M., J. Walker, C.M. Doland, K. Winer, and R.A. Street, "Hydrogen Incorporation in Silicon Thin Films Deposited with a Remote Hydrogen Plasma," Appl. Phys. Lett., Vol. 54, 1989, pp. 1872-1874.

Jones, D.I., W.E. Spear, and P.G. LeComber, "Transport Properties of Amorphous Germanium Prepared by the Glow Discharge Techniques," J. Non-Cryst. Solids, Vol. 20, 1976, pp. 259-270.

Jones(1), S.J., S.M. Lee, W.A. Turner, and W. Paul, "Substrate Temperature Dependence of the Structural Properties of Glow Discharge Produced a-Ge:H," Mat. Res. Soc. Symp. Proc., Vol. 149, 1989, pp. 45-50.

Jones(2), S.J., Y. Chen, D.L. Williamson, and G.D. Mooney, "Small-Angle X-Ray Scattering Studies of Glow-Discharge-Produced a-SiGe:H Alloys," Mat. Res. Soc. Symp. Proc., Vol. 258, 1992, pp. 229-234.

Kakkad, R., S.J. Fonash, and S. Weideman, "Highly Conductive Ultrathin Crystalline Si Layers by Thermal Crystallization of Amorphous Si," Appl. Phys. Lett., Vol. 59, 1991, pp. 3309-3311.

Kamisako, K., K. Aota, and Y. Tarui, "Analysis of Deposition Rate Distribution in the Photo-CVD of a-Si by a Unified Reactor with a Lamp," Jpn. J. Appl. Phys., Vol. 23, 1984, pp. L776-L778.

Kamiya, T., M. Kishi, A. Ushirokawa, and T. Katoda, "Observation of the Amorphous-to-Crystalline Transition in Silicon by Raman Scattering," Appl. Phys. Lett., Vol. 38, 1981, pp. 377-379.

Kampas(1), F.J., "Reactions of Atomic Hydrogen in the Deposition of Hydrogenated Amorphous Silicon by Glow Discharge and Reactive Sputtering," J. Appl. Phys., Vol. 53, 1982, pp. 6408-6412.

Kampas(2), F.J., "An Optical Emission Study of the Glow-Discharge Deposition of a-Si:H from Argon-Silane Mixture," J. Appl. Phys., Vol. 54, 1983, pp. 2276-2280.

Kampas(3), F.J., "Chemical Reactions in Plasma Deposition," Semiconductors and Semimetals, Vol. 21, Part A, 1984, pp. 153-177.

Kampas(4), F.J., "Gas-Phase Free Radical Reactions in the Glow-Discharge Deposition of Hydrogenated Amorphous Silicon from Silane and Disilane," J. Appl. Phys., Vol. 57, 1985, pp. 2290-2291.

Kampas(5), F.J., and M.J. Kushner, "Effect of Silane Pressure on Silane-Hydrogen RF Glow Discharges," IEEE Transactions on Plasma Science, Vol. PS-14, 1986, pp. 173-178.

Kane, P.F., and G.B. Larrabee, "Characterization of Solid Surfaces," Plenum Press, N.Y., 1978.

Kanicki(1), J., "Contact Resistance to Undoped and Phosphorus-Doped Hydrogenated Amorphous Silicon Films," Appl. Phys. Lett., Vol. 53, 1988, pp. 1943-1945.

Kanicki(2), J., F.R. Libsch, J. Griffith, and R. Polastre, "Performance of Thin Hydrogenated Amorphous Silicon Thin-Film Transistors," J. Appl. Phys., Vol. 69, 1991a, pp. 2339-2345.

Kanicki(3), J., "Amorphous and Microcrystalline Semiconductor Devices: Optoelectronic Devices," Artech House, MA; 1991b.

Kaplan, D., N. Sol, G. Velasco, and P.A. Thomas, "Hydrogenation of Evaporated Amorphous Silicon Films by Plasma Treatment," Appl. Phys. Lett., Vol. 33, 1978, pp. 440-442.

Kasper, W., H. Bohm, and B. Hirschauer, "The influence of electrode areas on radio-frequency glow discharge," J. Appl. Phys. Vol. 71, 1992, pp. 4168-4172.

Kato(1), I., S. Wakana, S. Hara, and H. Kezuka, "Microwave Plasma CVD System for the Fabrication of Thin Solid Films," Jpn. J. Appl. Phys., Vol. 21, 1982, pp. L470-L472.

Kato(2), I., "Microwave Sputtering System for the Fabrication of Thin Solid Films," Rev. Sci. Instrum., Vol. 53, 1982, pp. 214-216.

Kato, S., and T. Aoki, "High-Rate Deposition of a-Si:H using Electron Cyclotron Resonance Plasma," J. Non-Crystal. Solids, Vols. 77&78, 1985, pp. 813-816.

Kausche, H., K. Prasad, and R. Plattner, "Wave Resonance Plasma for Deposition of Thin Films for Amorphous Silicon Solar Cells," Proc. of 9th Euro. Comm. PV Solar Energy Conf., 1989, pp. 595-598.

Kawai, Y., and K. Sakamoto, "Production of a Large Diameter Hot-Electron Plasma by Electron Cyclotron Resonance Heating," Rev. Sci. Instrum., Vol. 53, 1982, pp. 606-609.

Kaya, H., T. Imura, T. Kusano, A. Hiraki, O. Nakamura, Y. Okayasu, & M. Matsumura, "Evaluation of Boron and Phosphorous Doping Microcrystalline Silicon Films," Jpn. J. Appl. Phys., Vol. 23, 1984, pp. L549-L551.

Kazmerski(1), L.L., "Analysis and Characterization of Thin Films: A Tutorial," Solar Cells, Vol. 24, 1988, pp. 387-418.

Kazmerski(2), L.L., "Spectroscopic Scanning Tunneling Microscope Studies of Hydrogenated Amorphous Silicon Solar Cells," Proc. 9th Euro. Comm. PV Solar Energy Conf., 1989, pp. 589-594.

Kenne, J., Y. Ohashi, T. Matsushita, M. Konagai, and K. Takahashi, "High Deposition Rate Preparation of Amorphous Silicon Solar Cells by RF Glow Discharge Decomposition of Disilane," J. Appl. Phys., Vol. 55, 1984, pp. 560-564.

Kim(1), W.Y., H. Tasaki, M. Hallerdt, M. Konagai, and K. Takahashi, "Carbon-Alloyed Graded-Bandgap Layers Prepared by Photo-CVD: Effect on the p/i Interface and Application to the i-Layer of Amorphous Silicon Solar Cells," Proc. 19th PV Specialists Conf., 1987a, pp. 306-311.

Kim(2), W.Y., H. Tasaki, M. Konagai, and K. Takahashi, "Use of Carbon-Alloyed Graded-Bandgap Layer at the p/i Interface to Improve the Photocharacteristics of Amorphous Silicon Alloyed p-i-n Solar Cells Prepared by Photochemical Vapor Deposition," J. Appl. Phys. Vol. 61, 1987b, pp. 3071-3076.

Kim(3), W.Y., S. Yamanaka, M. Hallerdt, M. Konagai, and K. Takahashi, "Control of Amorphous Silicon Solar Cell Interfaces to Obtain High Conversion Efficiencies," Proc. 3rd PV Sci. & Eng. Conf., 1987c, pp. 175-178.

Kim(4), W.Y., M. Konagai, and K. Takahashi, "High Quality Amorphous Silicon Films Prepared by Atmospheric-Pressure Photo-CVD," Proc. 20th IEEE PV Specialists Conf., 1988, pp. 277-281.

Kishi, Y., H. Inoue, H. Tanaka and Y Kuwano, "New-Type of Ultralight Flexible a-Si Solar Cell and its Application on an Airplane," Proc. 22nd IEEE PV Specialists Conf., 1991, pp. 1213-1218.

Kitagawa(1), M., S. Ishihara, K. Setsune, Y. Manabe, and T. Hirao, "Low Temperature Preparation of Hydrogenated Amorphous Silicon by Microwave

Electron-Cyclotron-Resonance Plasma CVD," Jpn. J. Appl. Phys., Vol. 26, 1987b, pp. L231-L233.

Kitagawa(2), M., and T. Hirao, "Low-temperature preparation of doped hydrogenated amorphous silicon films by ac-biased microwave ECR plasma CVD method," Jpn. J. Appl. Phys., Vol. 29, 1990, pp. L1753-L1756.

Knights(1), J.C., "Substitutional Doping in Amorphous Silicon," AIP Conf. Proc., Vol. 31, Amer. Inst. of Physics, New York, 1976, pp. 296-300.

Knights(2), J.C., G. Lucovsky, and R.J. Nemanich, "Defects in Plasma-Deposited a-Si:H," J. Non-Crystal. Solids, Vol. 32, 1979a, pp. 393-403.

Knights(3), J.C., "Characterization of Plasma-Deposited Amorphous Si:H Thin Films," Proc. 10th Conf. on Solid State Devices, Tokyo, Japan, 1978; also in Jpn. J. Appl. Phys., Vol. 18, Supplement 18-1, 1979b, pp. 101-108.

Knights(4), J.C., and R.A. Lujan, "Microstructure of Plasma-Deposited a-Si:H Films," Appl. Phys. Lett., Vol. 35, 1979c, pp. 244-246.

Knights(5), J.C., R.A. Lujan, M.P. Rosenblum, R.A. Street, D.K. Biegelsen, and J.A. Reimer, "Effects of Inert Gas Dilution of Silane on Plasma-Deposited a-Si:H Films," Appl. Phys. Lett., Vol 38, 1981, pp. 331-333.

Knights(6), J.C., "Structure and Chemical Characterization," in Topics in Applied Physics, Vol. 55; The Physics of Hydrogenated Amorphous Silicon I, Structure, Preparation and Devices, J. D. Joannopoulos and G. Lucovsky, eds., Berlin: Springer Verlag, 1984, pp. 5-62.

Knights(7), J.C., "Plasma Deposition of a-Si:H," Mat. Res. Soc. Symp. Proc., Vol. 38, Chang and Abeles, eds., 1985, pp. 371-381.

Kobayashi, K., M. Hayama, S. Kawamoto, and H. Miki, "Characteristics of Hydrogenated Amorphous Silicon Films Prepared by Electron Cyclotron Resonance Microwave Plasma Chemical Vapor Deposition Method and their Application to Photodiodes," Jpn. J. Appl. Phys., Vol. 26, 1987, pp. 202-208.

Komiya, T., A. Kamo, H. Kujirai, I. Shimizu, and J-I. Hanna, "Preparation of Crystalline Si Thin Films by Spontaneous Chemical Deposition," Mat. Res. Soc. Symp. Proc., Vol. 164, 1990, pp. 63-68.

Komuro, S., Y. Aoyagi, Y. Segawa, S. Namba, A. Masuyama, H. Okamoto, and Y. Hamakawa, "Study of Optically Induced Degradation of Conductivity in Hydrogenated Amorphous Silicon by Transient Grating Method," Appl. Phys.

Lett., Vol. 42, 1983, pp. 807-809.

Konagai(1), M., H. Takai, W.Y. Kim, and K. Takahashi, "Preparation of Amorphous Silicon and Related Semiconductors by Photochemical Vapor Deposition and their Application to Solar Cells," Proc. 18th IEEE PV Specialists Conf., 1985, pp. 1372-1377.

Konagai(2), M., "Status of Amorphous Silicon and Related Alloys Prepared by Photochemical Vapor Deposition," Mat. Res. Soc. Symp., Vol. 70, 1986, pp. 257-268.

Konagai(3), M., "Photochemical Vapor Deposition for a-Si Solar Cells," Technical Digest of the Int'l PVSEC-III, Tokyo, Japan, 1987, pp. 15-20.

Kondo, M. H. Nishio, H. Yamagashi, M. Yamaguchi, K. Tsuge, & Y. Tawada, "Effects of Low Level Doping of i-Layer in a-SiC:H/a-Si:H Heterojunction Solar Cells," Proc. 19th PV Specialists Conf., 1987, pp. 604-609.

Konuma, M., "Film Deposition by Plasma Techniques," Springer Verlag, Berlin, Springer Series on Atoms and Plasmas No. 10, 1992.

Koo, Y.C., R. Perrin, K.T. Aust, S. Zukotynski, and R.V. Kruzelecky, "The Effects of the Glass Substrate on the Properties of RF Glow Discharge a-Si:H Thin Films," J. Appl. Phys., Vol. 63, 1988, pp. 2443-2445.

Kragler, G., H. Berner, J. Meier, K. Loble, E. Bucher, and H. Curtins, Proc. 9th Euro. Comm. PV Conf., 1989, pp. 1013-1016.

Kumar, S., and S.C. Agarwal, "Study of Surface Photovoltage in Hydrogenated Amorphous Silicon," J. Appl. Phys., Vol. 58, 1985, pp. 3798-3808.

Kumata, K., U. Itoh, Y. Toyoshima, N. Tanaka, H. Anzai, and A. Matsuda, "Photochemical Vapor Deposition of Hydrogenated Amorphous Silicon Films from Disilane and Trisilane using a Low Pressure Mercury Lamp," Appl. Phys. Lett., Vol. 48, 1986, pp. 1380-1382.

Kumeda, M., H. Komatsu, T. Shimizu, N. Fukuda, and N. Kitagawa, "NMR and Electron Spin Resonance Studies on a-Si:H Prepared by Glow Discharge Decomposition of Si_2H_6," Jpn. J. Appl. Phys., Vol. 24, 1985, pp. L495-L497.

Kurata, H., M. Hirose, and Y. Osaka, "Wide optical-gap photoconductive a-SiN:H," Japanese J. Appl. Phys. 20 (11), 1981, pp. L811-L813.

Kuriyama, H., S. Kiyama, S. Noguchi, T. Kuwahara, S. Ishida, T. Nohda, K.

Sano, H. Iwata, H. Kawata, M. Osumi, S. Tsuda, S. Nakano, and Y. Kuwano, "Enlargement of Poly-Si Film Grain Size by Excimer Laser Annealing and Its Application to High-Performance Poly-Si Thin Film Transistor," Jpn. J. Appl. Phys., Vol. 30, 1991, pp. 3700-3703.

Kurtz, S.R., Y.S. Tsuo, and R. Tsu, "Correlation of Stress with Light-Induced Defects in Hydrogenated Amorphous Silicon Films," Appl. Phys. Lett., Vol. 49, 1986, pp. 951-953.

Kushner(1), M.J. "A Plasma Chemistry and Surface Model for the Deposition of a-SiH from RF Glow Discharge: A Study of Hydrogen Content," Mat. Res. Soc. Symp. Proc., Vol. 68, 1986a, pp. 49-53.

Kushner(2), M.J., "Mechanisms for Power Deposition in Ar/SiH_4 Capacitively Coupled RF Discharges," IEEE Transactions on Plasma Science, Vol. PS-14, April 1986b, pp. 188-196.

Kushner(3), M.J., "A Phenomenological Model for Surface Deposition Kinetics during Plasma and Sputter Deposition of Amorphous Hydrogenated Silicon," J. Appl. Phys, Vol. 62, 1987a, pp. 4763-4772.

Kushner(4), M.J., "On the Balance between Silylene and Silyl Radicals in RF Glow Discharges in Silane: The Effect on Deposition Rates of a-Si:H," J. Appl. Phys., Vol. 62, 1987b, pp. 2803-2811.

Kuwano(1), Y., 1982, Technological Applications of Tetrahedral Amorphous Solids, Palo Alto, unpublished, as shown in Madan, A., "Amorphous Silicon Solar Cells," Chapter 9 of Silicon Processing for Photovoltaics, edited by C. P. Khattak and K. V. Ravi, North-Holland Physics Publisher, 1985.

Kuwano(2), Y., "Photovoltaic Structures by Plasma Deposition," In Plasma Deposited Thin Films, CRC Press, 1986, pp. 161-186.

Kuwano(3), Y. and S. Tsuda, "High Quality P-Type a-SiC Film Doped with $B(CH_3)_3$ and its Application to a-Si solar Cells," Mat. Res. Soc. Symp. Proc., Vol. 118, 1988, pp. 557-567.

Kuwano(4), Y., S. Nakano, M. Tanaka, T. Takahama, T. Matsuyama, M. Isomura, N. Nakamura, H. Haku, M. Nishikuni, H. Nishiwaki, and S. Tsuda, "A More Than 18% Efficiency HIT Structure a-Si/c-Si Solar Cell Using Artificially Constructed Junction," Mat. Res. Soc. Symp. Proc. Vol. 258, 1992, pp. 857-868.

La Marche, P.H. and R. Levi-Setti, "Amorphous Silicon as an Inorganic Resist," Proc. Soc. Photoelectric Instrument Engineers, Vol. 471, 1984, pp. 60-65.

Langford(1), A., B. Stafford, and Y.S. Tsuo, "Maintaining Window Transparency in Photo-Chemical Vapor Deposition: A Simultaneous Etch/Deposition Method," Proc. 19th IEEE PV Specialists Conf., 1987, pp. 573-576.

Langford(2), A.A., M.L. Fleet, and A.H. Mahan, "Correction for Multiple Reflections in Infrared Spectra of Amorphous Silicon," Solar Cells, Vol. 27, 1989a, pp. 373-383.

Langford(3), A.A., M.L. Fleet, A.J. Nelson, S.E. Asher, J.P. Goral, and A. Mason, "Determination of the Fluorine Content in a-Si:H:F by Infrared Spectroscopy, Electron Probe Microanalysis, X-Ray Photoelectron Spectroscopy, and Secondary Ion Mass Spectrometry," J. Appl. Phys., Vol. 65, 1989b, pp. 5154-5160.

Langford(4), A.A., B.P. Nelson, M.L. Fleet, and R.S. Crandall, "Identification of Bonded Species in Hydrogenated Fluorinated Amorphous Silicon," Phys. Rev. B, Vol. 42, 1990, pp. 7245-7248.

Langford(5), A.A., M.L. Fleet, B.P. Nelson, W.A. Lanford, and N. Maley, "Infrared Absorption Strength and Hydrogen Content of Hydrogenated Amorphous Silicon," Phys. Rev. B, Vol. 45, 1992, pp. 13367-13377.

Lannin, J.S.,"Local Structural Order in Amorphous Semiconductors," Physics Today, Vol. 41, July 1988, pp. 28-35.

Le Comber(1), P.G., A. Madan, and W.E. Spear, "Electronic Transport and State Distribution in Amorphous Si Films," J. Non-Cryst. Solids, Vol. 11, 1972, pp. 219-233.

Le Comber(2), P.G., R.J. Loveland, W.E. Spear, and R.A. Vaughan, "The Effect of Preparation Conditions and of Oxygen and Hydrogen on the Properties of Amorphous Silicon," in Amorphous and Liquid Semiconductors, edited by J. Stuke & W. Brenning, Taylor & Francis, London, 1974, pp. 245-250.

Le Comber(3), P.G., "Present and Future Applications of Amorphous Silicon and Alloys," Journal of Non-Crystal. Solids, Vol. 114, 1989, pp. 1-13.

Lechner, P., M. Gorn, H. Rubel, B. Scheppat, and N. Kniffler, "Material Properties of p-Type a-SiC Layers using Either Diborane or Trimethylboron as a Doping Gas," Mat. Res. Soc. Symp. Proc., Vol. 192, 1990, pp. 81-86.

Lee, Y.H., and M.M. Chen, "Silicon Doping Effects in Reactive Plasma Etching," J. Vac. Sci. Technol. B, Vol. 4, 1986, pp. 468-475.

Lee, T.C., and G.W. Neudeck, "Effects of Controlled Hydrogenation by Ion Implantation on the Localized States in Vacuum Evaporated Amorphous Silicon," J. Appl. Phys., Vol. 54, 1983, pp. 3977-3982.

Lehmann, V., and U. Gösele, "Porous Silicon: Quantum Sponge Structures Grown via a Self-Adjusting Etching Process," Advanced Materials, Vol. 4, 1992, pp. 114-116.

Leopold, D.J., P.A. Fedders, R.E. Norberg, J.B. Boyce, and J.C. Knights, "Deuteron and Proton Magnetic Resonance in Amorphous Silicon," Phys. Rev. B, Vol.31, 1985, pp. 5642-5656.

Ley, L., "Photoemission and Optical Properties," Topics in Applied Physics, Vol. 56, 1984, pp. 61-168.

Li, Y-M. and B. F. Fieselmann, "Improvement of the Optical and Photoelectric Properties of a-SiC:H by using Trisilylmethane as a Feedstock," Appl. Phys. Lett. Vol. 59, 1991, pp. 1720-1722.

Lin(1), G.H., J.R. Doyle, M. He, and A. Gallagher, "Argon Sputtering Analysis of the Growing Surface of Hydrogenated Amorphous Silicon Films," J. Appl. Phys., Vol. 64, 1988, pp. 188-194.

Lin(2), G.H., M. Kapur, R.C. Kainthla, and J. O'M Bockris, "One-step method to produce hydrogen by a triple-stack amorphous silicon solar cell", Appl. Phys. Lett. Vol. 55, 1989, pp. 386-387.

Little, T.W., K. Takahara, H. Koike, T. Nakazawa, I. Yudasaka, and H. Ohshima, "Low Temperature Poly-Si TFTs Using Solid Phase Crystallization of Very Thin Films and an Electron Resonance Chemical Vapor Deposition Gate Insulator," Jpn. J. Appl. Phys., Vol. 30, 1991, pp. 3724-3728.

Liu, J.Z., X. Li, P. Roca i Cabarrocas, J.P. Conde, A. Maruyama, H. Park, S. Wagner, and A.E. Delahoy, "Ambipolar Diffusion Length in a-Si:H(F) and a-Si,Ge:H,F Measured with the Steady State Photocarrier Grating Technique," 21st IEEE PV Spec. Conf. Proc., 1990, pp. 1606-1609.

Lloret, A., E. Bertran, J.L. Andujar, A. Canillas, and J.L. Morenza, "Ellipsometric Study of a-Si:H Thin Films Deposited by Square Wave Modulated RF Glow Discharge," J. Appl. Phys., Vol. 69, 1991, pp. 632-638.

Longeway(1), P.A., H.A. Weakliem, and R.D. Estes, "Mechanism of the Direct Current Plasma Discharge Decomposition of Disilane," J. Phys. Chem., Vol. 88, 1984a, pp. 3282-3287.

Longeway(2), P.A., "Plasma Kinetics," Semiconductors and Semimetals, Vol. 21, 1984b, pp. 179-193.

Longeway(3), P.A., H.A. Weakliem, and R.D. Estes, "Analysis of a Flowing Silane DC Discharge in the Presence of a Hot Surface," J. Appl. Phys. Vol. 57, 1985, pp. 5499-5505.

Lu, H.-Y., and M. A. Petrich, "Effects of temperature and bias on the microstructure of plasma-deposited amorphous silicon carbide," J. Appl. Phys. Vol. 72, 1992, pp. 2054-2056.

Lu, Z.H., E. Sacher, and A. Yelon, "Kinetics of the Room-Temperature Air Oxidation of Hydrogenated Amorphous Silicon and Crystalline Silicon," Phil. Mag. B, Vol. 58, 1988, pp. 385-388.

Lucovsky(1), G., R.J. Nemanich, and J.C. Knights, "Structural Interpretation of the Vibrational Spectra of a-Si:H Alloys," Phys. Rev. B, Vol. 19, 1979, pp. 2064-2073.

Lucovsky(2), G., and W.B. Pollard, "Vibrational Spectra of Defect and Alloy Atom Complexes in Amorphous Silicon Films," Physica, Vol. 117B & 118B, 1983, pp. 865-867.

Lucovsky(3), G., G.N. Parsons, C. Wang, B.N. Davidson, and D.V. Tsu, "Low Temperature Deposition of Hydrogenated Amorphous Silicon (a-Si:H): Control of Polyhydride Incorporation and Its Effects on Thin Film Properties," Solar Cells, Vol. 27, 1989, pp. 121-136.

Lucovsky(4), G., C. Wang, and Y.L. Chen, "Barrier-Limited Transport in μc-Si and μc-Si,C Thin Films Prepared by Remote Plasma-Enhanced Chemical-Vapor Deposition," J. Vac. Sci. Technol. A, Vol. 10, 1992, pp. 2025-2031.

Luft(1), W., "Report on Hydrogenated Amorphous Silicon-Germanium Alloys," Proc. 20th IEEE PV Specialists Conf., 1988a, pp. 218-223.

Luft(2), W., and Y.S. Tsuo, "Plasma Deposition of Hydrogenated Amorphous Silicon Films," Appl. Phys. Comm., Vol. 8, 1988b, pp. 1-74.

Luft(3), W., Editor, "Photovoltaic Safety," Amer. Inst. of Physics Conf. Proc., Vol. 166, 1988c.

Luft(4), W., "High-Rate Deposition of Hydrogenated Amorphous Silicon Films and Devices," Appl. Phys. Comm., Vol. 8, 1988d, pp. 239-298.

Luft(5), W., "Characteristics of Hydrogenated Amorphous Silicon-Germanium Alloys," Appl. Phys. Comm., Vol. 9, Nos. 1&2, 1989, pp. 43-63.

Luft(6), W., "Photovoltaic Applications for Silicon Carbon Films and Microcrystalline Silicon Films," Applied Physics Communications, Vol. **9**, No. 4, 1989-1990, pp. 329-357.

Luft(7), W., B. Stafford, and B. von Roedern, "The Amorphous Silicon Program Needs with Respect to Stabilized Module Performance," *Solar Cells*, Vol. 30, Nos. 1-4, 1991, pp. 195-205.

Maa, J. and S. Lin, "Low Temperature Crystallization of Amorphous Silicon Films," Thin Solid Films, Vol. 64, 1979, pp. 63-64.

Mackenzie(1), K.D., P.G. Le Comber, and W.E. Spear, "The Density of States in Amorphous Silicon Determined by Space-Charge-Limited Current Measurements," Phil. Mag. B, Vol. 46, 1982, pp. 377-389.

Mackenzie(2), K.D., J.R. Eggert, D.J. Leopold, Y.M. Li, S. Lin, and W. Paul, "Structural, Electrical, and Optical Properties of a-SiGe:H and an Infrared Electronic Band Structure," Phys. Rev. B, Vol. 31, 1985a, pp. 2198-2212.

Mackenzie(3), K.D., J. Hannna, J.R. Eggert, Y.M. Li, Z. Sun, and W. Paul, "Properties of a-SiGe:H and a-SiGe:H:F Alloys," J. Non-Crystal. Solids, Vols. 77&78, 1985b, pp. 881-884.

Mackenzie(4), K.D., J.H. Burnett, J.R. Eggert, Y.M. Li, and W. Paul, "Comparison of the structural, electrical, and optical properties of amorphous silicon-germanium alloys produced from hydrides and fluorides," Phys. Rev. B, Vol. 38, 1988, pp 6120-6136.

Madan(1), A., S.R. Ovshinsky, and E. Benn, "Electrical and Optical Properties of Amorphous Si:F:H Alloys," Phil. Mag. B, Vol. 40, 1979, pp. 259-277.

Madan(2), A. and M.P. Shaw, "The Physics and Applications of Amorphous Semiconductors," Academic Press, San Diego, California, 1988.

Magarino, J., D. Kaplan, A. Friederich, and A. Deneuville, "Doping Effects on Post-Hydrogenated Chemical-Vapour-Deposited Amorphous Silicon," Phil. Mag. B, Vol. 45, 1982, pp. 285-306.

Mahan(1), A.H., B. von Roedern, and A. Madan, "Assessment of the suitability of a-SiSn:H alloys for solar cell purposes," Internal SERI report, 1984a.

Mahan(2), A.H., D.L. Williamson, and A. Madan, "Properties of Amorphous Silicon Tin Alloys Produced using the Radio Frequency Glow Discharge Technique," Appl. Phys. Lett., Vol. 44, 1984b. pp. 220-222,

Mahan(3), A.H., P. Raboisson, and R. Tsu, "Influence of Microstructure on the Photoconductivity of Glow Discharge Deposited Amorphous SiC:H and Amorphous SiGe:H Alloys," Appl. Phys. Lett., Vol. 50, 1987, pp. 335-337.

Mahan(4), A.H., A. Mascarenhas, D. L. Williamson, & R. S. Crandall, "Microstructure and the Urbach Edge in Glow Discharge Deposited a-SiC:H," Mat. Res. Soc. Symp. Proc. Vol. 118, 1988. pp. 641-646.

Mahan(5), A.H., Private communication 1989.

Mahan(6), A.H., Y. Chen, D.L. Williamson, and G.D. Mooney, "The Structure of a-Si:H by Small Angle X-Ray Scattering," J. Non-Crystal. Solids, Vol. 137&138, 1991a, pp. 65-70.

Mahan(7), A.H., J. Carapella, B.P. Nelson, R.S. Crandall, and I. Balberg, "Deposition of device-quality, low-H content amorphous silicon," J. Appl Phys, Vol. 69, 1991b, pp. 6728-6730.

Maley(1), N. and J.S. Lennin, "Influence of Hydrogen on Vibrational and Optical Properties of a-$Si_{1-x}H_x$ Alloys," Phys. Rev. B, Vol. 36, 1987, pp. 1146-1152.

Maley(2), N., "Critical Investigation of the Infrared Transmission Data Analysis of Hydrogenated Amorphous Silicon Alloys," Phys. Rev. B, Vol. 46, 1992, pp. 2078-2085.

Marshall, J.M., R.A. Street, and M.J. Thompson, "Localized States in Compensated a-Si:H," Phys. Rev. B, Vol. 29, 1984, pp. 2331-2333.

Martin, P.J., R.P. Netterfield, W.G. Sainty, and D.R. McKenzie, "Optical Properties of Thin Amorphous Silicon and Amorphous Hydrogenated Silicon Films Produced by Ion Beam Techniques," Thin Solid Films, Vol. 100, 1983, pp. 141-147.

Matsuda(01), A., S. Yamasaki, K. Nakagawa, H. Okushi, K. Tanaka, S. Izima, M. Matsumura, and H. Yamamoto, "Electrical and Structural Properties of Phosphorous-Doped Glow-Discharge Si:F:H and Si:H Films," Jpn. J. Appl. Phys., Vol. 19, 1980a, pp. L305-L308.

Matsuda(02), A., K. Nakagawa, K. Tanaka, M. Matsumura, S. Yamasaki, H. Okushi, and S. Iizima, "Plasma Spectroscopy—Control and Analysis of a-Si:H

Deposition," J. Non-Cryst. Solids, Vol. 35 & 36, 1980b, pp. 183-188.

Matsuda(03), A., and K. Tanaka, "Plasma Spectroscopy—Glow Discharge Deposition of Hydrogenated Amorphous Silicon," Thin Solid Films, Vol. 92, 1982, pp. 171-187.

Matsuda(04), A., T. Kaga, H. Tanaka, L. Malhotra, and K. Tanaka, "Glow Discharge Deposition of a-Si:H from Pure Si_2H_6 and Pure SiH_4," Jpn. J. Appl. Phys., 22, 1983a, pp. L115-L117.

Matsuda(05), A., "Formation Kinetics and Control of Microcrystallite in uc-Si:H from Glow Discharge Plasma," J. Non-Cryst. Solids, Vol.59 & 60, 1983b, pp. 767-774.

Matsuda(06), A., K. Yagii, T. Kaga, and K. Tanaka, "Glow-Discharge Deposition of Amorphous Silicon from SiH_3F," Jpn. J. Appl. Phys., Vol. 23, 1984a, pp. L576-L578.

Matsuda(07), A., T. Kaga, H. Tanaka, and K. Tanaka, "Influence of Power-Source Frequency on the Properties of GD a-Si:H," Jpn. J. Appl. Phys., Vol. 23, 1984b, pp. L567-569.

Matsuda(08), A., K. Yagii, M. Koyama, M. Toyama, Y. Imanishi, N. Ikuchi, and K. Tanka, "Preparation of Highly Photosensitive Hydrogenated Amorphous Si-Ge Alloys using a Triode Plasma Reactor," Appl. Phys. Lett., Vol. 47, 1985, pp. 1061-1063.

Matsuda(09), A., M. Koyama, N. Ikushi, Y. Imanishi, and K. Tanaka, "Guiding Principle in the Preparation of High-Photosensitive Hydrogenated Amorphous Si-Ge Alloys from Glow-Discharge Plasma," Jpn. J. Appl. Phys., Vol. 25, 1986a, pp. L54-L56.

Matsuda(10), A., T. Yamaoka, S. Wolff, M. Koyama, Y. Imanishi, H. Kataoka, H. Matsuura, and K. Tanaka, "Preparation of Highly Photosensitive Hydrogenated Amorphous Si-C Alloys from a Glow-Discharge Plasma," J. Appl. Phys., Vol. 60, 1986b, pp. 4025-4027.

Matsuda(11), A. and K. Tanaka, "Guiding Principle for Preparing Highly Photosensitive Si-Based Amorphous Alloys," J. Non-Cryst. Solids, Vol. 97 & 98, 1987, pp 1367-1374.

Matsuda(12), A., and T. Goto, "Role of Surface and Growth-Zone Reactions in the Formation Process of μc-Si:H," Mat. Res. Soc. Symp. Proc., Vol. 164, 1990, pp. 3-14.

Matsumura(1), H. and S. Furukawa, "A New Hydro-Fluorinated Amorphous Silicon Produced by Using Intermediate Species SiF_2," J. Non-Crystal. Solids, Vols. 59 & 60, 1983, pp. 739-742.

Matsumura(2), H., T. Uesugi, and H. Ihara, "Photoconductive Amorphous Silicon-Carbide Produced by Intermediate Species SiF_2 and CF_4 Mixture," Jpn. J. Appl. Phys., Vol. 24, 1985a, pp. L24-L26.

Matsumura(3), H., and H. Tachibana, "Amorphous Silicon Produced by a New Thermal Chemical Vapor Deposition Method using Intermediate Species SiF_2," Appl. Phys. Lett., Vol. 47, 1985b, pp. 833-835.

Matsumura(4), H., "Catalytic Chemical Vapor Deposition (CTL-CVD) Method to Obtain High Quality Amorphous Silicon Alloys," Mat. Res. Soc. Symp. Proc., Vol. 118, 1988, pp. 43-48.

Matsumura(5), H., "Study on Catalytic Chemical Vapor Deposition Method to Prepare Hydrogenated Amorphous Silicon," J. Appl. Phys., Vol. 65, 1989, pp. 4396-4402.

Matsumura(6), H., M. Yamaguchi, and K. Morigaki, "Properties of Catalytic CVD Amorphous Silicon Germanium (a-SiGe:H)", Mat. Res. Soc. Symp. Proc., Vol. 192, 1990, pp. 499-504.

Matsunami, H, T. Shirafuji, and M. Yoshimoto, "a-Si:H Deposited by Direct Photo-CVD Using a Microwave-Excited Xe Lamp," Mat. Res. Soc. Symp. Proc., Vol. 192, 1990, pp. 505-510.

Matsuo(1), S. and M. Kiuchi, "Low Temperature Chemical Vapor Deposition Method Utilizing an Electron Cyclotron Resonance Plasma," Jpn. J. Appl. Phys., Vol. 22, 1983, pp. L210-L212.

Matsuo(2), S., M. Ueda, T. Imura, & Y. Osaka, "Effect of Hydrogen Dilution on Structure of a-Si:H Prepared by Substrate Impedance Tuning Technique", Jpn. J. Appl. Phys., Vol.27, 1988, pp. 475-479.

Matsushita, T., K. Komori, M. Konagai, and K. Takahashi, "High Performance Hydrogenated Amorphous Si Solar Cells with Graded Boron-Doped Intrinsic Layers Prepared from Disilane at High Deposition Rates," Appl. Phys. Lett., Vol. 44, 1984, pp. 1092-1094.

Matsuyama, T., M. Taguchi, M. Tanaka, T. Matsuoka, S. Tsuda, S. Nakano, Y. Kuwano, "High-Quality p-Type μc-Si Films Prepared by the Solid Phase Crystallization Method," Jpn. J. Appl. Phys., Vol. 29, 1990, pp. 2690-2693.

Mayo, M.J., "Photodeposition: Enhancement of Deposition Reactions by Heat and Light," Solid State Technology, April 1986, pp. 141-144.

McCaughey, M. J., and M. J. Kushner, "Simulation of the bulk and surface properties of amorphous hydrogenated silicon deposited from silane plasmas," J. Appl. Phys. Vol. 65, 1988, pp. 186-195.

McCurdy, R.J. and R.G. Gordon, "Effects of Substrate Temperature and Gas Phase Chemistry on the APCVD of a-Si:H Films from Disilane," Mat. Res. Soc. Symp. Proc., Vol. 118, 1988, pp. 97-102.

McMahon(1), T.J., and R. Konenkamp, "Diffusion Length and Collection Width for the Evaluation of a-Si," Proc. 16th IEEE PV Specialists Conf., 1982, pp. 1389-1393.

McMahon(2), T.J. and R. Tsu, "Equilibrium Temperature and Related Defects in Intrinsic Glow Discharge Amorphous Silicon," Appl. Phys. Lett., Vol. 51, 1987, pp. 412-414.

McMahon(3), T.J. and R.S. Crandall, "Hole trapping, light-soaking, and secondary photocurrent transitions in amorphous silicon," Phys. Rev. B, Vol. 39, 1989, pp. 1766-1771.

McMahon(4), T.J., "Defect Equilibration in Device Quality a-Si:H and its Relation to Light-Induced Defects," Proc. Amer. Inst. of Phys., Vol. 234, 1991, pp. 83-90.

Meassen, K.M.H., M.J.M. Pruppers, J. Bezemer, F.H. Habraken, and W.F. von der Weg, "Hydrogen Content and the Optical Bandgap in Amorphous Silicon," Mat. Res. Soc. Symp. Proc., Vol. 95, 1987, pp. 201-205.

Meiling(1). H., W. Lenting, J. Bezemer, and W.F. van der Weg, "The influence of electrode shape and hydrogen dilution of SiH_4 on the optical properties of glow-discharge hydrogenated amorphous silicon," Phil. Mag. B, 1990a, Vol. 62, pp. 19-28.

Meiling(2), H., M.J. van den Boogaard, R.E.I. Schropp, J. Bezemer, and W.F. van der Weg, "Hydrogen dilution of Silane: Correlation between the Structure and Optical Band Gap in GD a-Si:H Films," Mat. Res. Soc. Symp. Proc., Vol. 192, 1990b, pp. 645-650.

Mejia(1), S.R., R.D. McLeod, K.C. Kao, and H.C. Card, "The Effects of Deposition Parameters on a-Si:H Films Fabricated by Microwave Glow Discharge Techniques," J. Non-Cryst. Solids, Vol. 59 & 60, 1983, pp. 727-730.

Mejia(2), S.R., R.D. McLeod, W. Pries, P. Shufflebotham, D.J. Thomson, J. White, J. Shellenberg, K.C. Kao, and H.C. Card, "Fabrication of a-Si:H Films by Microwave Plasmas under Electron Cyclotron Resonance Conditions," J. Non-Cryst. Solids, Vol. 77 & 78, 1985, pp. 765-768.

Mejia(3), S.R., R.D. McLeod, K.C. Kao, and H.C. Card, "Electron-Cyclotron-Resonant Microwave Plasma System for Thin-Film Deposition," Rev. Sci. Instrum., Vol. 57, 1986, pp. 493-496.

Menna, P., A.H. Mahan, and R. Tsu, "Determination of the Void Fraction in Hydrogenated Amorphous Silicon Carbon", Proc. 19th PV Specialists Conf., 1987, pp. 832-834.

Meunier(1), M., J.H. Flint, J.S. Haggerty, and D. Adler, "Laser-Induced Chemical Vapor Deposition of Hydrogenated Amorphous Silicon. I. Gas-Phase Process Model," J. Appl. Phys., Vol. 62, 1987a, pp. 2812-2821.

Meunier(2), M., J.H. Flint, J.S. Haggerty, and D. Adler, "Laser-Induced Chemical Vapor Deposition of Hydrogenated Amorphous Silicon. II. Film Properties," J. Appl. Phys., Vol. 62, 1987b, pp. 2822-2829.

Miller, D.L., H. Lutz, H. Weismann, E. Rock, A.K. Ghosh, S. Ramamoorthy, and M. Strongin, "Some Properties of Evaporated Amorphous Silicon Made with Atomic Hydrogen," J. Appl. Phys., Vol. 49, 1978, pp. 6192-6193.

Mitchell, K., Y-H. Shing, V. Grosvenor, and P. Yan, "Thin Film SiGe:H Single-Junction and Tandem Solar Cells and Collorary Plasma Diagnostics and Thin Film Properties", Proc. 18th IEEE PV Specialists Conf. 1985, pp. 894-899.

Miyachi, K., H. Tanaka, T. Igarashi, Y. Ohashi, and N. Fukuda, "High Performance Solar Cells Fabricated by Using Disilane," Tech. Digest 3rd Int'l PV Sci. & Eng. Conf., 1987, pp. 179-182.

Moddel, G., D. A. Anderson, and W. Paul, "Derivation of the Low-Energy Optical-Absorption Spectra of a-Si:H from Photoconductivity," Phys. Rev. Vol. B22, 1980, pp. 1918-1925.

Mohring, H.-D., G. Schumm, G. H. Bauer, "Mobility-Lifetime Product for Electrons and Holes in 1.75 eV - 1.95 eV a-SiC:H Films," Proc. 22nd IEEE PV Spec. Conf., 1991, Vol. II, 1991, pp. 1357-1362.

Monroe(1), D., J. Orenstein, and M. Kastner, "Density of states in the gap of a-As_2Se_3 by Photocurrent Transient Spectroscopy," J. de Phys., Vol. 42-C4, 1981, pp. 559-562.

Monroe(2), D., "Hopping in Exponential Band Tails," Phys. Rev. Lett., Vol.54, 1985, pp. 146-149.

Moore, A.R., "Collection Length of Holes in a-Si:H by Surface Photovoltage Using a Liquid Schottky Barrier," Appl. Phys. Lett., Vol. 45, 1982, pp. 403-405.

Moore, C.A., G.P. Davis, and R.A. Gottscho, "Sensitive, Nonintrusive, *In-Situ* Measurement of Temporally and Spatially Resolved Plasma Electric Fields," Phys. Rev. Lett., Vol. 52, 1984, pp. 538-541.

Morimoto(1), A., M. Matsumoto, M. Kumeda, and T. Shimizu, "Effect of reduction in impurity content for a-Si:H films," Jpn. J. Appl. Phys., Vol. 29, 1990, L1747-L1749.

Morimoto(2), A., S. Oozora, M. Kumeda, and T. Shimizu, "Annealing Behavior of Hydrogenated Amorphous Silicon-Nitrogen Alloy Films Prepared by Sputtering," Phys. Stat. Sol. (b), Vol. 119, 1983, pp. 715-720.

Morin, P. A., N. W. Wang, and S. Wagner, "The farthest reaches of a-Si,Ge:H,F parameter space...where no one has gone before," Mat. Res. Soc. Symp. Proc., Vol. 258, 1992, pp. 523-528.

Mort, J. and F. Jansen, "Plasma Deposited Thin Films," CRC Press, Inc., Boca Raton, Florida, 1986.

Moustakas(1), T.D., "Sputtering," Semiconductors and Semimetals, Vol. 21A, 1984, pp. 55-82.

Moustakas(2), T.D., H.P. Maruska, and R. Friedman, "Properties anf Photovoltaic Applications of Microcrystalline Silicon Films Prepared by RF Reactive Sputtering," J. Appl. Phys., Vol. 58, 1985, pp. 983-986.

Muramatsu(1), S., N. Nakamura, S. Matsubara, H. Itoh, and T. Shimada, "Amorphous Silicon Solar Cells Fabricated Using a Hot-Wall Type Symmetric Plasma CVD Reactor," Tech. Digest 3rd Int'l PV Sci. & Eng. Conf., 1987, pp. 383-386.

Muramatsu(2), S., H. Kajiyama, H. Itoh, S. Matsubara, and T. Shimada, "Structural Investigation of Hydrogenated Amorphous Silicon Germanium Alloys," Proc. 20th IEEE PV Specialists Conf., 1988, pp. 61-66.

Murnick, D. E., R. B. Robinson, and D. Stoneback, "Optogalvanic signals from argon metastables in rf glow discharge," Appl. Phys. Lett. Vol. 54, 1989, pp. 792-794.

Nagahara, T., K. Fujimoto, N. Kono, Y. Kashiwagi, and H. Kakinoki, "*In-Situ* Chemically Cleaning Poly-Si Growth at Low Temperature," Jpn. J. Appl. Phys., Vol. 31, Dec. 1992.

Najar, S., B. Equer, and N. Lakhoua, "Electronic Transport Analysis by Electron-Beam-Induced Current at Variable Energy of Thin-Film Amorphous Semiconductors," J. Appl. Phys., Vol. 69, 1991, pp. 3975-3985.

Nakamura, G., K. Sato, and Y. Yukimoto, "High Performance Tandem Type Amorphous Solar Cells," Proc. 16th IEEE PV Spec. Conf., 1982, pp. 1331-1337.

Nakamura, N., T. Takahama, M. Isomura, M. Nishikuni, K. Yoshida, S. Tsuda, S. Nakano, M. Ohnishi, and Y. Kuwano, "The Influence of the Si-H2 Bond on the Light-Induced Effect in a-Si Films and a-Si Solar Cells," Jpn. J. Appl. Phys. 28, 1989, pp. 1762-1768.

Nakano(1), S., H. Tarui, H. Haku, T. Takahama, T. Matsuyama, M. Isomura, and Y. Kuwano, "Materials Investigations for High-Efficiency and High-Reliability a-Si Solar Cells," Proc. 19th IEEE PV Specialists Conf., 1987, pp. 678-683.

Nakano(2), S., K. Wakisaka, M. Kameda, M. Isomura, T. Matsuyama, N. Nakamura, S. Tsuda, M. Ohnishi, and Y. Kuwano, "High-Quality a-Si Films Prepared by the Direct Photo-CVD Method," Mat. Res. Soc. Symp. Proc., Vol. 149, 1989, pp. 417-422.

Nakashita(1), T., Y. Osaka, M. Hirose, T. Imura, and A. Hiraki, "Defect States and Electronic Properties of Post-Hydrogenated CVD Amorphous Silicon," Jpn. J. Appl. Phys., Vol. 22, 1983, pp. 1766-1770.

Nakashita(2), T., M. Hirose, and Y. Osaka, "Localized-State-Density Distribution in Post-Hydrogenated CVD Amorphous Silicon," Jpn. J. Appl. Phys., Vol. 23, 1984a, pp. 146-149.

Nakashita(3), T., K. Kohno, T. Imura, and Y. Osaka, "Electronic Properties of Post-Hydrogenated Lightly-Boron-Doped CVD Amorphous Silicon," Jpn. J. Appl. Phys., Vol. 23, 1984b, pp. 1547-1550.

Nakatani(1), K., M. Yano, K. Suzuki, & H. Okaniwa, "Properties of microcrystalline P-Doped Si:H Films," J. Non-Crystal. Solids, Vol. 59 & 60, 1983, pp. 827-830.

Nakatani(2), K., M. Ogasawara, K. Suzuki, H. Okaniwa, K. Hamamoto, and H. Ozaki, "Electrical Properties of Hydrogenated Amorphous Silicon Layers on a Polymer Film Substrate under Tensile Stress," Appl. Phys. Lett., Vol. 54, 1989,

pp. 1678-1680.

Nakayama(1), Y., K. Wakimura, S. Takahashi, H. Kita, and T. Kawamura, "Plasma Deposition of a-Si:H:F Films from SiH_2F_2 and SiF_4-SiH_4," J. Non-Crystal. Solids, Vol. 77&78, 1985, pp. 797-800.

Nakayama(2), Y., S. Akita, K. Wakita, and T. Kawamura, "Synthesis of Highly Photosensitive a-SiC:H Films at High Deposition Rate by Plasma Decomposition of SiH_4 and C_2H_2," Mat. Res. Soc. Symp. Proc. Vol. 118, 1988, pp. 73-78.

Nakazawa, K. and K. Tanaka, "Effect of Substrate Temperature on Recrystallization of Plasma Chemical Vapor Deposition Amorphous Silicon Films," J. Appl. Phys., Vol. 68, pp. 1029-1032, 1990.

Nath, P., C. Vogeli, K. Hoffman, and T. Laarman, "Amorphous Silicon Alloy Tandem Devices and Modules on Thin Polyimide Substrates," Tech. Digest 5th Int'l PV Sci. & Eng. Conf., 1990, pp. 627-629.

Neudeck, G.W. and T.C. Lee, "Reduction in the Localized Band-Gap States in Amorphous Silicon by Annealing and Hydrogen Implantation," Appl. Phys. Lett., Vol. 43, 1983, pp. 680-682.

Nevin, W.A., H. Yamagishi, K. Asaoka, and Y. Tawada, "Wide-bandgap Hydrogenated Amorphous Silicon Carbide Prepared from an Aromatic Carbon Source," Proc. 22nd IEEE PV Specialists Conf., 1991, pp. 1347-1351.

Newton, J.L., and K. Kritikson, "Electrical and Structural Properties of a-Ge:H Alloys," 21st IEEE PV Specialists Conf., 1990, pp. 1662-1666.

Nguyen, H.V., I. An, Y. Li, C.R. Wronski, and R.W. Collins, "Optical Properties and Structure of Microcrystalline Silicon," Materials Res. Soc. Symp. Proc. Vol. 258, 1992, pp. 235-240.

Nishida, A., K. Nakagawa, H. Kakibayashi, and T. Shimada, "Microstructure of Visible Light Emitting Porous Silicon," Jpn. J. Appl. Phys., Vol. 31, 1992, pp. L1219-L1222.

Nishida(1), S., H. Tasaki, M. Konagai, & K. Takahashi, "Highly Conductive and Wide Band Gap Amorphous-Microcrystalline Mixed-Phase Silicon Films Prepared by Photochemical Vapor Deposition," J. Appl. Phys., Vol. 58, 1985, pp. 1427-1431.

Nishikawa, S., H. Kakinuma, T. Watanabe, and K. Nihei, "Influence of Deposition Conditions on Properties of Hydrogenated Amorphous Silicon

Prepared by RF Glow Discharge," Jpn. J. Appl. Phys., Vol. 24, 1985, pp. 639-645.

Nishikuni, M., K. Ninomiya, S. Okamoto, T. Takahama, S. Tsuda, M. Ohnishi, S. Nakano, and Y. Kuwano, "Amorphous Silicon-Carbon Alloy Prepared by the (CMP) Controlled Plasma Magnetron Method," Jpn. J. Appl. Phys., Vol. 30, 1991, pp. 7-12.

Oda(1), S., S. Ishihara, N. Shibata, S. Takagi, H. Shirai, A. Miyauchi, and I. Shimizu, "Properties of a-Si Based Alloys Prepared from Fluorides and Hydrogen," J. Non-Crystal. Solids, Vol. 77&78, 1985, pp. 877-880.

Oda(2), S., S. Ishihara, N. Shibata, H. Shirai, A. Miyauchi, K. Fukuda, A. Tanabe, H. Ohtoshi, J. Hanna, and I. Shimizu, "The Role of Hydrogen Radicals in the Growth of a-Si and Related Alloys," Jpn. J. Appl. Phys., Vol. 25, 1986, pp. L188-L190.

Oda(3), S., J. Noda, and M. Matsumura, "Preparation of a-Si:H Films by VHF Plasma CVD," Mat. Res. Soc. Symp. Proc., Vol. 118, 1988, pp. 117-122.

Oda(4), S., J. Noda, and M. Matsumura, "Diagnostic study of VHF plasma and deposition of hydrogenated amorphous silicon films," Jpn. J. Appl. Phys., Vol. 29, 1990, pp. 1889-1895.

Ogawa, K., I. Shimizu, and E. Inoue, "Preparations of a-Si:H from Higher Silanes with the High Growth Rate," Jpn. J. Appl. Phys., Vol. 20, 1981, pp. L639-L642.

Ohagi, H., M. Yamazaki, J. Nakata, J. Shirafuji, K. Fujibayashi, and Y. Inuishi, "Optical Degradation of a-Si:H Films with Different Morphology and Preparation of Degradation-Resistive Films," Tech. Digest 3rd Int'l PV Sci. and Eng., 1987, pp. 671-674.

Ohnishi(1), M., H. Nishiwaki, E. Enomoto, Y. Nakashima, S. Tsuda, T. Takahama, H. Tarui, M. Tanaka, H. Dojo, and Y. Kuwano, "Preparation and Properties of Amorphous Silicon Produced by a Consecutive, Separated Reaction Chamber Method," J. of Non-Cryst. Solids, Vol. 59 & 60, 1983, pp. 1107-1110.

Ohnishi(2), M., N. Nishiwaki, M. Tanaka, N. Nakamara, S. Tsuda, S. Nakano, and Y. Kuwano, "Preparation and Properties of a-Si Films Deposited with a High Deposition Rate," Proc. 1st Int'l PV Sci. and Eng. Conf., Kobe, Japan, Nov. 13-16, 1984, pp. 719-722.

Ohnishi(3), M., K. Uchihashi, H. Nishiwaki, H. Shibuya, M. Tanaka, K.

Ninomiya, M. Nishikuni, N. Nakamura, S. Tsuda, S. Nakano, and Y. Kuwano, "High-Quality a-Si Films Prepared at High Deposition Rate by Controlled Plasma Magnetron (CPM) Method," Tech. Digest 3rd Int'l PV Sci. and Eng. Conf., 1987, pp. 25-28.

Ohnishi(4), M., H. Nishiwaki, K. Uchihashi, K. Yoshida, M. Tanaka, K. Ninomiya, M. Nishikuni, N. Nakamura, S. Tsuda, S. Nakano, T. Yazaki, and Y. Kuwano, "Preparation and Properties of a-Si Films Deposited at a High Deposition Rate under a Magnetic Field," Jpn. J. Appl. Phys., Vol. 27, 1988, pp. 40-46.

Okada(1), Y., D. Slobodin, S. F. Chou, R. Schwarz, and S. Wagner, "Infrared Spectroscopy of Deuterated a-Si,Ge:D,F Alloys Prepared by DC Glow Discharge Deposition," Mat. Res. Soc. Symp. Proc., Vol. 70, 1986, pp. 289-293.

Okada(2), Y., and S. Wagner, "Etching Selectivity of SiF_4 and H_2 Plasmas for c-Si, a-Si:H, and SiO_2," Mat. Res. Soc. Symp. Proc. Vol. 192, 1990a, pp. 541-546.

Okada(3), Y., J. Chen, I.H. Campbell, P.M. Fauchet, and S. Wagner, "Mechanisam of the Growth of Amorphous and Microcrystalline Silicon from Silicon Tetrafluoride and Hydrogen," J. Appl. Phys., Vol. 67, 1990b, pp. 1757-1760.

Okamoto, H., H. Kida, T. Kamada, and Y. Hamakawa, "Below-gap Primary Photocurrent Associated with Correlated Defects in Hydrogenated Amorphous Silicon," Phil. Mag. B, Vol. 52, 1985, pp. 1115-1133.

Orton, J.W., "On the Analysis of Space-Charge-Limited Current-Voltage Characteristics and the Density of States in Amorphous Silicon," Phil. Mag. B, Vol. 49, 1984, pp. L1-L7.

Oswald, R., and J. O'Dowd, "Large Area Triple-Junction a-Si Alloy Production Scale-Up," Semi-Annual Technical Progress Report, Sept. 1992, National Renewable Energy Lab. Subcontract No. ZM-2-11040-2.

Overhof, H., and W. Beyer, "Electronic Transport in Hydrogenated Amorphous Silicon," Phil. Mag. B, Vol. 47, 1983, pp. 377-392.

Palik, E.D., J.W. Faust, H.F. Gray, and R.F. Green, "Study of the Etch-Stop Mechanism in Silicon," J. Electrochem. Soc., Vol. 129, 1982, pp. 2051-2059.

Pankove(1), J.I., "Photoluminescence Recovery in Rehydrogenated Amorphous Silicon," Appl. Phys. Lett., Vol. 32, 1978, pp. 812-813.

Pankove(2), J.I., M.A. Lampert, and M.L.Tarng, "Hydrogenation and Dehydrogenation of Amorphous and Crystalline Silicon," Appl. Phys. Lett., Vol. 32, 1978, pp. 439-441.

Pankove(3), J.I., Editor, "Hydrogenated Amorphous Silicon," Semiconductors and Semimetals, Vol. 21, Parts A, B, C, and D, 1984.

Pantelides, S.T., "Defects in Amorphous Silicon: A New Perspective," Phys. Rev. Lett., Vol. 57, 1986, pp. 2979-2982.

Park, H.R., J.Z, Liu, P. Roca i Cabarrocas, A. Maruyama, M. Isomura, S. Wagner, J.R. Abelson, and F. Finger, "Dependence of the saturated Light-Induced Defect Density on macroscopic properties of Hydrogenated Amorphous Silicon," Appl. Phys. Lett. Vol. 57, 1990, pp. 1440-1442.

Parsons(1), G.N., D.V. Tsu, and G. Lucovsky, "Growth of a-Si:H Films by Remote Plasma Enhanced CVD," Mat. Res. Soc. Symp. Proc., Vol. 118, 1988a, pp. 37-42.

Parsons(2), G.N., D.V. Tsu, and G. Lucovsky, "Properties of Intrinsic and Doped a-Si:H Deposited by Remote Plasma Enhanced Chemical Vapor Deposition," J. Vac. Sci. Technol. A, Vol. 6, 1988b, pp. 1912-1916.

Patel(1), R.I., J. Shirck, D.J. Olsen, N.T. Tran, and K.A. Epstein, "High Deposition Rate Studies of Hydrogenated Amorphous Silicon," Proc. 18th IEEE PV Specialists Conf., 1985, pp. 499-502.

Patel(2), R.I., D.J. Olsen, J.R. Shirck, and N.T. Tran, "Comparison Studies of Hydrogenated Amorphous Silicon Films Prepared from Silane-Hydrogen and Silane-Helium Mixtures," Mat. Res. Soc. Symp. Proc., Vol. 70, 1986, pp. 55-58.

Paul, D.K., B. von Roedern, S. Oguz, J. Blake, and W. Paul, "Structural Properties of Glow Discharge a-SiGe:H Alloys as Revealed by Infrared Absorption and Hydrogen Evolution Techniques," J. Phys. Soc. Jpn., (Suppl. A) Vol. 49, 1980, p. 1261.

Paul(1), W., A.J. Lewis, G.A.N. Connell, and T.D. Moustakas, "Doping, Schottky Barrier and p-n Junction Formation in Amorphous Germanium and Silicon by rf Sputtering," Solid State Comm., Vol. 20, 1976, pp. 969-972.

Paul(2), W., and D.A. Anderson, "Properties of Amorphous Hydrogenated Silicon, with Special Emphasis on Preparation by Sputtering," Solar Energy Materials, Vol. 5, 1981, pp. 229-316.

Paul(3), W., D.K. Paul, B. von Roedern, J. Blake, and S. Oguz, "Preferential Attachment of H in Amorphous Hydrogenated Binary Semiconductors and Consequent Inferior Reduction of Pseudogap State Density," Phys. Rev. Lett., Vol. 46, 1981, pp. 1016-1020.

Paul(4), W., and K. D. Mackenzie, "Amorphous Hydrogenated Silicon-Germanium Alloys Suitable for Photoelectronic Applications," Annual Subcontract Report, 1 July 1985-30 June 1986, SERI/STR-211-3107, Golden, CO: Solar Energy Research Inst., 1987.

Perez-Mendez, V., G. Cho, I. Fujieda, S.N. Kaplan, S. Qureshi, and R.A. Street, "The Application of Thick Hydrogenated Amorphous Silicon Layers to Charged Particle and X-Ray Detection," Mat. Res. Soc. Symp. Proc. Vol. 149, 1989, pp. 621-630.

Pernisz, U.C., L. Tarhay, J.J. D'Errico, and K.G. Sharp, "Fluorinated Silane Precursors to Amorphous Silicon," Proc. 19th IEEE PV Specialists Conf., 1987, pp. 582-587.

Perry, J.W., Y.H. Shing, and C.E. Allevato, "Diagnostics of Silane and Germane Radio Frequency Plasmas by Coherent Anti-Stokes Raman Spectroscopy," Appl. Phys. Lett., 1988, Vol. 52, pp. 2022-2024.

Peters, J.W. and F.L. Gebhart, "Mobile Transparent Window Apparatus and Method for Photochemical Vapor Deposition," U.S. Patent No. 4,265,932, May 5, 1981.

Phillips, J.C., "Structure of Amorphous Semiconductors," in *Disordered Semiconductors*, M.C. Kastner, G.A. Thomas, and S.R. Ovshinsky, eds. Plenum Press, New York, 1987, pp.257-258.

Photovoltaic Energy Technology Division, Office of Solar Electric Technologies, U.S. Department of Energy, National Photovoltaics Program, Five-Year Research Plan, 1987-1991, Washington, DC: U.S. DOE.

Pinarbasi(1), M., M.J. Kushner, and J.R. Abelson, "Effect of Hydrogen Content on the Light Induced Defect Generation in Direct Current Magnetron Reactively Sputtered Hydrogenated Amorphous Silicon Thin Films," J. Appl. Phys., Vol. 68, 1990, pp. 2255-2264.

Pinarbasi(2), M., N. Maley, J.R. Abelson, V. Chu, and S. Wagner, "Carrier Transport Properties of DC Magnetron Reactive Sputtered a-Si:H Films," Mat. Res. Soc. Symp. Proc., Vol. 149, 1989, pp. 205-210.

Pochet, T., J. Dubeau, L.A. Hamel, B. Equer, and A. Karar, "Charge Collection in a-Si:H Particle Detectors," Mat. Res. Soc. Symp. Proc., Vol.149, 1989, pp. 661-666.

Pointu(1), A.M., "A Model of Radio Frequency Planar Discharges," J. Appl. Phys., Vol. 60, 1986, pp. 4113-4118.

Pointu(2), A.M., "Model of RF Discharges at Frequencies Greater than the Ionic Plasma Frequency," Appl. Phys. Lett., Vol. 50, 1987, pp. 1047-1049.

Pollak, F.H., and R. Tsu, "Raman Characterization of Semiconductors Revisited," Proc. Int'l Soc. for Optical Engineering, SPIE (Society of Photoelectric Instrument Engineers) Spectroscopic Characterization Techniques for Semiconductor Technology, Vol. 452, 1983, pp. 26-43.

Ponpon, J.P., and B. Bourdon, "Oxidation of Glow Discharge a-Si:H," Solid State Electronics, Vol. 25, 1982, pp. 875-876.

Postol, T.A., C.M. Falco, R.T. Kampwirth, I.K. Schuller, and W.B. Yellon, "Structure of Amorphous Silicon and Silicon Hydrides," Phys. Rev. Lett., Vol. 45, 1980, pp. 648-652.

Potts, J.E., J.A. McMillan, and E.M. Peterson, "Effects of RF Power and Reactant Gas Pressure on Plasma Deposited a-Si:H," J. Appl. Phys., Vol. 52, 1981, pp. 6665-6667.

Powell, M.J., "The Physics of Amorphous-Silicon Thin-Film Transistors," IEEE Transactions on Electron Devices, Vol.36, 1989, pp. 2753-2763.

Pramanik, D., and J. Glanville, "Aluminum Film Analysis with Focused Ion Beam Microscope," Solid State Tech., May 1990, pp. 77-80.

Pratt, B., and R. Weil, "Post-Hydrogenation of Evaporated a-Si," J. Non-Crystal. Solids, Vol. 89, 1987, pp. 9-12.

Properties of Amorphous Silicon, EMIS Datareviews Series No. 1, published by INSPEC, The Inst. of Electrical Engineers, London and New York, 1985.

Putley, E.H., Hall Effect and Related Phenomena, Butterworth & Co., Ltd., London, 1960.

Qian, Z.M., A. Van Ammel, H. Michiel, J. Nijs, and R. Mertens, "Preparation of Hydrogenated Amorphous Silicon with Tunable Gap by Homogeneous Chemical Vapor Deposition," J. Appl. Phys., Vol. 68, 1990, pp. 143-155.

Qiao, J., Z. Jiang, and Z. Ding, "Effect of Some Process Parameters on the Deposition Mechanism of a-Si:H Film," J. Non-Crystal. Solids, Vol. 77 & 78, 1985, pp. 829-832.

Rajeswaran, G., F.J. Kampas, P.E. Vanier, R.L. Sabatini, and J. Tafto, "Substrate Temperature Dependence of Microcrystallinity in Plasma-Deposited, Boron-Doped Hydrogenated Silicon Alloys," Appl. Phys. Lett., Vol. 43, 1983, pp. 1045-1047.

Rao, C.N.R., Chemical Applications of Infrared Spectroscopy, Academic Press, New York, 1963.

Ray(1), S., P. Chaudhuri, A.K. Batabyal, and A.K. Barua, "Electronic and Optical Properties of Boron Doped Hydrogenated Amorphous Silicon Thin Films," Solar Energy Materials, Vol. 10, 1984, pp. 335-347.

Ray(2), S., D. Das, and A.K. Barua, "Infrared Vibrational Spectra of Hydrogenated Amorphous Silicon Carbide Thin Films Prepared by Glow Discharge," Solar Energy Materials, Vol. 15, 1987, pp. 45-57.

Ready, S.E., J.B. Boyce, N.M. Johnson, J. Walker, and K.S. Stevens, "Hydrogen Bonding in a-Si:H Prepared by Remote Hydrogen Plasma Deposition", Mat. Res. Soc. Symp. Proc., Vol. 192, 1990, pp. 127-132.

Redfield, D., and R.H. Bube, "Interpretation of Degradation Kinetics of Amorphous Silicon", Appl. Phys. Lett., Vol. 54, 1989, p. 1037 and "Identification of Defects in Amorphous Silicon," Phys. Rev. Lett. Vol. 65, 1990, p. 464.

Reimer(1), J.A., B.A. Scott, D.J. Wolford, and H. Nijs, "Low Spin Density Amorphous Hydrogenated Germanium Prepared by Homogeneous Chemical Vapor Deposition," Appl. Phys. Lett., Vol. 46, 1985, pp. 369-371.

Reimer(2), J.A., M.J. McCarthy, K.K. Gleason, and P.W. Morrison, "Identification of Chemical Growth Mechanisms in Amorphous Semiconductors," Mat. Res. Soc. Symp. Proc., Vol. 95, 1987, pp. 209-217.

Ristein, J., J. Hautala, and P.C. Taylor, "Excitation-Energy Dependence of Optically Induced ESR in a-Si:H," Phys. Rev. B, Vol. 40, 1989, pp. 88-92.

Ritter, D., K. Weiser, and E. Zeldov, "Steady-State Photocarrier Grating Technique for Diffusion-Length Measurement in Semiconductors: Theory and Experimental Results for Amorphous Silicon and Semi-Insulating GaAs," J. Appl. Phys., Vol. 62, 1987, pp. 4563-4570.

Robertson, J., "Electronic structure of silicon nitride," Phil. Mag. B, Vol. 63, 1991, pp. 47-77.

Robertson, P.A., Y.K. Bhatnagar, and W.I. Milne, "A Comparative Study of the Photo-CVD of a-Si:H from Internal and External Lamp Systems," Technical Digest of the Int'l PVSEC-3, Tokyo, Japan, 1987, pp. 271-274.

Robertson(1), R., D. Hils, H. Chatham, and A. Gallagher, "Radical Species in Argon-Silane Discharges," Appl. Phys. Lett. Vol.43, 1983, pp. 544-546.

Robertson(2), R., and A. Gallagher, "Reaction Mechanism and Kinetics of Silane Pyrolysis on a Hydrogenated Amorphous Silicon Surface," J. Chem. Phys., Vol. 85, 1986a, pp. 3623-3630.

Robertson(3), R., and A. Gallagher, "Mono- and Disilicon Radicals in Silane and Silane-Argon DC Discharges," J. Appl. Phys., Vol. 59, 1986b, pp. 3402-3411.

Roca i Cabarrocas(1), P., "Detailed Study of Ion Bombardment in RF Glow Discharge Deposition Systems: The Role of Helium Dilution," Mat. Res. Soc. Symp. Proc., Vol. 149, pp. 33-38, 1989a.

Roca i Cabarrocas(2), P., S. Kumar, and B. Drevillon, "In-Situ Study of the Thermal Decomposition of (B_2H_6:SiH_4) by Combining Spectroscopic Ellipsometry and Kelvin Probe Measurements," Proc. 9th E.C. PV Solar Energy Conf., 1989b, pp. 263-266.

Roca i Cabarrocas(3), P., P. Morin, J. Conde, V. Chu, J.Z. Liu, H.R. Park, and S. Wagner, " Does Ion Bombardment induce a Degradation of the Electronic Properties of a-Si:H Films?," Mat. Res. Soc. Symp. Proc., Vol. 192, 1990, pp. 745-750.

Roca i Cabarrocas(4), P., P. Morin, V. Chu, J.P. Conde, J.Z. Liu, H.R. Park, and S. Wagner, "Optoelectronic properties of hydrogenated amorphous silicon films deposited under negative substrate bias", J. Appl. Phys, Vol. 69, 1991, pp. 2942-2950.

Rocheleau(1), R.E., S.S. Hegedus, and W.N. Baron, "Properties of Intrinsic a-Si Films Deposited from Higher Order Silanes by Chemical Vapor Deposition," Mat. Res. Soc. Symp. Proc., Vol. 49, 1985, pp. 15-20.

Rocheleau(2), R.E., R.M. Tullman, D.E. Albright, "a-Si Materials and a-Si Solar Cells," J. Non-Crystal. Solids, Vol. 87, 1986a, pp. 46-63.

Rocheleau(3), R.E., S.C. Jackson, S.S. Hedegus, and B.N. Baron, "Properties of a-Si:H and a-SiGe:H Films Deposited by Photo-Assisted CVD," Mat. Res. Soc. Symp. Proc., Vol. 70, 1986b, pp. 37-42.

Rocheleau(4), R.E., S.S. Hegedus, W.A. Buchanan, and R.M. Tullman, "Effects of Impurities on Film Quality and Device Performance in a-Si:H Deposited by Photo-assisted CVD," Proc. 19th IEEE PV Specialists Conf., 1987, pp. 699-704.

Rocheleau(5), R.E., R.M. Tullman, D.E. Albright, and S.S. Hegedus, "Amorphous Silicon-Germanium Deposited by Photo-CVD: Effects of Hydrogen Dilution and Substrate Temperature," Mat. Res. Soc. Symp. Proc., Vol. 118, 1988, pp. 653-658.

Rosan, K., "Hydrogenated Amorphous-Silicon Image Sensors," IEEE Transactions on Electron Devices, Vol. 36, 1989, pp. 2923-2927.

Ross(1), R.C., I.S.T. Tsong, R. Messier, W.A. Lanford, and C. Burman, "Quantification of Hydrogen in a-Si:H Films by IR Spectrometry, ^{15}N Nuclear Reaction, and SIMS," J. Vac. Sci. Technol., Vol. 20, 1982, pp. 406-409.

Ross(2), R.C., and J. Jaklik, "Plasma Polymerization and Deposition of Amorphous Hydrogenated Silicon from rf and dc Silane Plasmas," J. Appl. Phys., Vol. 55, 1984, pp. 3785-3794.

Roth, R.M., K.G. Spears, and G. Wong, "Spatial Concentrations of Silicon Atoms by Laser-Induced Fluorescence in a Silane Glow Discharge," Appl. Phys. Lett., Vol. 45, 1984, pp. 28-30.

Rothwarf, A., "A Mechanism for Enhanced Recombination at the p-i Junction of a-Si:H Solar Cells," Proc. 20th IEEE PV Specialists Conf., 1988, pp. 166-170.

Rudder(1), R.A., J.W. Cook, Jr., and G. Lucovsky, "High Photoconductivity in Dual Magnetron Sputtered Amorphous Hydrogenated Silicon and Germanium Alloy Films," Appl. Phys. Lett., Vol. 45, 1984, pp. 887-889.

Rudder(2), R.A., J.W. Cook, Jr., and G. Lucovsky, "Thin Films of a-$Si_{1-x}Ge_x$:H Alloys by Dual Magnetron Sputtering in a UHV Chamber," J. Vac. Sci. Technol. A, Vol. 3, 1985, pp. 567-571.

Ryners, S. W., A. Scheeline, and P. W. Bohn, "Structure evolution in a-SiC:H films prepared from tetramethylsilane," J. Appl. Phys. Vol. 69, 1991, pp. 2951-2960.

Sabisky, E.S., Z. Kiss, F. Ellis, E. Eser, S. Gau, F. Kampas, J. Van Dine, H. Weakliem, and T. Varvar, "Eureka - A 10 MW_p a-Si:H Module Processing

Line," Proc. 9th Euro. Comm. PV Conf., 1989, pp. 27-29.

Sah, W.-J., S.-C. Lee, H-K. Tsai, and J-H. Chen, "Amorphous Silicon Edge Detectors for Application to Neural Network Image Sensors," Appl. Phys. Lett., Vol. 56, 1990, pp. 2539-2541.

Saito, N., K. Aoki, H. Sannomiya, and T. Yamaguchi, "Optical and Electrical Properties of Hydrogenated Amorphous $Si_{1-x}Ge_x$ Alloy Thin Films Prepared by Planar Magnetron Sputtering," Thin Solid Films, Vol. 115, 1984, pp. 253-262.

Sakai, H., T. Yoshida, T. Hama, and Y. Ichikawa, "Effects of Surface Morphology of Transparent Electrode on the Open-Circuit Voltage in a-Si:H Solar Cells," Jpn. J. Appl. Phys., Vol. 29, 1990, pp. 630-635.

Sakamoto, Y. "Measurement of Power Transfer Efficiency from Microwave Field to Plasma under ECR Conditions," Jpn. J. Appl. Phys., Vol. 16, 1977, pp. 1993-1998.

Sakata(1), I., S. Okazaki, M. Yamanaka, Y. Hayashi, "Raman Scattering Studies on Hydrogenated Amorphous Silicon Prepared under High Deposition Rate Conditions," Jpn. J. Appl. Phys., Vol. 24, 1985, pp. L428-L430.

Sakata(2), I., M. Yamanaka, and Y. Hayashi, "Properties of Plasma-Deposited Hydrogenated Amorphous Silicon Prepared under Visible Light Illumination," J. Appl. Phys., Vol. 67, 1990, pp. 3737-3743.

Sakata(3), I., M. Yamanaka, and Y. Hayashi, "Defects in Plasma-Deposited Hydrogenated Amorphous Silicon Prepared under Visible Light Illumination," J. Appl. Phys., Vol. 69, 1991, pp. 2561-2567.

Sakka, T., K. Toyoda, and M. Iwasaki, "Evolution of Hydrogen from Plasma-Deposited Amorphous Hydrogenated Silicon Films Prepared from a SiH_4/H_2 Mixture," Appl. Phys. Lett., Vol. 55, 1989, pp. 1068-1070.

Sasaki, H., M. Aiga, M. Usui, K. Kawabata, T. Ishihara, S. Terazono, K. Sato, K. Okaniwa, T. Itagaki, G. Nakamura, Y. Yukimoto, and K. Fujikawa, "Improvements of Interfaces in Tandem-Type Amorphous Silicon Alloy Solar Cells," Proc. 2nd Int'l PV Sci. and Eng. Conf., 1986, pp. 467-470.

Sato, K., K. Kawabata, S. Terazono, T. Ishihara, H. Sasaki, M. Deguchi, T. Itagaki, H. Morikawa, M. Aiga, and K. Fujikawa, "Preparation of High Quality a-SiGe:H Films and its Application to the High Efficiency Triple-Junction Amorphous Solar Cells," Proc. 20th IEEE PV Specialists Conf., 1988, pp. 73-78.

Scapple, R.Y., J.W. Peters, J.F. Linder, and E.M. Yee, "Method for Photochemical Vapor Deposition," U.S. Patent No. 4,597,986, July 1, 1986.

Schade(1), H., Z E. Smith, and A. Catalano, "Correlation between Bulk p-Layer Properties of a-SiC:H and Performance of a-SiC:H/a-Si:H Heterojunction Solar Cells," Solar Energy Materials, Vol. 10, 1984, pp. 317-328.

Schade(2), H., "Study of Microcrystalline Silicon-Carbon p-Layers Prepared by Photo-CVD and Glow Discharge," STR-211-3527, SERI Subcontractor Report, 1989

Scher, H., M.F. Shlesinger, and J.T. Bendler, "Time-Scale Invariance in Transport and Relaxation," Physics Today, Vol. 44, 1991, pp. 26-34.

Schiff, E.A., "Defects, Metastability, and Photocarrier Process in a-Si:H and a-SiGe:H," Proc. 1989 a-Si Subcontractors' Review Meeting, SERI/CP-211-3514, June 1989, pp. 211-220.

Schmitt, J.P.M., "Fundamental Mechanisms in Silane Plasma Decomposition and Amorphous Silicon Deposition," J. Non-Cryst. Solids, Vol. 59 & 60, 1983, pp. 649-658.

Schubert, M.B. and G.H. Bauer "Studies of Substrate Induced Disorder in a-Si:H by In-Situ-Raman Backscattering During Film Growth," Proc. 21st IEEE PV Specialists Conf., 1990, pp. 1595-1599.

Schumm, G., E. Lotter, and G.H. Bauer, "Charged Dangling Bonds in Undoped Amorphous Silicon," Appl. Phys. Lett., Vol. 60, 1992, pp. 3262-3264.

Scott(1), B.A., M.H. Brodsky, D.C. Green, P.B. Kirby, R.M. Plecenik, and E.E. Simonyi, "Glow Discharge Preparation of Amorphous Hydrogenated Silicon from Higher Silanes," Appl. Phys. Lett., Vol. 37, 1980, pp. 725-727.

Scott(2), B.A., J.A. Reimer, and P.A. Longeway, "Growth and Defect Chemistry of Amorphous Hydrogenated Silicon," J. Appl. Phys., Vol. 54, 1983, pp. 6853-6863.

Scott(3), B.A., "Homogeneous Chemical Vapor Deposition," Semiconductors and Semimetals, Vol. 21A, 1984, pp. 123-149.

Selwyn, G.S., "Optical Diagnostic Techniques for RIE," Proc. 6th Symp. Plasma Processing, Oct. 1986, Electrochemical Soc., Vol. 87-6, Mathad, Schwarts, and Gottscho, eds., 1986, pp. 220-253.

Shanks, H., C.J. Fang, L. Ley, M. Cardona, F.J. Demond, and S. Kalbitzer, "Infrared Spectrum and Structure of Hydrogenated Amorphous Silicon," Phys. Stat. Sol. (b), 100, 1980, pp. 43-56.

Shen, D.S., R.E.I. Schropp, H. Chatham, R.E. Hollingsworth, P.K. Bhat, and J. Xi, "Improving Tunneling Junction in Amorphous Silicon Tandem Solar Cell," Appl. Phys. Lett., Vol. 56, 1990, pp. 1871-1873.

Sheng, T.Y., Z.Q. Yu, and G.J. Collins, "Disk Hydrogen Plasma Assisted Chemical Vapor Deposition of Aluminum Nitride," Appl. Phys. Lett., Vol. 52, 1988, pp. 576-578.

Sherman, A., "Chemical Vapor Deposition for Microelectronics Principles, Technology, and Applications," Noyes Publications, New Jersey, 1987.

Shibata, A., Y. Kazama, K. Seki, W.Y. Kim, S. Yamanaka, M. Konagai, and K. Takahashi, "High Efficiency Amorphous Silicon Solar Cells with 'Delta-Doped' P-Layer," 20th IEEE PV Specialists Conf., 1988, 317-319.

Shibata(1), N., A. Miyauchi, A. Tanabe, J. Hanna, S. Oda, and I. Shimizu, "Hole Transport in a-Si:H(F) Prepared by Hydrogen-Radical-Assited Chemical Vapor Deposition," Jpn. J. Appl. Phys., Vol. 25, 1986a, pp. 1783-1787.

Shibata(2), N., A. Tanabe, J. Hanna, S. Oda, and I. Shimizu, "Preparation of Highly Photoconductive a-SiGe from Fluorides by Controlling Reactions with Atomic Hydrogen," Jpn. J. Appl. Phys., Vol. 25, 1986b, pp. L540-L543.

Shibata(3), N., K. Fukuda, H. Ohtoshi, J. Hanna, S. Oda, and I. Shimizu, "Growth of Amorphous and Crystalline Silicon by HR-CVD (Hydrogen Radical Enhanced CVD)," Mat. Res. Soc. Symp. Proc., Vol. 95, 1987a, pp. 225-235.

Shibata(4), N., K. Fukuda, H. Ohtoshi, J. Hanna., S. Oda, and I. Shimuzu, "Preparation of Polycrystalline Silicon by Hydrogen-Radical-Enhanced Chemical Vapor Deposition," Jpn. J. Appl. Phys., Vol. 26, 1987b, pp. L10-L13.

Shibata(5), N., S. Oda, and I. Shimizu, "Hole Transport in Silicon Thin Films with Variable Hydrogen Content," Jpn. J. Appl. Phys., Vol. 26, 1987c, pp. L276-L279.

Shimada(1), T., N. Nakamura, S. Matsubara, H. Itoh, S. Muramatsu, and M. Migitaka, "High-Rate Deposition of a-Si:H Films from Monosilane by Hot-Wall Type Symmetric-Plasma CVD Reactor," Proc. 1st Int'l PV Sci. & Eng. Conf., 1984, pp. 445-448.

Shimada(2), T., S. Matsubara, H. Itoh, and S. Muramatsu, "Characteristics of p-a-SiC:H Films and Cells with Multi-Layered p," First Sunshine Workshop on Solar Cells, Japan, 1990.

Shimizu, K., H. Hosoya, O. Sugiura, and M. Matsumura, "High-Mobility Bottom-Gate Thin-Film Transistors with Laser-Crystallized and Hydrogen-Radical-Annealed Polysilicon Films," Jpn. J. Appl. Phys., Vol. 30, 1991, pp. 3704-3709.

Shimizu, I., "Properties of a-Si Based Alloys Prepared from Fluorides and Hydrogen," J. Non-Crystal. Solids, Vol. 77 & 78, 1985, pp. 877-880.

Shimizu, I., J.-I. Hanna, and H. Shirai, "Control of Chemical Reactions for Growth of Crystalline Si at Low Substrate Temperature," Mat. Res. Soc. Symp. Proc., Vol. 164, 1990, pp. 195-204.

Shimizu(1), T., K. Nakazawa, M. Kumeda, and S. Ueda, "Incorporation Scheme of Reducing Defects in a-Si Studied by NMR and ESR," Physica, Vol. 117B & 118B, 1983, pp. 926-928.

Shimizu(2), T., M. Kumeda, A. Morimoto, Y. Tsujimura, and I. Kobayashi, "NMR and ESR Studies on a-SiGe:H Films Prepared by Glow Discharge and Magnetron Sputtering," Mat. Res. Soc. Symp. Proc., Vol. 70, 1986, pp. 313-318.

Shimizu(3), T., X. Xu, H. Kidoh, A. Morimoto, and M. Kumeda, "Surface and Bulk Defects in Hydrogenated Amorphous Silicon and Silicon-Based Alloy Films," J. Appl. Phys., Vol. 64, 1988, pp. 5045-5049.

Shimozuma, M., G. Tochitani, H. Ohno, H. Tagashira, and J. Nakahara, "Hydrogenated Amorphous Carbon Films Deposited by Low-Frequency Plasma Chemical Vapor Deposition at Room Temperature," J. Appl. Phys., Vol. 66, 1989, pp. 447-449.

Shinar, J., R. Shinar, S. Mitra, M. L. Albers, H. R. Shanks, and T. D. Moustakas, "Porous Morphology and Oxydation Kinetics in a-Si:H RF Sputtered by He/H_2," Mat. Res. Soc. Symp. Proc., Vol. 95, 1987, pp. 183-189.

Shindo, M., S. Sato, I. Myokan, S. Mano, and T. Shibata, "High Rate Preparation of a-Si:H by Reactive Evaporation Method," Jpn. J. Appl. Phys., Vol. 23, 1984, pp. 273-276.

Shing(1), Y.H., J.W. Perry, D.R. Coulter, and G. Radhakrishnan, "Amorphous Silicon Deposition Diagnostics using Coherent Anti-Stokes Raman Spectroscopy," Mat. Res. Soc. Symp. Proc., Vol. 95, 1987a, pp. 237-242.

Shing(2), Y.H., J.W. Perry, and A.M. Hermann, "In Situ Process Diagnostics of Silane Plasma for Device-Quality a-Si:H Deposition", Conf. Record 19th IEEE PV Specilists Conf., 1987b, pp. 577-581.

Shing(3), Y.H., C.E. Allevato, C.L. Yang, F.S. Pool, "Electron Cyclotron Resonance and Glow Discharge of a-Si:H," Proc. 1989 Amorphous Silicon Subcontractors' Review Meeting, June 19-20, 1989, SERI report # CP-211-3514, pp. 59-73.

Shirafuji, J., S. Nagata, and M. Kuwagaki, "Effect of Hydrogen Dilution of Silane on Optoelectronic Properties in Glow-Discharged Hydrogenated Silicon Films," J. Appl. Phys., Vol. 58, 1985, pp. 3661-3663.

Shirai, K., T. Iizuka, and S. Gonda, "Electric Probe Measurements in an ECR Plasma CVD Apparatus," Jpn. J. Appl. Phys., Vol. 28, 1989, pp. 897-902.

Shirai, H., D. Das, J.-I. Hanna, and I. Shimizu, "The Concept of 'Chemical Annealing' for the Improvement of the Stability in Si-Network," Techn. Digest 5th Int'l PV Sci. and Eng., 1990, pp. 59-62.

Shufflebotham, P.K., D.J. Thomson, and H.C. Card, "Behavior of Downstream Plasmas Generated in a Microwave Plasma Chemical-Vapor Deposition Reactor," J. Appl. Phys., Vol. 64, 1988, pp. 4398-4403.

Sichanugrist(1), P., M. Konagai, and K. Takahashi, "High Performance a-Si:H Solar Cells Prepared from SiH_4 at High Deposition Rates," Proc. 1st Int'l PV Sci. & Eng. Conf., 1984, pp. 187-190.

Sichanugrist(2), P., M. Konagai, and K. Takahashi, "High Quality Hydrogenated Amorphous Germanium Films Prepared by Photochemical Vapor Deposition," Extended Abstracts 17th Conf. Solid State Devices and Materials, Tokyo, Japan, 1985, pp. 103-106.

Sichanugrist(3), P., H. Suzuki, M. Konagai, and K. Takahashi, "High-Rate Preparation of Amorphous-Silicon Solar Cells with Monosilane," Jpn. J. Appl. Phys., Vol. 25, 1986, pp. 440-443.

Silver, M., N.C. Giles, E. Snow, M.P. Shaw, V. Canella, and D. Adler, "Study of the Electronic Structure of Amorphous Silicon Using Reverse-Recovery Techniques," Appl. Phys. Lett., Vol. 41, 1982, pp. 935-937.

Slobodin, D., S. Aljishi, Y. Okada, D-S. Shen, V. Chu, and S. Wagner, "a-(Si,Ge):H,F Alloys Prepared from SiH_4 and GeF_4," Mat. Res. Soc. Symp. Proc., Vol. 70, 1986, pp. 275-281.

Smith, E.B., Y.S. Tsuo, and R.S. Crandall, "Light-Induced Metastable Defects in Rehydrogenated and Posthydrogenated a-Si:H," Proc. Industry-University Advanced Materials Conf. II, 1989, pp. 1-189.

Smith, J.F., and D.C. Hinson, "Thin Film Characterization," Solid State Tech., Nov. 1986, pp. 135-140.

Smith, R.A., "Semiconductors," 2nd Ed., Cambridge Univ. Press, N. Y., 1978.

Smith, Z E. and S. Wagner, "Band Tails, Entropy, and Equilibrium Defects in Hydrogenated Amorphous Silicon," Phys. Rev. Lett., Vol. 59, 1987, pp. 688-691.

Snell, A.J., W.E. Spear, and P.G. LeComber, "The Lifetime of Injected Carriers in Amorphous Silicon p-n Junctions," Phil. Mag. B, Vol. 43, 1981, pp. 407-417.

Solanki, R., C.A. Moore, and G.J. Collins, "Laser-Induced Chemical Vapor Deposition," Solid State Technology, June 1985, pp. 220-227.

Soule, D.E., G.T. Reedey, E.M. Peterson, and J.A. McMillan, "Relation Between Silicon-Hydrogen Complexes and Microvoids in Amorphous Silicon Films from IR Absorption," Amer. Inst. Phys. Conf. Proc., No. 73, 1981, pp. 89-94.

Spear(1), W.E., and P.G. LeComber, "Electronic Properties of Substitutionally Doped Amorphous Si and Ge," Phil. Mag., Vol. 33, 1976, pp. 935-949.

Spear(2), W.E. G. Willeke, P. G. Le Comber, & A. G. Fitzgerald, "Electronic Properties of Microcrystalline Silicon Films Prepared in a Glow Discharge Plasma," J. De Physique, 4C, Supp. No.10, Vol. 42, 1981, pp. 257-260.

Staebler(1), D.L., and C.R. Wronski, "Reversible Conductivity Changes in Discharge-Produced Amorphous Silicon," Appl. Phys. Lett., Vol. 31, 1977, pp. 292-294.

Staebler(2), D.L., "Laser-Beam Annealing of Discharge-Produced Amorphous Silicon," J. Appl. Phys., Vol. 50, 1979, pp. 3648-3652.

Staebler(3), D.L., and C.R. Wronski, "Optically Induced Conductivity Changes in Discharge-Produced Hydrogenated Amorphous Silicon," J. Appl. Phys., Vol. 51, 1980, pp. 3262-3268.

Stafford(1), B.L., and E. Sabisky, editors, "Stability of Amorphous Silicon Alloy Materials and Devices," Amer. Inst. of Phys. Conf. Proc., Vol.157, 1987.

Stafford(2), B.L., editor, "Amorphous Silicon Materials and Solar Cells," Amer.

Inst. of Physics Conf. Proc., Vol. 234, 1991.

Sterling, H.F., and R.C.G. Swann, "Chemical Vapour Deposition Promoted by R.F. Discharge," Solid-State Electronics, Vol. 8, 1965, pp. 653-654.

Stevens, K.S., and N.M. Johnson, "Intrinsic stress in hydrogenated amorphous silicon deposited with a remote hydrogen plasma," J. Appl. Phys. Vol. 71, 1992, pp. 2628-2631.

Stoddart, H.A., Z. Vardeny, and J. Tauc, "Transient-photomodulation-spectroscopy studies of Carrier Thermalization and Recombination in a-Si:H," Phys. Rev. B, Vol. 38, 1988. pp. 1362-1377,

Stone(1), J.L., "Progress of Amorphous Silicon Based Solar Cells and Modules," Tech. Digest 3rd Int'l PV Sci. Eng. Conf., Tokyo, Japan, 1987, pp. 161-166.

Stone(2), J.L., "Recent Advances in Thin-Film Solar Cells," Proc. 5th Int'l PV Sci. and Eng. Conf., 1990, pp. 227-232.

Street(1), R.A., J.C. Knights, and D.K. Biegelsen, "Luminescence Studies of Plasma-Deposited Hydrogenated Silicon," Phys. Rev. B, Vol. 18, 1978, pp. 1880-1891.

Street(2), R.A., "Photoconductivity and Related Measurements of the Conduction Band Tail in a-Si:H," J. de Physic., Vol. 42-C4, 1981, pp. 575-578.

Street(3), R.A., D.K. Biegelsen, and J. Zesch, "Spin-Dependent Recombination at Dangling Bonds in a-Si:H," Phys. Rev. B, Vol. 25, 1982a, pp. 4334-4337.

Street(4), R.A., "Doping and Fermi Energy in Amorphous Silicon," Phys. Rev. Lett., Vol. 49, 1982b, pp. 1187-1190.

Street(5), R.A., and M.J. Thompson, "Electronic states at the hydrogenated amorphous silicon/silicon nitride interface," Appl. Phys. Lett., Vol. 45, 1984, pp. 769-771.

Street(6), R.A., J. Kakalios, and T.M. Hayes, "Thermal Equilibration in Doped Amorphous Silicon," Phys. Rev. B, Vol. 34, 1986, pp. 3030-3033.

Street(7), R.A., J. Kakalios, C.C. Tsai, and T.M. Hayes, "Thermal-Equilibrium Processes in Amorphous Silicon," Phys. Rev. B, Vol. 35, 1987, pp. 1316-1333.

Street(8), R.A., and K. Winer, "Defect Equilibria in Undoped a-Si:H," Phys. Rev. B, Vol. 40, 1989, pp. 6236-6249.

Street(9), R.A., "Hydrogenated Amorphous Silicon," Cambridge University Press, New York; 1991.

Stutzmann(1), M., W.B. Jackson, and C.C. Tsai, "Light-Induced Metastable Defects in Hydrogenated Amorphous Silicon: A Systematic Study", Phys. Rev. B, Vol. 32, 1985a, pp. 23-47.

Stutzmann(2), M., "The Role of Mechanical Stress in the Light-Induced Degradation of Hydrogenated Amorphous Silicon," Appl. Phys. Lett., Vol. 47, 1985b, pp. 21-23.

Stutzmann(3), M., D.K. Biegelsen, and R.A. Street, "Detailed Investigation of Doping in Hydrogenated Amorphous Silicon and Germanium," Phys. Rev. B, Vol. 35, 1987, pp. 5666-5701.

Stutzmann(4), M., R.A. Street, C.C. Tsai, J.B. Boyce, and S.E. Ready, "Structural, Optical, and Spin Properties of a-SiGe:H Alloys," J. Appl. Phys., Vol. 66, 1989, pp. 569-592.

Stutzmann(5), M., "The defect density in amorphous silicon," Phil. Mag. B, Vol. 60, 1989, pp. 531-546.

Sugai, H., H. Toyoda, A. Yoshida, and T. Okuda, "Ion and Radical Contributions to Hydrogenated Amorphous Silicon Film Formation in a DC Toroidal Discharge," Appl. Phys. Lett., Vol. 46, 1985, pp. 1048-1050.

Sze, S.M., "Physics of Semiconductor Devices," 2nd Edition, John Wiley & Sons, New York, 1981.

Szydlo, N., J. Magarino, and D. Kaplan, "Post-Hydrogenated Chemical Vapor Deposited Amorphous Silicon Schottky Diodes," J. Appl. Phys., Vol. 53, 1982, pp. 5044-5051.

Takagi, T., K. Matsubara, H. Takaoka, and I. Yamada, "New Developments in Ionized-Cluster Beam and Reactive Ionized-Cluster Beam Deposition Techniques," Thin-Solid Films, Vol. 63, 1979, pp. 41-51.

Takeda, T. and S. Sano, "Amorphous Silicon Position Sensor for Telephone Terminal," Mat. Res. Soc. Symp. Proc., Vol. 118, 1988, pp. 399-405.

Tanabe, H., M. Azuma, T. Uematsu, H. Shirai, J. Hanna, I. Shimizu, "Growth of crystalline silicon, microcrystalline and epitaxial at low substrate temperature," Mat. Res. Soc. Symp. Proc., Vol 149, 1989, pp. 17-22.

Tanaka(1), K., S. Yamasaki, K. Nakagawa, A. Matsuda, H. Okushi, M. Matsumura, and S. Iizima, "The Role of Argon Involved in Plasma-Deposited Amorphous Si:H Films," J. Non-Crystal. Solids, 35 & 36, 1980, pp. 475-480.

Tanaka(2), K., K. Nakagawa, A. Matsuda, M. Matsumura, H. Yamamoto, S. Yamasaki, H. Okushi, and S. Izima, "(Invited) Optical, Electrical, and Structural Properties of Plasma-Deposited Amorphous Silicon," Jpn. J. Appl. Phys, Vol. 20, Supplement 20-1, 1981, pp. 267-273.

Tanaka(3), K., editor, "Glow Discharge Hydrogenated Amorphous Silicon," KTK Scientific Publishers, Tokyo, 1989.

Tanaka, M., K. Ninomiya, N. Nakamura, S. Tsuda, S. Nakano, M. Ohnishi, and Y. Kuwano, "High-Rate Deposition a-Si Film with a Separated Plasma Triode Method," Jpn. J. Appl. Phys., Vol. 27, 1988, pp. 14-19.

Tanielian, M., "Adsorbate Effects on the Electrical Conductance of a-Si:H," Phil. Mag. B, Vol. 45, 1982, pp. 435-462.

Tanaguchi, M., M. Hirose, T. Hamasaki, and Y. Osaka, "Novel Effects of Magnetic Field on the Silane Glow Discharge," Appl. Phys. Lett., Vol. 37, 1980, pp. 787-788.

Tarui, H., T. Matsuyama, S. Okomoto, Y. Hishikawa, H. Dohjo, N. Nakamura, S. Tsuda, S. Nakano, M. Ohnishi, and Y. Kuwano, "Preparation and Properties of High-Quality p-type a-SiC Materials for a-Si Solar Cells," Tech. Digest 3rd Int'l PV Sci. and Eng. Conf., 1987, pp. 41-44.

Tauc, J., In Optical Properties of Solids, F. Abeles, ed., North-Holland Publ. Co., 1972, p. 279; Also in Amorphous and Liquid Semiconductors, Chapter 4, J. Tauc, ed., New York: Plenum Press, 1974.

Tawada(1), Y., K. Tsuge, M. Kondo, H. Okamoto, and Y. Hamakawa, "Properties and Structure of a-SiC:H for High-Efficiency a-Si Solar Cell," J. Appl. Phys., Vol. 53, 1982a, pp. 5273-5281.

Tawada(2), Y., M. Kondo, H. Okamoto, & Y. Hamakawa, "Window Effects of Hydrogenated Amorphous Silicon Carbide in a p-i-n a-Si Solar Cell," Jpn. J. Appl. Phys., Vol. 21, 1982b, Supplement 21-1, pp. 297-303.

Taylor, P.C., and S.G. Bishop, "Optical Effects in Amorphous Semiconductors," Amer. Inst. of Physics Conf. Proc., No. 120, 1984.

The Physics of Hydrogenated Amorphous Silicon II, Electronic and Vibrational

Properties, Topics in Applied Physics, Vol. 56, J. D. Joannopoulos and G. Lucovsky, eds., Berlin: Springer Verlag, 1984.

Thomas, J.P., M. Fallavier, K. Affolter, W. Lüthy, and M. Dupuy, "CW Laser Annealing of Hydrogenated Amorphous Silicon Obtained by RF Sputtering," J. Appl. Phys., Vol. 52, 1981, pp. 476-479.

Thomas, P.A. and J.C. Flachet, "Effect of Hydrogenation on the Conductivity of UHV-Deposited Amorphous Silicon," Phil. Mag. B, Vol. 51, 1985, pp. 55-66.

Thompson(1), M.J., "Sputtered Material," In "The Physics of Hydrogenated Amorphous Silicon I," J.D. Joannopoulos and G. Lucovsky, eds., Springer-Verlag, New York, 1984, pp. 119-175.

Thompson(2), M.J., "The Materials Issues and Applications of Amorphous Silicon Thin Film Transistors," Mat. Res. Soc. Symp. Proc., Vol. 70, 1986, pp. 613-624.

Thornton, J.A. and A.S. Penfold, "Cylindrical Magnetron Sputtering," In "Thin-Film Process," J.L. Vossen and W. Kern, eds., Academic Press, New York, 1978, pp. 75-113.

Tobin, M.C., "Laser Raman Spectroscopy," Wiley-Interscience, New York, 1971.

Tomikawa, T., H. Itozaki, and N. Fujita, "Properties of a-SiGe:H Prepared by a Mercury Sensitized Photo CVD," Tech. Digest 3rd Int'l PV Sci. and Eng. Conf., 1987, pp. 267-270.

Tong(1), B.Y., P.K. John, S.K. Wong, and K.P. Chik, "Highly Stable, Photosensitive Evaporated Amorphous Silicon Films," Appl. Phys. Lett., Vol. 38, 1981, pp. 789-790.

Tong(2), B.Y., P.K. Gogna, P.K. John, S.K. Wong, and K.P. Chik, "Absence of Staebler-Wronski Instability in Amorphous Silicon Films Hydrogenated in a Theta-Pinch Plasma," Thin Solid Films, Vol. 105, 1983, L79-L81.

Tonouchi, M., F. Moriyama, and T. Miyasato, "Characterization of μc-Si:H Films Prepared by H_2 Sputtering," Jpn. J. Appl. Phys., Vol. 29, 1990, pp. L385-L387.

Toyoda, H., H. Sugai, K. Kato, A. Yoshida, and T. Okuda, "Hydrogenated Amorphous Silicon Formation by Flux Control and Hydrogen Effects on the Growth Mechanism," Appl. Phys. Lett., Vol. 48, 1986, pp. 1648-1650.

Toyoshima, Y., K. Kumata, U. Itoh, K. Arai, A. Matsuda, N. Washida, G. Inoue, and K. Katsuumi, "Ar ($^3P_2$) Induced Chemical Vapor Deposition of Hydrogenated Amorphous Silicon," Appl. Phys. Lett., Vol. 46, 1985, pp. 584-586.

Trevor, D.J., N. Sadeghi, T. Nakano, J. Derouard, R.A. Gottscho, P.D. Foo, and J.M. Cook, "Spatially Resolved Ion Velocity Distributions in a Diverging Field Electron Cyclotron Resonance Plasma Reactor," Appl. Phys. Lett., Vol. 57, 1990, pp. 1188-1190.

Triska, A., D. Dennison, and H. Fritzsche, "Hydrogen Content in Amorphous Ge and Si Prepared by RF Decomposition of GeH_4 and SiH_4," Bull. Am. Phys. Soc., Vol. 20, 1975, p. 392.

Troxell, J.R., M.I. Harrington, and R.A. Miller, "Laser-Recrystallized Silicon Thin-Film Transistors on Expansion-Matched 800°C Glass," IEEE Electron Device Letters, Vol. EDL-8, 1987, pp. 576-580.

Tsai(1), C.C., M. Stutzmann, and W.B. Jackson, "The Staebler-Wronski Effect in Undoped a-Si:H: Its Intrinsic Nature and the Influence of Impurities," Amer. Inst. of Physics Conf. Proc. No. 120, 1984, pp. 242-249.

Tsai(2), C.C., J.C. Knights, G. Chang, and B. Wacker, "Film Formation Mechanisms in the Plasma Deposition of Hydrogenated Amorphous Silicon," J. Appl. Phys., Vol. 59, 1986, pp. 2998-3001.

Tsai(3), C.C., J.G. Shaw, B. Wacker, and J.C. Knights, "Film Growth Mechanisms of Amorphous Silicon in Diode and Triode Glow Discharge Systems," Mat. Res. Soc. Symp. Proc., Vol. 95, 1987, pp. 219-224.

Tsai(4), C.C., R. Thompson, C. Doland, F.A. Ponce, G.B. Anderson, and B. Wacker, "Transition from Amorphous to Crystalline Silicon: Effect of Hydrogen on Film Growth," MRS Symp. Proc., Vol. 118a, 1988a, pp. 49-54.

Tsai(5), C.C., "Plasma Deposition of Amorphous and Crystalline Silicon: The Effect of Hydrogen on the Growth, Structure and Electronic Properties," Amorphous Silicon and Related Materials, H. Fritzsche, ed., World Scientific Publishing Company, pp. 123-147, 1988b.

Tsai(6), C.C., G.B. Anderson, R. Thompson, B. Wacker, and C. Doland, "Temperature Dependence of Structure, Transport and Growth of Microcrystalline Silicon: Does Grain Size Correlate with Transport?" Mat. Res. Soc. Symp. Proc., Vol. 149, 1989, pp. 297-302.

Tsai(7), C.C., G.B. Anderson, and R. Thompson, "Growth of Amorphous

Microcrystalline and Epitaxial Silicon in Low Temperature Plasma Deposition", Mat. Res. Soc. Symp. Proc., Vol. 192, 1990, pp. 475-480.

Tsai, H.-K., W.-L. Lin, W.-J. Sah, and S-C. Lee, "The Characteristics of Amorphous Silicon Carbide Hydrogen Alloy", J. Appl. Phys. Vol. 64, 1988, pp. 1910-1915.

Tscharner, R., D. Fischer, H. Keppner, A.V. Shah, A.A. Howling, J.L. Dorier, Ch. Hollenstein, "Fast Deposition of High Quality Amorphous Silicon Solar Cells without Powder Formation", Proc. 6th Int'l PV Sci. and Eng. Conf., 1992, pp. 311-316.

Tsu(1), R., J. Gonzalez-Hernandez, S.S. Chao, S.C. Lee, and K. Tanaka, "Critical Volume Fraction of Crystallinity for Conductivity Percolation in Phosphorous-Doped Si:F:H Alloys," Appl. Phys. Lett., Vol. 40, 1982, pp. 534-535.

Tsu(2), R., J. Gonzalez-Hernandez, and F.H. Pollak, "Determination of Energy Barrier for Structural Relaxation in a-Si and a-Ge by Raman Scattering," J. Non-Cryst. Solids, Vol. 66, 1984, pp. 109-114.

Tsu(3), R., J. Gonzalez-Hernandez, S.S. Chao, and D. Martin, "Dependence of Grain Size on the Substrate Temperature of Si and Ge Films Prepared by Evaporation under Ultrahigh Vacuum," Appl. Phys. Lett., Vol. 48, 1986, pp. 647-649.

Tsu(4), R., P. Menna, and A. H. Mahan, "Optical Absorption and Disorder in Hydrogenated Amorphous Si-Ge and Si-C Alloy Systems," Solar Cells, Vol. 21, 1987, pp.189-194.

Tsuda(1), S., H. Tarui, H. Haku, Y. Nakashima, Y. Hishikawa, S. Nakano, and Y. Kuwano, "High Quality a-SiGe Alloys Prepared by New Methods," J. Non-Cryst. Solids, Vol. 77&78, 1985, pp. 845-848.

Tsuda(2), S., H. Haku, H. Tarui, T. Martsuyama, K. Sayama, Y. Nakashima, S. Narano, M. Ohnishi, and Y. Kuwano, "High-Quality a-Si Based Alloys: a-SiGe Films Fabricated in a Super Chamber and Superlattice Structure a-Si Films Prepared by a Photo-CVD Method," Mat. Res. Soc. Symp. Proc., Vol. 95, 1987a, pp. 311-316.

Tsuda(3), S., T. Takahama, M. Isomura, H. Tarui, Y. Nakashima, Y. Hishikawa, N. Nakamura, T. Matsuoka, H. Nishiwaki, S. Nakano, M. Ohnishi, and Y. Kuwano, "Preparation and Properties of High-Quality a-Si Films with a Super Chamber (Separated Ultra-High Vacuum Reaction Chamber)," Jpn. J. Appl. Phys., Vol. 26, 1987b, pp. 33-38.

Tsuo(01), Y.S., J.L. Hurd, R.J. Matson, and T.F. Ciszek, "Electron Channeling and EBIC Studies of Edge-Supported Pulling Silicon Sheets," IEEE Trans. Electron Devices, Vol. ED-31, 1984, pp. 614-618.

Tsuo(02), Y.S., E.B. Smith, and S.K. Deb, "Ion Beam Hydrogenation of Amorphous Silicon," Appl. Phys. Lett., Vol. 51, 1987a, pp. 1436-1438.

Tsuo(03), Y.S., R. Weil, S. Asher, A. Nelson, Y. Xu, and R. Tsu, "Deposition and Etching of a-Si:H:F using Silane and Xenon Difluoride," Proc. 19th IEEE PV Specialists Conf., 1987b, pp. 705-708.

Tsuo(04), Y.S., X.J. Deng, E.B. Smith, Y. Xu, and S.K. Deb, "Ion-Beam Rehydrogenation and Posthydrogenation of a-Si:H," J. Appl. Phys, Vol. 64, 1988a, pp. 1604-1607.

Tsuo(05), Y.S., E.B. Smith, X.J. Deng, Y. Xu, and S.K. Deb, "Ion-Beam Hydrogenation of Amorphous Silicon," Solar Cells, Vol. 24, 1988b, pp. 249-256.

Tsuo(06), Y.S., Y. Xu, R.S. Crandall, H.S. Ullal, and K. Emery, "Hydrogen-Plasma Recative Flushing for a-Si:H P-I-N Solar Cell Fabrication," Mat. Res. Soc. Symp. Proc., Vol. 149, 1989a, pp. 471-476.

Tsuo(07), Y.S. and A.L. Langford, "Method and Apparatus for Removing and Preventing Window Deposition During Photochemical Vapor Deposition Processes," U.S. Patent No. 4,816,294, March 28, 1989b.

Tsuo(08), Y.S. and S.K. Deb, "Hydrogen Ion Microlithography," U.S. Patent No. 4,960,675, Oct. 2, 1990.

Tsuo(09), Y. S., and W. Luft, "Alternative Deposition Processes for Hydrogenated Amorphous Silicon and Related Alloys," Appl. Phys. Comm., Vol. 10, 1990, pp. 71-141.

Tsuo(10), Y. S., Y. Xu, E.A. Ramsay, R.S. Crandall, S.J. Salamon, I. Balberg, B.P. Nelson, Y. Xiao, and Y. Chen, "Methods of Improving Glow-Discharge-Deposited a-SiGe:H," Mat. Res. Soc. Symp. Proc., Vol. 219, 1991a, pp. 769-774.

Tsuo(11), Y. S., Y. Xu, I. Balberg, and R. Crandall, "Effect of Helium Dilution on Glow Discharge Deposition of a-SiGe:H Alloys," Conf. Record 22nd IEEE PV Specialists Conf., Vol. II, 1991b, pp. 1334-1337.

Tsuo(12), Y.S., Y. Xu, D.W. Baker, and S.K. Deb, "Etching Properties of Hydrogenated Amorphous Silicon," Mat. Res. Soc. Symp. Proc., Vol. 219, 1991c, pp. 805-810

Tsuo(13), Y.S., Y. Xu, E.A. Ramsay, R.S. Crandall, S.J. Salamon, I. Balberg, B.P. Nelson, Y. Xiao, and Y. Chen, "Methods of Improving Glow Discharge-Deposited a-$Si_{1-x}Ge_x$:H," Mat. Res. Soc. Symp. Proc., Vol. 219, 1991d, pp. 769-774.

Turner, D.P., I.P. Thomas, J. Allison, M.J. Thompson, A.J. Rhodes, I.G. Austin, and T.M. Searle, "The Growth and Properties of Bias-Sputtered a-Si:H," Amer. Inst. of Phys. Conf. Proc. No. 73, 1983, pp. 47-51.

U.S. Department of Energy, Five-Year Research Plan, 1987-1991, Photovoltaics: USA's Energy Opportunity, National Photovoltaics Program, May 1987, Photovoltaic Energy Technology Division, Office of Solar Electric Technologies, U. S. Department of Energy, DOE/CH10093-7.

Uchida, H., K. Takechi, S. Nishida, and S. Kaneko, "High-Mobility and High-Stability a-Si:H Thin Film Transistors with Smooth SiN_x/a-Si Interface," Jpn. J. Appl. Phys., Vol. 30, 1991, pp. 3691-3694.

Uchida(1), Y., H. Kanoh, O. Sugiura, and M. Matsumura, "Hydrogen-radical annealing of chemical vapor-deposited amorphous silicon films," Jpn. J. Appl. Phys., Vol. 29, 1990, pp. L2171-L2173.

Uchida(2), Y., S. C Deane, and W. I. Milne, "Posthydrogenation of low-pressure chemicalvapor-deposited amorphous silicon using a novel internal lamp system and its application to thin-film transistor fabrication," J. Appl. Phys. Vol. 72, 1992, pp. 3150-3154.

Ueda(1), M., A. Chayahara, T. Hamasaki, M. Hirose, and Y. Osaka, "A New Mode of Plasma Deposition in a Cylindrical Drum Type Reactor," Extended Abstracts 16th Int. Conf. Solid State Devices and Materials, 1984, pp. 543-546.

Ueda(2), M., T. Imura, and Y. Osaka, "Effect of Substrate Temperature on Properties of a-Si:H Prepared at High Deposition Rate," Proc. 10th Symp. on Ion Sources and Ion-Assisted Technology, 1986.

Urbach, F., "The Long-Wavelength Edge of Photographic Sensitivity and of the Electronic Absorption of Solids," Phys. Rev., Vol. 92, 1953, p.1324.

van Oort, R.C., M.J. Geerts, J.C. van den Heuvel, and J.W. Metselaar, "Hydrogen Plasma Etching of Amorphous and Microcrystalline Silicon," Electronics Lett., Vol. 23, 1987, pp. 967- 968.

Vanecek(1), M., J. Kocka, J. Stuchlik, Z. Kozisek, O. Stika, and A. Triska, "Density of the Gap States in Undoped and Doped Glow Discharge a-Si:H,"

Solar Energy Materials, Vol. 8, 1983, pp. 411-423.

Vanecek(2), M., A. Abraham, O. Stika, J. Stuchlik, and J. Kocka, "Gap States Density in a-Si:H Deduced from Subgap Optical Absorption Measurement on Schottky Solar Cells," Phys. Stat. Sol. (a), Vol. 88, 1984, pp. 617-623.

Vanier, P. E., Deposition of Amorphous Silicon Solar Cells at High Rates by Glow Discharge of Disilane, SERI/STR-211-3016, Solar Energy Research Inst., September 1986.

Venugopalan, M., "Aim a Laser at Your Plasma and Expose Its Secrets," Research & Development, June 1989, pp. 62-68.

Veprek, S., M. Heintze, R. Bayer, and M. Jurčik-Rajman, "From the Understanding of the Reaction Mechanism towards Optimizing the Deposition Rate and Optoelectronic Properties of a-Si:H," Mat. Res. Soc. Symp. Proc., Vol. 149, 1989, pp. 3-9.

Verdeyen, J.T., J. Beberman, and L. Overzet, "Modulated Discharges: Effect on Plasma Parameters and Deposition," J. Vac. Sci. and Tech., Vol. 8, 1990, pp. 1851-1856.

Vèrie, C., "Recent Progress in Studies of a-$Si_{1-x}Sn_x$:H for Novel Amorphous Silicon-Based Solar Cells," Proc. 17th IEEE PV Specialists Conf., 1984. pp. 245-247,

Vieira, M., R. Martins, A. Macarico, I. Baia, F. Soares, and L. Guimarães, "Role of Power Density, U.V. Light and Hydrogen Dilution on Transition of Amorphous to Microcrystalline Structure on Films Produced by a TCDDC System," Mat. Res. Soc. Symp. Proc., Vol. 118, 1988, pp. 113-116.

Viturro, R.E. and K. Weiser, "Some Properties of Hydrogenated Amorphous Silicon Produced by Direct Reaction of Silicon and Hydrogen Atoms," Phil. Mag. B, Vol. 53, 1986, pp. 93-103.

von Roedern(1), B., D.K. Paul, J. Blake, R.W. Collins, G. Moddel, and W. Paul, "Optical Absorption, Photoconductivity, and Photoluminescence of Glow Discharge Amorphous SiGe Alloys," Phys. Rev. B, Vol. 25, 1982, pp. 7678-7687.

von Roedern(2), B., A.H. Mahan, T.J. McMahon, and A. Madan, "An Assessment of a-SiGe:H Alloys with Band Gap of 1.5 eV as to their Suitability for Solar Cell Applications," Mat. Res. Soc. Symp. Proc., Vol. 49, 1985, pp. 167-172.

von Roedern(3), B., "Shortfall of Defect Models for Amorphous Silicon Solar Cell Performance," J. Appl. Phys., Vol. 62, 1993.

Wagner, S., private communication, 1992

Walsh, P.J. and N. Bottka, "Internal Photolysis Reactor," U.S. Patent No. 4,454,835, June 19, 1984.

Wang, K., D. Han, M. Kemp, and M. Silver, "Time Resolved Electroluminescence in Hydrogenated Amorphous Silicon," Appl. Phys. Lett., Vol. 62, 1993, pp. 157-159.

Wang, Q., H. Antoniadis, and E.A. Schiff, "Electron Drift Mobility Measurements on Annealed and Light-Soaked a-Si:H," Appl. Phys. Lett., Vol. 60, 1992, pp. 2791-2793.

Watanabe(1), H., K. Katoh, and M. Yasui, "Properties of amorphous films prepared from SiH4-N2-H2 gas mixture," Jpn. J. Appl. Phys., Vol. 21, 1982, pp. L341-L343.

Watanabe(2), T., K. Azuma, M. Nakatani, K. Suzuki, T. Sonobe, and T. Shimada, "Chemical Vapor Deposition of a-Si:H Films Utilizing a Microwave Excited Ar Plasma Stream," Jpn. J. Appl. Phys., Vol. 25, 1986, pp. 1805-1810.

Watanabe(3), T., K. Azuma, M. Tanaka, M. Nakatani, T. Sonobe, and T. Shimada, "IR Study of a-SiGe:H Films Prepared by a Microwave-Excited Plasma CVD Method," Tech. Digest 3rd Int'l PV Sci. and Eng. Conf. 1987a, pp. 57-60.

Watanabe(4), T., M. Tanaka, K. Azuma, M. Nakatani, T. Sonobe, and T. Shimada, "Chemical Vapor Deposition of a-SiGe:H Films Utilizing a Microwave-Excited Plasma," Jpn. J. Appl. Phys., Vol. 26, 1987b, pp. L288-L290.

Weakliem, H.A., R.D. Estes, and P.A. Longeway, "Depomposition Kinetics of Silane Glow Discharges," Proc. 6th Int'l Symp. on Plasma Chemistry, Vol. 3, 1983, pp. 838-842.

Weisfield(1), R.L., "Space-Charge-Limited Currents: Refinements in Analyses and Applications to a-Si,Ge Alloys," Ph.D. Thesis, Harvard University Press, 1983.

Weisfield(2), R.L., "Amorphous-Silicon Linear-Array Device Technology: Applications in Electronic Copying," IEEE Trans. Electr. Devices, Vol. 36, 1989, pp. 2935-2939.

Weitzel, I., R. Primig, and K. Kempter, "Preparation of Glow Discharge Amorphous Silicon for Passivation Layers," Thin Solid Films, Vol. 75, 1981, pp. 143-150.

Weller, H.C., S.M. Paasche, C.E. Nebel, F. Kessler, and G.H. Bauer, "Highly Photoconductive 1.4-1.5 eV Amorphous Silicon-Germanium Alloys Prepared by Glow Discharge," Proc. 19th IEEE PV Specialists Conf., 1987, pp. 872-877.

Wertheimer, M.R. and M. Moisan, "Comparison of Microwave and Lower Frequency Plasmas for Thin Film Deposition and Etching," J. Vac. Sci. Tech. A, Vol. 3, 1985, pp. 2643-2649.

Wiesendanger, R., L. Rosenthaler, H.R. Hidber, H.J. Güntherodt, A.W. McKinnon, and W.E. Spear, "Hydrogenated Amorphous Silicon Studied by Scanning Tunneling Microscopy," J. Appl. Phys., Vol. 63, 1988, pp. 4515-4517.

Wiesmann(1), H., A.K. Ghosh, T. McMahon, and M. Strongin, "a-Si:H Produced by High-Temperature Thermal Decomposition of Silane," J. Appl. Phys., Vol. 50, 1979, pp. 3752-3754.

Wiesmann(2), H., J. Dolan, G. Fricano, and V. Danginis, Research on High-Efficiency Single-Junction Monolithic Thin Film Amorphous Silicon Cells, STR-211-3097, Solar Energy Research Inst., January 1987.

Willeke, G., and R. Martins, "Structural Properties of Weakly Absorbing Highly Conductive SiC Thin Films Prepared in a TCDDC System", Proc. 20th IEEE PV Specialists Conf., 1988, pp. 320-327.

Williamson, D.L., A.H. Mahan, B.P. Nelson, and R.S. Crandall, "Microvoids in a-$Si_{1-x}C_x$:H Alloys Studied by Small Angle X-Ray Scattering," Appl. Phys. Lett., Vol. 55, 1989, pp. 783-786.

Wilson, B.A., A.M. Sergent, K.W. Wecht, A.J. Williams, T.P. Kerwin, C.M. Taylor, and J.P. Harbison, "Effects of Annealing on Plasma-Deposited a-Si:H Films Grown under Optimal Conditions," Phys. Rev. B, Vol. 30, 1984, pp. 3320-3332.

Winer, K., I. Hirabayashi, and L. Ley, "Distribution of Occupied Near-Surface Band-Gap States in a-Si:H," Phys. Rev. B, Vol. 38, 1988, pp. 7680-7693.

Winters, H.F., and J.W. Coburn, "The Etching of Silicon with XeF_2 Vapor," Appl. Phys. Lett., Vol. 34, 1979, pp. 70-73.

Wolf, S., and R.N. Tauber, "Silicon Processing for the VLSI Era," Lattice Press,

Sunset Beach, California, 1987.

Wormhoudt, J., A.C. Stanton, and J. Silver, "Techniques for Characterization of Gas-Phase Species in Plasma Etching and Vapor Deposition Processes," Proc. of SPIE - Int'l Soc. Optical Engineering, Spectroscopic Characterization Techniques for Semiconductor Technology, Vol. 452, 1983, pp. 88-89.

Wronski(1), C.R., "Amorphous Silicon and Its Applications," Solid State Technology, June 1988, pp. 113-117.

Wronski(2), C.R., "Instabilities in a-Si:H Solar Cells: Materials and Device Issues," Proc. 21st IEEE PV Specialists Conf., 1990, pp.1487-1492.

Wronski(3), C.R., "Review of Direct Measurements of Mobility Gaps in a-Si:H using Internal Photoemission," J. Non-Cryst. Solids, Vol. 141, 1992, pp. 16-23.

Wu, Z. Q., C.Y. Xu, W.P. Zhang, Z.B. Zheng, and R.C. Fang, "Infrared Spectra of a-Si:H,Cl Films," J. Non-Cryst. Solids, Vol. 59&60, 1983, pp. 217-220.

Wu(1), Z-Y., A. Lloret, J. M. Siefert, P. Roca i Cabarrocas, and B. Equer, "Is B(CH3)3 an Interesting Alternative to B_2H_6 for a-Si:H p Doping?" Mat. Res. Soc. Symp. Proc., Vol. 149, 1989a, pp. 291-296.

Wu(2), Z., A. Lloret, B. Equer, J.-M. Siefert, and H. Tran Quoc, "A Comparison Between Diborane and Trimethylboron for p-type Doping in a-Si:H Solar Cells," Proc. 9th E.C. PV Solar Energy Conf., 1989b, pp. 1035-1037.

Xi, J., R.E. Hollingsworth, R. Buitrago, D. Oakley, J.P. Cumalot, U. Nauenberg, J. McNeil, D.F. Anderson, and V. Perez-Mendez, "Minimum Ionizing Particle Detection using Amorphous Silicon Diodes," Nuclear Instruments and Methods in Physics Research, Vol. A301, 1991, pp. 219-222.

Xiao, Y., M.J. Heben, J.M. McCullough, Y.S. Tsuo, J.I. Pankove, and S.K. Deb, "Enhancement and Stabilization of Porous Silicon Photoluminescence by a Remote-Plasma Treatment," Appl. Phys. Lett., Vol. 62, 1993, pp. 1152-1154.

Xu, X., A. Morimoto, M. Kumeda, and T. Shimizu, "ESR and Constant Photocurrent Studies of Surface and Bulk Defects in a-Si:H," Jpn. J. Appl. Phys., Vol. 26, 1987, pp. L1818-L1820.

Xu, Y., Y.S. Tsuo, and R.S. Crandall, unpublished results on PED-CVD using hydrogen plasma for the etching cycles, 1988.

Yacobi, B.G., "Electron-Beam-Induced Information Storage in Hydrogenated

Amorphous Silicon Devices," Appl. Phys. Lett., Vol. 44, 1984, pp. 695-697.

Yamada, H., and Y. Torii, "Low-Temperature Film Growth of Si by Reactive Ion Beam Deposition," Appl. Phys. Lett., Vol. 50, 1987, pp. 386-388.

Yamada(1), I., I. Nagai, M. Horie, and T. Takagi, "Preparation of Doped Amorphous Silicon Films by Ionized-Cluster Beam Deposition," J. Appl. Phys., Vol. 54, 1983, pp. 1583-1587.

Yamada(2), I., "Ionized-Cluster Beam Deposition," Semiconductors and Semimetals, Vol. 21A, 1984, pp. 83-107.

Yamamoto(1), Y., et al., "Germanium and Carbon Composition Graded Layer in P/I Interface of a-SiGe:H Solar Cell", Proc. 19th PV Specialists Conf., 1987a, pp. 302-305.

Yamamoto(2), Y., K. Nomoto, T. Okuno, S. Moriuchi, Y. Nakata, T. Inoguchi, "A Role of Composition Graded Layer in P/I Interface of Amorphous Silicon Solar Cell", Proc. 19th PV Specialists Conf., 1987b, pp. 901-904.

Yamanaka(1), S., M. Konagai, & K. Takahashi, "Theoretical Investigation of the Optimum Design for Amorphous Silicon Based Solar Cells," Proc. 20th IEEE PV Specialists Conf., 1988, pp. 160-165.

Yamanaka(2), S., M. Konagai, and K. Takahashi, "Numerical Study of Amorphous Silicon Based Solar Cell Performance Toward 15% Conversion Efficiency," Jpn. J. Appl. Phys., Vol. 28, 1989, pp. 1178-1184.

Yamano, M. and H. Takesada, "Full Color Liquid Crystal Television Addressed by Amorphous Silicon TFTs," J. Non-Crystal. Solids, Vol.77&78, 1985, pp. 1383-1388.

Yang, D., K. S. Ambo, and J. W. Holm-Kennedy, "Four-Color Discriminating Sensor Using Amorphous Silicon Drift-Type Photodiode," Sensors and Actuators, Vol. 14, 1988, pp. 69-77.

Yang, J., R. Ross, T. Glatfelter, R. Mohr, G. Hammond, C. Bernotaitis, EA. Chen, J. Burdick, M Hopson, and S. Guha, "High Efficiency Multi-Junction Solar Cells using Amorphous Silicon and Amorphous Silicon-Germanium Alloys", 20th IEEEE PV Specialists Conf., 1988, pp. 241-246.

Yang(1), L., J. Newton, and B. Fieselmann, "Raman Spectroscopy of a-SiGe:H Alloys," Mat. Res. Soc. Symp. Proc., Vol. 149, 1989, pp. 497-502.

Yang(2), L., I. Balberg, A. Catalano, and M. Bennett, "Strong Thickness Dependence of Photoelectronic Properties in Hydrogenated Amorphous Silicon," Mat. Res. Soc. Symp. Proc., Vol. 192, 1990, pp. 243-248.

Yang(3), L., L. Chen, and A. Catalano, "Intensity and Temperature Dependence of Photodegradation of Amorphous Silicon Solar Cells under Intense Illumination," Appl. Phys. Lett., Vol. 59, 1991, pp. 840-842.

Yang(4), L., L. Chen, S. Wiedeman, and A. Catalano, "Microcrystalline Silicon in a-Si:H-based Multijunction Solar Cells", Mat. Res. Soc. Symp. Proc., Vol. 283, 1992. (in press)

Ye, Y.J., W.A. Anderson, and Y.S. Tsuo, "Thickness Dependence of Photoelectrical Properties of Intrinsic Amorphous Silicon," Solar Cells, Vol. 23, 1988, pp. 191-199.

Yokota(1), K., T. Kageyama, and S. Katayama, "Behavior of Oxygen Adsorbed in Annealed Hydrogenated Amorphous Silicon," Jpn. J. Appl. Phys., Vol. 22, 1983, p. 370.

Yokota(2), K., T. Kageyama, and S. Katayama, "Oxidation and Stress Relief in Air at Room Temperature of Amorphous Silicon Hydrogenated in a Glow Discharge," Solid State Electronics, Vol. 28, 1985, pp. 893-901.

Yokota(3), K., M. Takada, Y. Ohno, and S. Katayama, "Deposition of hydrogenated amorphous silicon by an inductively coupled glow discharge reactor with shield electrodes," J. Appl. Phys., Vol. 72, 1992, pp. 1188-1190.

Yoshida, A., K. Inoue, H. Ohashi, and Y. Saito, "High Quality Hydrogenated Amorphous Silicon Films by Windowless Hydrogen Discharge," Appl. Phys. Lett., Vol. 57, 1990, pp. 484-486.

Yoshida(1), T., S. Fujikake, H. Shimabukuro, Y. Ichikawa, & H. Sakai, "Open-Circuit Voltage of p-i-n a-Si Based Solar Cells," 20th IEEE PV Specialists Conf., 1988a, pp. 335-339.

Yoshida(2), T., T. Ihara, S. Fujikake, Y. Ichikawa, & H. Sakai, "Multijunction a-SiC:H/a-Si:H Solar Cells", Proc. 8th E.C. PV Solar Energy Conf., 1988b, pp. 893-897.

Yoshida(3), T., Y. Ichikawa, and H. Sakai, "A Novel PD-CVD Technique and Its Application to a-Si Solar Cells," Proc. 9th Euro. Comm. PV Conf., 1989, pp. 1006-1009.

Yoshida(4), T., T. Hokaya, Y. Ichikawa, and H. Sakai, "Pulse discharge CVD and its application to a-Si solar cells," Techn. Digest 5th Int'l PVSEC, 1990, p. 537.

Zelikson, M., J. Salzman, K. Weiser, and J. Kanicki, "Enhanced Electro-Optic Effect in Amorphous Hydrogenated Silicon Based Waveguides," Appl. Phys. Lett., Vol. 61, 1992, pp. 1664-1666.

Zhang, G., Z. Xi, J. Yan, and Z. Shi, "a-Si:H films prepared with center finger electrode hot-wall plasma CVD," Techn. Digest Int'l PVSEC-5, 1990, pp. 795-798.

Ziegler, Y., H. Curtins, J. Baumann, and A. Shah, "VHF-GD a-Si:H Films Prepared at Very Low Temperature," Mat. Res. Soc. Symp. Proc., Vol. 149, 1989, pp. 81-86.

SUBJECT INDEX

I

N

O

P

Q

R

S

W

X

Y

Z